Texts and Monographs in Physics

T0224101

Johannes Voit

The Statistical Mechanics of Financial Markets

Third Editon

With 99 Figures

 Springer

Dr. Johannes Voit
Deutscher Sparkassen-und Giroverband
Charlottenstraße 47
10117 Berlin
Germany
E-mail: johannes.voit@dsgv.de

ISBN-13 978-3-642-06578-1   e-ISBN-13 978-3-540-26289-3

Springer is a part of Springer Science+Business Media
springeronline.com
© Springer-Verlag Berlin Heidelberg 2010
Printed in The Netherlands

Cover design: *design & production* GmbH, Heidelberg

Printed on acid-free paper

One must act on what has not happened yet.

Lao Zi

Preface to the Third Edition

The present third edition of *The Statistical Mechanics of Financial Markets* is published only four years after the first edition. The success of the book highlights the interest in a summary of the broad research activities on the application of statistical physics to financial markets. I am very grateful to readers and reviewers for their positive reception and comments. Why then prepare a new edition instead of only reprinting and correcting the second edition?

The new edition has been significantly expanded, giving it a more practical twist towards banking. The most important extensions are due to my practical experience as a risk manager in the German Savings Banks' Association (DSGV): Two new chapters on risk management and on the closely related topic of economic and regulatory capital for financial institutions, respectively, have been added. The chapter on risk management contains both the basics as well as advanced topics, e.g. coherent risk measures, which have not yet reached the statistical physics community interested in financial markets. Similarly, it is surprising how little research by academic physicists has appeared on topics relating to Basel II. Basel II is the new capital adequacy framework which will set the standards in risk management in many countries for the years to come. Basel II is responsible for many job openings in banks for which physicists are extemely well qualified. For these reasons, an outline of Basel II takes a major part of the chapter on capital.

Feedback from readers, in particular Guido Montagna and Glenn May, has led to new sections on American-style options and the application of path-integral methods for their pricing and hedging, and on volatility indices, respectively. To make them consistent, sections on sensitivities of options to changes in model parameters and variables ("the Greeks") and on the synthetic replication of options have been added, too. Chin-Kun Hu and Bernd Kälber have stimulated extensions of the discussion of cross-correlations in financial markets. Finally, new research results on the description and prediction of financial crashes have been incorporated.

Some layout and data processing work was done in the Institute of Mathematical Physics at the University of Ulm. I am very grateful to Wolfgang Wonneberger and Ferdinand Gleisberg for their kind hospitality and generous support there. The University of Ulm and Academia Sinica, Taipei, provided opportunities for testing some of the material in courses.

My wife, Jinping Shen, and my daughter, Jiayi Sun, encouraged and supported me whenever I was in doubt about this project, and I would like to thank them very much.

Finally, I wish You, Dear Reader, a good time with and inspiration from this book.

Berlin, July 2005   *Johannes Voit*

Preface to the First Edition

This book grew out of a course entitled "Physikalische Modelle in der Finanzwirtschaft" which I have taught at the University of Freiburg during the winter term 1998/1999, building on a similar course a year before at the University of Bayreuth. It was an experiment.

My interest in the statistical mechanics of capital markets goes back to a public lecture on self-organized criticality, given at the University of Bayreuth in early 1994. Bak, Tang, and Wiesenfeld, in the first longer paper on their theory of self-organized criticality [Phys. Rev. A **38**, 364 (1988)] mention Mandelbrot's 1963 paper [J. Business **36**, 394 (1963)] on power-law scaling in commodity markets, and speculate on economic systems being described by their theory. Starting from about 1995, papers appeared with increasing frequency on the Los Alamos preprint server, and in the physics literature, showing that physicists found the idea of applying methods of statistical physics to problems of economy exciting and that they produced interesting results. I also was tempted to start work in this new field.

However, there was one major problem: my traditional field of research is the theory of strongly correlated quasi-one-dimensional electrons, conducting polymers, quantum wires and organic superconductors, and I had no prior education in the advanced methods of either stochastics and quantitative finance. This is how the idea of proposing a course to our students was born: learn by teaching! Very recently, we have also started research on financial markets and economic systems, but these results have not yet made it into this book (the latest research papers can be downloaded from my homepage http://www.phy.uni-bayreuth.de/btp314/).

This book, and the underlying course, deliberately concentrate on the main facts and ideas in those physical models and methods which have applications in finance, and the most important background information on the relevant areas of finance. They lie at the interface between physics and finance, not in one field alone. The presentation often just scratches the surface of a topic, avoids details, and certainly does not give complete information. However, based on this book, readers who wish to go deeper into some subjects should have no trouble in going to the more specialized original references cited in the bibliography.

Despite these shortcomings, I hope that the reader will share the fun I had in getting involved with this exciting topic, and in preparing and, most of all, actually teaching the course and writing the book.

Such a project cannot be realized without the support of many people and institutions. They are too many to name individually. A few persons and institutions, however, stand out and I wish to use this opportunity to express my deep gratitude to them: Mr. Ralf-Dieter Brunowski (editor in chief, Capital – Das Wirtschaftsmagazin), Ms. Margit Reif (Consors Discount Broker AG), and Dr. Christof Kreuter (Deutsche Bank Research), who provided important information; L. A. N. Amaral, M. Ausloos, W. Breymann, H. Büttner, R. Cont, S. Dresel, H. Eißfeller, R. Friedrich, S. Ghashghaie, S. Hügle, Ch. Jelitto, Th. Lux, D. Obert, J. Peinke, D. Sornette, H. E. Stanley, D. Stauffer, and N. Vandewalle provided material and challenged me in stimulating discussions. Specifically, D. Stauffer's pertinent criticism and many suggestions signficantly improved this work. S. Hügle designed part of the graphics. The University of Freiburg gave me the opportunity to elaborate this course during a visiting professorship. My students there contributed much critical feedback. Apart from the year in Freiburg, I am a Heisenberg fellow of Deutsche Forschungsgemeinschaft and based at Bayreuth University. The final correction were done during a sabbatical at Science & Finance, the research division of Capital Fund Management, Levallois (France), and I would like to thank the company for its hospitality. I also would like to thank the staff of Springer-Verlag for all the work they invested on the way from my typo-congested LATEX files to this first edition of the book.

However, without the continuous support, understanding, and encouragement of my wife Jinping Shen and our daughter Jiayi, this work would not have got its present shape. I thank them all.

Bayreuth,
August 2000   *Johannes Voit*

Contents

1. Introduction

1.1 Motivation

The public interest in traded securities has continuously grown over the past few years, with an especially strong growth in Germany and other European countries at the end of the 1990s. Consequently, events influencing stock prices, opinions and speculations on such events and their consequences, and even the daily stock quotes, receive much attention and media coverage. A few reasons for this interest are clearly visible in Fig. 1.1 which shows the evolution of the German stock index DAX [1] over the two years from October 1996 to October 1998. Other major stock indices, such as the US Dow Jones Industrial Average, the S&P500, or the French CAC40, etc., behaved in a similar manner in that interval of time. We notice three important features: (i) the continuous rise of the index over the first almost one and a half years which

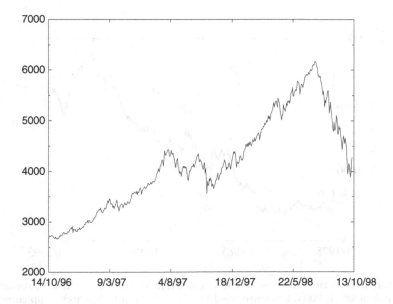

Fig. 1.1. Evolution of the DAX German stock index from October 14, 1996 to October 13, 1998. Data provided by Deutsche Bank Research

was interrupted only for very short periods; (ii) the crash on the "second black Monday", October 27, 1997 (the "Asian crisis", the reaction of stock markets to the collapse of a bank in Japan, preceded by rumors about huge amounts of foul credits and derivative exposures of Japanese banks, and a period of devaluation of Asian currencies). (iii) the very strong drawdown of quotes between July and October 1998 (the "Russian debt crisis", following the announcement by Russia of a moratorium on its debt reimbursements, and a devaluation of the Russian rouble), and the collapse of the Long Term Capital Management hedge fund.

While the long-term rise of the index until 2000 seemed to offer investors attractive, high-return opportunities for making money, enormous fortunes of billions or trillions of dollars were annihilated in very short times, perhaps less than a day, in crashes or periods of extended drawdowns. Such events – the catastrophic crashes perhaps more than the long-term rise – exercise a strong fascination.

To place these events in a broader context, Fig. 1.2 shows the evolution of the DAX index from 1975 to 2005. Several different regimes can be distinguished. In the initial period 1975–1983, the returns on stock investments were extremely low, about 2.6% per year. Returns of 200 DAX points, or 12%, per year were generated in the second period 1983–1996. After 1996, we see a marked acceleration with growth rates of 1200 DAX points, or 33%, per year. We also notice that, during the growth periods of the stock market, the losses incurred in a sudden crash usually persist only over a short

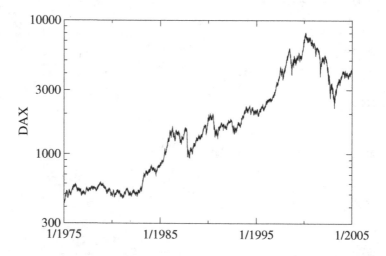

Fig. 1.2. Long-term evolution of the DAX German stock index from January 1, 1975 to January 1, 2005. Data provided by Deutsche Bank Research supplemented by data downloaded from Yahoo, http://de.finance.yahoo.com

time, e.g. a few days after the Asian crash [(ii) above], or about a year after the Russian debt crisis [(iii) above]. The long term growth came to an end, around April 2000 when markets started sliding down. The fourth period in Fig. 1.2 from April 2000 to the end of the time series on March 12, 2003, is characterized by a long-term downward trend with losses of approximately 1400 DAX points, or 20% per year. The DAX even fell through its long-term upward trend established since 1983. Despite the overall downward trend of the market in this period, it recovered as quickly from the crash on September 11, 2001, as it did after crashes during upward trending periods. Finally, the index more or less steadily rose from its low at 2203 points on March 12, 2003 to about 4250 points at the end of 2004. Only the future will show if a new growth period has been kicked off.

This immediately leads us to a few questions:

- Is it possible to earn money not only during the long-term upward moves (that appears rather trivial but in fact is not) *but also during the drawdown periods?* These are questions for investors or speculators.
- What are the factors responsible for long- and short-term price changes of financial assets? How do these factors depend on the type of asset, on the investment horizon, on policy, etc.?
- How do the three growth periods of the DAX index, discussed in the preceding paragraph, correlate with economic factors? These are questions for economists, analysts, advisors to politicians, and the research departments of investment banks.
- What statistical laws do the price changes obey? How smooth are the changes? How frequent are jumps? These problems are treated by mathematicians, econometrists, but more recently also by physicists. The answer to this seemingly technical problem is of great relevance, however, also to investors and portfolio managers, as the efficiency of stop-loss or stop-buy orders [2] directly depends on it.
- How big is the risk associated with an investment? Can this be measured, controlled, limited or even eliminated? At what cost? Are reliable strategies available for that purpose? How big is any residual risk? This is of interest to banks, investors, insurance companies, firms, etc.
- How much fortune is at risk with what probability in an investment into a specific security at a given time?
- What price changes does the evolution of a stock price, resp. an index, imply for "financial instruments" (derivatives, to be explained below, cf. Sect. 2.3)? This is important both for investors but also for the writing bank, and for companies using such derivatives either for increasing their returns or for hedging (insurance) purposes.
- Can price changes be predicted? Can crashes be predicted?

1.2 Why Physicists? Why Models of Physics?

This book is about financial markets from a physicist's point of view. Statistical physics describes the complex behavior observed in many physical systems in terms of their simple basic constituents and simple interaction laws. Complexity arises from interaction and disorder, from the cooperation and competition of the basic units. Financial markets certainly are complex systems, judged both by their output (cf., e.g., Fig. 1.1) and their structure. Millions of investors frequent the many different markets organized by exchanges for stocks, bonds, commodities, etc. Investment decisions change the prices of the traded assets, and these price changes influence decisions in turn, while almost every trade is recorded.

When attempting to draw parallels between statistical physics and financial markets, an important source of concern is the complexity of human behavior which is at the origin of the individual trades. Notice, however, that nowadays a significant fraction of the trading on many markets is performed by computer programs, and no longer by human operators. Furthermore, if we make abstraction of the trading volume, an operator only has the possibility to buy or to sell, or to stay out of the market. Parallels to the Ising or Potts models of Statistical Physics resurface!

More specifically, take the example of Fig. 1.1. If we subtract out long-term trends, we are left essentially with some kind of random walk. In other words, the evolution of the DAX index looks like a random walk to which is superposed a slow drift. This idea is also illustrated in the following story taken from the popular book "A Random Walk down Wall Street" by B. G. Malkiel [3], a professor of economics at Princeton. He asked his students to derive a chart from coin tossing.

> "For each successive trading day, the closing price would be determined by the flip of a fair coin. If the toss was a head, the students assumed the stock closed 1/2 point higher than the preceding close. If the flip was a tail, the price was assumed to be down 1/2. ... The chart derived from the random coin tossing looks remarkably like a normal stock price chart and even appears to display cycles. Of course, the pronounced 'cycles' that we seem to observe in coin tossings do not occur at regular intervals as true cycles do, but neither do the ups and downs in the stock market. In other simulated stock charts derived through student coin tossings, there were head-and-shoulders formations, triple tops and bottoms, and other more esoteric chart patterns. One of the charts showed a beautiful upward breakout from an inverted head and shoulders (a very bullish formation). I showed it to a chartist friend of mine who practically jumped out of his skin. "What is this company?" he exclaimed. "We've got to buy immediately. This pattern's a classic. There's no question the stock will be up 15 points next week." He did not respond kindly to me when I told him the chart had been produced by flipping a coin." Reprinted from B. G. Malkiel: *A Random Walk down Wall Street*, ©1999 W. W. Norton

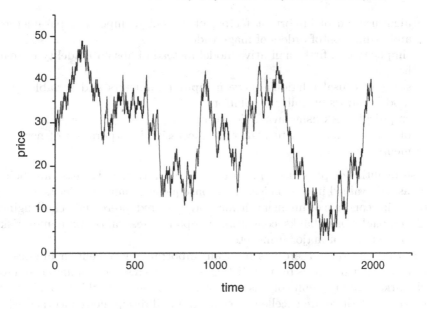

Fig. 1.3. Computer simulation of a stock price chart as a random walk

The result of a computer simulation performed according to this recipe, is shown in Fig. 1.3, and the reader may compare it to the DAX evolution shown in Fig. 1.1. "THE random walk", usually describing Brownian motion, but more generally any kind of stochastic process, is well known in physics; so well known in fact that most people believe that its first mathematical description was achieved in physics, by A. Einstein [4].

It is therefore legitimate to ask if the description of stock prices and other economic time series, and our ideas about the underlying mechanisms, can be improved by

- the understanding of parallels to phenomena in nature, such as, e.g.,
    - diffusion
    - driven systems
    - nonlinear dynamics, chaos
    - formation of avalanches
    - earthquakes
    - phase transitions
    - turbulent flows
    - stochastic systems
    - highly excited nuclei
    - electronic glasses, etc.;
- the associated mathematical methods developed for these problems;
- the modeling of phenomena which is a distinguished quality of physics. This is characterized by

- identification of important factors of causality, important parameters, and estimation of orders of magnitude;
- simplicity of a first qualitative model instead of absolute fidelity to reality;
- study of causal relations between input parameters and variables of a model, and its output, i.e. solutions;
- empirical check using available data;
- progressive approach to reality by successive incorporation of new elements.

These qualities of physicists, in particular theoretical physicists, are being increasingly valued in economics. As a consequence, many physicists with an interest in economic or financial themes have secured interesting, challenging, and well-paid jobs in banks, consulting companies, insurance companies, risk-control divisions of major firms, etc.

Rather naturally, there has been an important movement in physics to apply methods and ideas from statistical physics to research on financial data and markets. Many results of this endeavor are discussed in this book. Notice, however, that there are excellent specialists in all disciplines concerned with economic or financial data, who master the important methods and tools better than a physicist newcomer does. There are examples where physicists have simply rediscovered what has been known in finance for a long time. I will mention those which I am aware of, in the appropriate context. As an example, even computer simulations of "microscopic" interacting-agent models of financial markets have been performed by economists as early as 1964 [5]. There may be many others, however, which are not known to me. I therefore call for modesty (the author included) when physicists enter into new domains of research outside the traditional realm of their discipline. This being said, there is a long line of interaction and cross-fertilization between physics and economy and finance.

1.3 Physics and Finance – Historical

The contact of physicists with finance is as old as both fields. Isaac Newton lost much of his fortune in the bursting of the speculative bubble of the South Sea boom in London, and complained that while he could precisely compute the path of celestial bodies to the minute and the centimeter, he was unable to predict how high or low a crazy crowd could drive the stock quotations.

Carl Friedrich Gauss (1777–1855), who is honored on the German 10 DM bill (Fig. 1.4), has been very successful in financial operations. This is evidenced by his leaving a fortune of 170,000 Taler (contemporary, local currency unit) on his death while his basic salary was 1000 Taler. According to rumors, he derived the normal (Gaussian) distribution of probabilities in

Fig. 1.4. Carl Friedrich Gauss on the German 10 DM bill (detail), courtesy of Deutsche Bundesbank

estimating the default risk when giving credits to his neighbors. However, I have failed to find written documentation of this fact.

His calculation of the pensions for widows of the professors of the University of Göttingen (1845–1851) is a seminal application of probability theory to the related field of insurance. The University of Göttingen, where Gauss was professor, had a fund for the widows of the professors. Its administrators felt threatened by ruin as both the number of widows, as well as the pensions paid, increased during those years. Gauss was asked to evaluate the state of the fund, and to recommend actions to save it. After six years of analysis of mortality tables, historical data, and elaborate calculations, he concluded that the fund was in excellent financial health, that a further increase of the pensions was possible, but that the membership should be restricted. Quite contrary to the present public discussion!

The most important date in the perspective of this book is March 29, 1900 when the French mathematician Louis Bachelier defended his thesis entitled "Théorie de la Spéculation" at the Sorbonne, University of Paris [6]. In his thesis, he developed, essentially correctly and comprehensively, the theory of the random walk – and that five years before Einstein. He constructed a model for exchange quotes, specifically for French government bonds, and estimated the chances of success in speculation with derivatives that are somewhat in between futures and options, on those bonds. He also performed empirical studies to check the validity of his theory. His contribution had been forgotten for at least 60 years, and was rediscovered independently in the financial community in the late 1950s [7, 8]. Physics is becoming aware of Bachelier's important work only now through the interface of statistical physics and quantitative finance.

More modern examples of physicists venturing into finance include M. F. M. Osborne who rediscovered the Brownian motion of stock markets in 1959 [7, 8], and Fisher Black who, together with Myron Scholes, reduced an option pricing problem to a diffusion equation. Osborne's seminal work was first presented in the Solid State Physics seminar of the US Naval Research Laboratory before its publication. Black's work will be discussed in detail in Chap. 4.

1.4 Aims of this Book

This book is based on courses on models of physics for financial markets ("Physikalische Modelle in der Finanzwirtschaft") which I have given at the Universities of Bayreuth, Freiburg, and Ulm, and at Academia Sinica, Taipei. It largely keeps the structure of the course, and the subject choice reflects both my taste and that of my students.

I will discuss models of physics which have become established in finance, or which have been developed there even before (!) being introduced in physics, cf. Chap. 3. In doing so, I will present both the physical phenomena and problems, as well as the financial issues. As the majority of attendees of the courses were physicists, the emphasis will be more on the second, the financial aspects. Here, I will present with approximately equal weight established theories as well as new, speculative ideas. The latter often have not received critical evaluation yet, in some cases are not even officially published and are taken from preprint servers [9]. Readers should be aware of the speculative character of such papers.

Models for financial markets often employ strong simplifications, i.e. treat idealized markets. This is what makes the models possible, in the first instance. On the other hand, there is no simple way to achieve above-average profits in such idealized markets ("there is no free lunch"). The aim of the course therefore is NOT to give recipes for quick or easy profits in financial markets. On the same token, we do not discuss investment strategies, if such should exist. Keeping in line with the course, I will attempt an overview only of the *most basic aspects* of financial markets and financial instruments. There is excellent literature in finance going much further, though away from statistical physics [10]–[16]. Hopefully, I can stimulate the reader's interest in some of these questions, and in further study of these books.

The following is a list of important issues which I will discuss in the book:

- Statistical properties of financial data. Distribution functions for fluctuations of stock quotes, etc. (stocks, bonds, currencies, derivatives).
- Correlations in financial data.
- Pricing of derivatives (options, futures, forwards).
- Risk evaluation for market positions, risk control using derivatives (hedging).

- Hedging strategies.
- Can financial data be used to obtain information on the markets?
- Is it possible to predict (perhaps in probabilistic terms) the future market evolution? Can we formulate equations of motion?
- Description of stock exchange crashes. Are predictions possible? Are there typical precursor signals?
- Is the origin of the price fluctuations exogenous or endogenous (i.e. reaction to external events or caused by the trading activity itself)?
- Is it possible to perform "controlled experiments" through computer simulation of microscopic market models?
- To what extent do operators in financial markets behave rationally?
- Can game-theoretic approaches contribute to the understanding of market mechanisms?
- Do speculative bubbles (uncontrolled deviations of prices away from "fundamental data", ending typically in a collapse) exist?
- The definition and measurment of risk.
- Basic considerations and tools in risk management.
- Economic capital requirements for banks, and the capital determination framework applied by banking supervisors.

The organization of this book is as follows. The next chapter introduces basic terminology for the novice, defines and describes the three simplest and most important derivatives (forwards, futures, options) to be discussed in more detail throughout this book. It also introduces the three types of market actors (speculators, hedgers, arbritrageurs), and explains the mechanisms of price formation at an organized exchange.

Chapter 3 discusses in some detail Bachelier's derivation of the random walk from a financial perspective. Though no longer state of the art, many aspects of Bachelier's work are still at the basis of the theories of financial markets, and they will be introduced here. We contrast Bachelier's work with Einstein's theory of Brownian motion, and give some empirical evidence for Brownian motion in stock markets and in nature.

Chapter 4 discusses the pricing of derivatives. We determine prices of forward and futures contracts and limits on the prices of simple call and put options. More accurate option prices require a model for the price variations of the underlying stock. The standard model is provided by geometric Brownian motion where the logarithm of a stock price executes a random walk. Within this model, we derive the seminal option pricing formula of Black, Merton, and Scholes which has been instrumental for the explosive growth of organized option trading. We also measures of the sensitivity of option prices with respect to the basic variables of the model ("The Greeks"), options with early-exercise features, and volatility indices for financial markets.

Chapter 5 discusses the empirical evidence for or against the assumptions of geometric Brownian motion: price changes of financial assets are uncorrelated in time and are drawn from a normal distribution. While the first assumption is rather well satisfied, deviations from a normal distribution will lead us to consider in more depth another class of stochastic process, stable Lévy processes, and variants thereof, whose probability distribution functions possess fat tails and which describe financial data much better than a normal distribution. Here, we also discuss the implications of these fat-tailed distributions both for our understanding of capital markets, and for practical investments and risk management. Correlations are shown to be an important feature of financial markets. We describe temporal correlations of financial time series, asset–asset correlations in financial markets, and simple models for markets with correlated assets.

An interesting analogy has been drawn recently between hydrodynamic turbulence and the dynamics of foreign exchange markets. This will be discussed in more depth in Chap. 6. We give a very elementary introduction to turbulence, and then work out the parallels to financial time series. This line of work is still controversial today. Multifractal random walks provide a closely related framework, and are discussed.

Once the significant differences between the standard model – geometric Brownian motion – and real financial time series have been described, we can carry on to develop improved methods for pricing and hedging derivatives. This is described in Chap. refchap:risk. An important step is the passage from the differential Black–Scholes world to an integral representation of the life scenarios of an option. Consequently, aside numerical procedures, path integrals which are well-known in physics, are shown to be important tools for option valuation in more realistic situations.

Chapter 8 gives a brief overview of computer simulations of microscopic models for organized markets and exchanges. Such models are of particular importance because, unlike physics, controlled experiments establishing cause–effect relationships are not possible on financial markets. On the other hand, there is evidence that the basic hypotheses underlying standard financial theory may be questionable. One way to check such hypotheses is to formulate a model of interacting agents, operating on a given market under a given set of rules. The model is then "solved" by computer simulations. A criterion for a "good" model is the overlap of the results, e.g., on price changes, correlations, etc., with the equivalent data of real markets. Changing the rules, or some other parameters, allows one to correlate the results with the input and may result in an improved understanding of the real market action.

In Chap. 9 we review work on the description of stock market crashes. We emphasize parallels with natural phenomena such as earthquakes, material failure, or phase transitions, and discuss evidence for and against the hyptothesis that such crashes are outliers from the statistics of "normal" price fluctuations in the stock market. If true, it is worth searching for characteristic patterns preceding market crashes. Such patterns have apparently been found in historical crashes and, most remarkably, have allowed the prediction of the Asian crisis crash of October 27, 1997, but also of milder events such as a reversal of the downward trend of the Japanese Nikkei stock index, in early 1999. On the other hand, bearish trend reversals predicted in many major stock indices for the year 2004 have failed to materialize. We discuss the controversial status of crash predictions but also the improved understanding of what may happen before and after major financial crashes.

Chapters 10 and 11 leave the focus of statistical physics and turn towards banking practice. This appears important because many job opportunities requiring strong quantitative qualifications have been (and continue to be) created in banks. On the other hand, both the basic practices and the hot topics of banking, regrettably, are left out of most presentation for physics audiences. Chapter 10 is concerned with risk management. We define risk and discuss various measures of risk. We classify various types of risk and discuss the basic tools of risk management.

Chapter 11 finally discusses capital requirements for banks. Capital is taken as a cushion against losses which a bank may suffer in the markets, and therefore is an important quantity to manage risk and performance. The first part of the chapter discusses economic capital, i.e. what a bank has to do under purely economic considerations. Regulatory authorities apply a different framework to the banks they supervise. This is explained in the second part of Chap. 11. The new Basel Capital Accord (Basel II) takes a significant fraction of space. On the one hand, it will set the regulatory capital and risk management standards for the decades to come, in many countries of the world. On the other hand, it is responsible for many of the employment opportunities which may be open to the readers.

There are excellent introductions to this field with somewhat different or more specialized emphasis. Bouchaud and Potters have published a book which emphasizes derivative pricing [17]. The book by Mantegna and Stanley describes the scaling properties of and correlations in financial data [18]. Roehner has written a book with emphasis on empirical investigations which include financial markets but cover a significantly vaster field of economics [19]. Another book presents computer simulation of "microscopic" market models [20]. The analysis of financial crashes has been reviewed in a book by one of its main protagonists [21]. Mandelbrot also published a volume summarizing his contributions to fractal and scaling behavior in financial time series [22]. The important work of Olsen & Associates, a Zurich-based company working on trading models and prediction of financial time series, is summarized in *High Frequency Finance* [23]. The application of stochastic processes and path integrals, respectively, to problems of finance is briefly discussed in two physics books [24, 25] whose emphasis, though, is on phyiscal methods and applications. Finally, there has been a series of conferences and workshops whose proceedings give an overview of the state of this rapidly evolving field of research at the time of the event [26]. More sources of information are listed in the Appendix.

2. Basic Information on Capital Markets

2.1 Risk

Risk and profit are the important drivers of financial markets. Briefly, risk is defined as deviation of the actual outcome of an investment from its expected outcome when this deviation is negative. An alternative definition would view risk as the negative changes of a future position with respect to the present position. The difference does not matter much until we define quantitative risk measures in Chap. 10.3. Taking risk, reducing risk, and managing risk are important motivations for many operations in financial markets.

An investor taking risk will expect a certain return as compensation, the more so the higher the risk. Risky assets therefore also possess, at least on the average, high expected growth rates. Investments in risky stocks should be rewarded by a high rate of growth of their price. Investments in risky bonds should be rewarded by a high interest coupon.

Almost all investments are risky. There are very few instances which, to a good approximation, can be considered riskless. An investment in US treasury notes and bonds is considered a riskless investement because there is no doubt that the US treasury will honor its payment obligation. The same applies to bonds emitted by a number of other states and a few corporations (the so-called "AAA-rated" states and corporations). The interest rate paid on these bonds is called the riskless interest rate r, and will play an important role in many theoretical arguments in our later discussion. Interest rates change with time, though, both nominally and effectively. The rate r paid on two otherwise identical bonds emitted at different dates may be different. And the effective return of a traded bond bought or sold at times between emission and maturity fluctuates as a result of trading. In line with neglecting this interest rate risk, we will assume the risk-free interest rate r to be constant over the time scale considered.

2.2 Assets

What are the objects we are concerned with in this book? Let us start by looking into the portfolio of assets of a bank, or into the financial pages of a major newspaper. The bank portfolio may contain stocks, bonds, currencies, commodities, (private) equity, real estate, loans, mutual funds, hedge funds, etc., and derivatives, such as futures, options, or warrants.

The financial pages of the major newspapers contain the quotations of the most important traded assets of this portfolio. In addition, they contain quotations of market indices. Indices measure the composite performance of national markets, industries, or market segments. Examples include (i) for stock markets the Dow Jones Industrial Average, S&P500, DAX, DAX 100, CAC 40, etc., for blue chip stocks in the US, Germany, and France, respectively, (ii) the NASDAQ or TECDAX indices measuring the US and German high-technology markets, (iii) the Dow Jones Stoxx 50 index measuring the performance of European blue chip stocks irrespective of countries, or their participation in the European currency system. (iv) Indices are also used for bond markets, e.g., the REX index in Germany, but bond markets are also characterized by the prices and returns of certain benchmark products [11].

There are several ways to classify these assets. Usually, the assets held by a bank are organized in different groups, called "books". A "trading book" contains the assets held for trading purposes, normally for a rather short time. A simple trading book may contain stocks, bonds, currencies, commodities, and derivatives. The "banking book" contains assets held for longer periods of time, and mostly for business motivations. Assets of the banking book often are loans, mortgage backed loans, real estate, private equity, stocks, etc.

Some assets are *securities*. Securities are normally traded on organized markets (in some cases *over the counter*, OTC, i.e. directly between a bank and its client) and include stock, bonds, currencies, and derivatives. Their prices are fixed by demand and supply in the trading process. The following assets in the bank portfolio are not securities: commodities, equity unless it is in stocks, real estate, loans. Prices of traded securities usually are available as time series with a reasonably high frequency. Market indices are not securities although investments products replicating market indices are securities, often with a hidden derivative element. On the statistical side, very good time series are available for market indices, as illustrated by Figs. 1.1 and 1.2, and many to follow. Good price histories are available, too, for commodities.

Mutual funds, hedge funds, etc., are portfolios of securities. A portfolio in an ensemble of securities held by an investor. Their price is fixed by trading their individual components. We shall explicitly consider portfolios of securities in Chap. 10 where we show that the return of such a portfolio can be maximized at given risk by buying the securities is specific quantities which can be calculated.

A special class of securities merits a general name and discussion of its own. A *derivative* (also derivative security, contingent claim) is a financial instrument whose value depends on other, more basic underlying variables [10, 12, 13]. Very often, these variables are the prices of other securities (such as stocks, bonds, currencies, which are then called "underlying securities" or, for short, just "the underlying") with, of course, a series of additional parameters involved in determining the precise dependence. There are also derivatives on commodities (oil, wheat, sugar, pork bellies [!], gold, etc.), on market indices (cf. above), on the volatility of markets and also on phenomena apparently exterior to markets such as weather. As indicated by the examples of commodities and market indices, the emission of a derivative on these assets produces an "artificial" security. Especially in the case of commodities and markets indicies, the existence of derivatives considerably facilitates investment in these assets. Recently, the related transformation of portfolios of loans into tradable securities, known as securitization, has become an important practice in banking.

Derivatives are traded either on organized exchanges, such as Deutsche Terminbörse, DTB, which has evolved into EUREX by fusion with its Swiss counterpart, the Chicago Board of Trade (CBOT), the Chicago Board Options Exchange (CBOE), the Chicago Mercantile Exchange (CME), etc., or *over the counter* (OTC). Derivatives traded on exchanges are standardized products, while over the counter trading is done directly between a financial institution and a customer, often a corporate client or another financial institution, and therefore allows the tailoring of products to the individual needs of the clients.

Here, we mostly focus on stocks, market indices, and currencies, and their respective derivatives. We do this for two main reasons: (i) much of the research, especially by physicists, has concentrated on these assets; (ii) they are conceptually simpler than, e.g., bonds and therefore more suited to explain the basic mechanisms. Bond prices are influenced by interest rates. The interest rates, however, depend on the maturity of the bond, and the time to maturity therefore introduces an additional variable into the problem. Notice, however, that bond markets typically are much bigger than stock markets. Institutional investors such as insurance companies invest large volumes of money on the bond market because there they face less risk than with investments in, e.g., stocks.

2.3 Three Important Derivatives

Here, we briefly discuss the three simplest derivatives on the market: forward and futures contracts, and call and put options. They are sufficient to illustrate the basic principles of operation, pricing, and hedging. Many more instruments have been and continue to be created. Pricing such instruments, and using them for speculative or hedging purposes may present formidable technical challenges. They rely, however, on the same fundamental principles which we discuss in the remainder of this book where we refer to the three basic derivatives described below. Readers interested in those more complex instruments, are referred to the financial literature [10]–[15].

2.3.1 Forward Contracts

A forward contract (or just: forward for short) is a contract between two parties (usually two financial institutions or a financial institution and a corporate client) on the delivery of an asset at a certain time in the future, the maturity of the contract, at a certain price. This delivery price is fixed at the time the contract is entered.

Forward contracts are not usually traded on exchanges but rather over the counter (OTC), i.e. between a financial institution and its counterparty. For both parties, there is an obligation to honor the contract, i.e., to deliver/pay the asset at maturity.

As an example, consider a US company who must pay a bill of 1 million pound sterling three months from now. The amount of dollars the company has to pay obviously depends on the dollar/sterling exchange rate, and its evolution over the next three months therefore presents a risk for the company. The company can now enter a forward over 1 million pounds with maturity three months from now, with its bank. This will fix the exchange rate for the company as soon as the forward contract is entered. This rate may differ from the spot rate (i.e., the present day rate for immediate delivery), and include the opinion of the bank and/or market on its future evolution (e.g., spot 1.6080, 30-day forward 1.6076, 90-day forward 1.6056, 180-day forward 1.6018, quoted from Hull [10] as of May 8, 1995) but will effectively fix the rate for the company three months from now to 1.6056 US$/£.

2.3.2 Futures Contract

A *futures contract* (futures) is rather similar to a forward, involving the delivery of an asset at a fixed time in the future (maturity) at a fixed price. However, it is standardized and traded on exchanges. There are also differences relating to details of the trading procedures which we shall not explore here [10]. For the purpose of our discussion, we shall not distinguish between forward and futures contracts.

The above example, involving popular currencies in standard quantities, is such that it could as well apply to a futures contract. The differences are perhaps more transparent with a hypothetical example of buying a car. If a customer would like to order a BMW car in yellow with pink spots, there might be 6 months delivery time, and the contract will be established in a way that assures delivery and payment of the product at the time of maturity. Normally, there will be no way out if, during the six months, the customer changes his preferences for the car of another company. This corresponds to the forward situation. If instead one orders a black BMW, and changes opinion before delivery, for a Mercedes-Benz, one can try to resell the contract on the market (car dealers might even assist with the sale) because the product is sufficiently standardized so that other people are also interested in, and might enter the contract.

2.3.3 Options

Options may be written on any kind of underlying assets, such as stocks, bonds, commodities, futures, many indices measuring entire markets, etc. Unlike forwards or futures which carry an *obligation* for both parties, options give their holder the *right* to buy or sell an underlying assets in the future at a fixed price. However, they imply an obligation for the writer of the option to deliver or buy the underlying asset.

There are two basic types of options: *call options* (calls) which give the holder the right to buy, and *put options* (puts) which give their holder the right to sell the underlying asset in the future at a specified price, the strike price of the option. Conversely, the writer has the obligation to sell (call) or buy (put) the asset. Options are distinguished as being of European type if the right to buy or sell can only be exercised at their date of maturity, or of American type if they can be exercised at any time from now until their date of maturity. Options are traded regularly on exchanges.

Notice that, for the holder, there is no obligation to exercise the options while the writer has an obligation. As a consequence of this asymmetry, there is an intrinsic cost (similar to an insurance premium) associated with the option which the holder has to pay to the writer. This is different from forwards and futures which carry an obligation for both parties, and where there is no intrinsic cost associated with these contracts.

Options can therefore be considered as insurance contracts. Just consider your car insurance. With some caveats concerning details, your insurance contract can be reinterpreted as a put option you bought from the insurance company. In the case of an accident, you may sell your car to the insurance company at a predetermined price, resp. a price calculated according to a predetermined formula. The actual value of your car after the accident is significantly lower than its value before, and you will address the insurance for compensation. Your contract protects your investment in your car against unexpected losses. Precisely the same is achieved by a put option on a capital market. Reciprocally, a call option protects its owner against unexpected rises of prices. As in our example, with real options on exercise, one often does not deliver the product (which is possible in simple cases but impossible, e.g., in the case of index options), but rather settles the difference in cash.

As another example, consider buying 100 European call options on a stock with a strike price (for exercise) of $X = $ DM 100 when the spot price for the stock is $S_t = $ DM 98. Suppose the time to maturity to be $T - t = 2m$.

- If at maturity T, the spot price $S_T < $ DM 100, the options expire worthless (it makes no sense to buy the stock more expensively through the options than on the spot market).
- If, however, $S_T > $ DM 100, the option should be exercised. Assume $S_T = $ DM 115. The price gain per stock is then DM 15, i.e., DM 1500 for the entire investment. However, the net profit will be diminished by the price

of the call option C. With a price of $C = $ DM 5, the total profit will be DM 1000.

- The option should be exercised also for DM 100 $< S_T <$ DM 105. While there is a net loss from the operation, it will be inferior to the one incurred $(-100\ C)$ if the options had expired.

The profile of profit, for the holder, versus stock price at maturity is given in Fig. 2.1. The solid line corresponds to the call option just discussed, while the dashed line shows the equivalent profile for a put.

When buying a call, one speculates on rising stock prices, resp. insures against rising prices (e.g., when considering future investments), while the holder of a put option speculates on, resp. insures, against falling prices.

For the holder, there is the possibility of unlimited gain, but losses are strictly limited to the price of the option. This asymmetry is the reason for the intrinsic price of the options. Notice, however, that in terms of practical, speculative investments, the limitation of losses to the option price still implies a total loss of the invested capital. It only excludes losses higher than the amount of money invested!

There are many more types of options on the markets. Focusing on the most elementary concepts, we will not discuss them here, and instead refer the readers to the financial literature [10]–[15]. However, it appears that much applied research in finance is concerned with the valuation of, and risk management involving, exotic options.

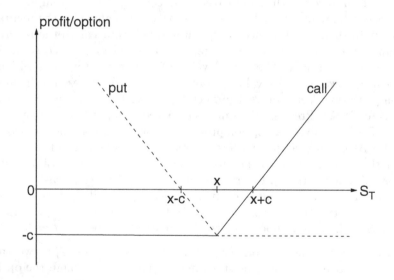

Fig. 2.1. Profit profile of call (*solid line*) and put (*dashed line*) options. S_T is the price of the underlying stock at maturity, X the strike price of the option, and C the price of the call or put

2.4 Derivative Positions

In every contract involving a derivative, one of the parties assumes the *long position,* and agrees to buy the underlying asset at maturity in case of a forward or futures contract, or, as the holder of a call/put option, has the right to buy/sell the underlying asset if the option is exercised. His partner assumes the *short position,* i.e., agrees to deliver the asset at maturity in a forward or futures or if a call option is exercised, resp. agrees to buy the underlying asset if a put option is exercised.

In the example on currency exchange rates in Sect. 2.3.1, the company took the long position in a forward contract on 1 million pounds sterling, while its bank went short. If the acquisition of a new car was considered as a forward or futures contract, the future buyer took the long position and the manufacturer took the short position.

With options, of course, one can go long or short in a call option, and in put options. The discussion of options in Sect. 2.3.3 above always assumed the long position. Observe that the profit profile for the writer of an option, i.e., the partner going short, is the inverse of Fig. 2.1 and is shown in Fig. 2.2. The possibilities for gains are limited while there is an unlimited potential for losses. This means that more money than invested may be lost due to the liabilities accepted on writing the contract.

Short selling designates the sale of assets which are not owned. Often there is no clear distinction from "going short". In practice, short selling is possible quite generally for institutional investors but only in very limited circumstances for individuals. The securities or derivatives sold short are

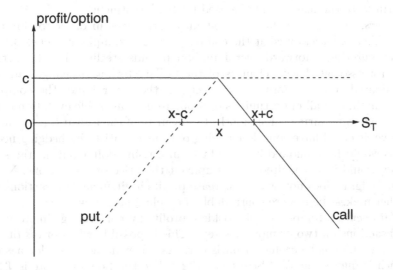

Fig. 2.2. Profit profile of call (*solid line*) and put (*dashed line*) options for the writer of the option (short position)

taken "on credit" from a broker. The hope is, of course, that their quotes will rise in the near future by an appreciable amount. We shall use short selling mainly for theoretical arguments.

Closing out an open position is done by entering a contract with a third party that exactly cancels the effect of the first contract. In the case of publicly traded securities, it can also mean selling (buying) a derivative or security one previously owned (sold short).

2.5 Market Actors

We distinguish three basic types of actors on financial markets.

- *Speculators* take risks to make money. Basically, they bet that markets will make certain moves. Derivatives can give extra leverage to speculation with respect to an investment in the underlying security. Reconsider the example of Sect. 2.3.3, involving 100 call options with $X = \text{DM } 100$ and $S_t = \text{DM } 98$. If indeed, after two months, $S_T = \text{DM } 115$, the profit of DM 1000 was realized with an investment of $100 \times C = \text{DM } 500$, i.e., amounts to a return of 200% in two months. Working with the underlying security, one would realize a profit of $100 \times (S_T - S_t) = \text{DM } 1700$ but on an investment of DM 9,800, i.e., achieve a return of "only" 17.34%. On the other hand, the risk of losses on derivatives is considerably higher than on stocks or bonds (imagine the stock price to stay at $S_T = \text{DM } 98$ at maturity). Moreover, even with simple derivatives, a speculator places a bet not only on the direction of a market move, but also that this move will occur before the maturity of the instruments he used for his investment.

- *Hedgers,* on the other hand, invest into derivatives in order to eliminate risk. This is basically what the company in the example of Sect. 2.3.1 did when entering a forward over 1 million pounds sterling. By this action, all risk associated with changes of the dollar/sterling exchange rate was eliminated. Using a forward contract, on the other hand, the company also eliminated all opportunities of profit from a favorable evolution of the exchange rate during three months to maturity of the forward. As an alternative, it could have considered using options to satisfy its hedging needs. This would have allowed it to profit from a rising dollar but, at the same time, would have required to pay upfront the price of the options. Notice that hedging does not usually increase profits in financial transactions but rather makes them more controllable, i.e., eliminates risk.

- *Arbitrageurs* attempt to make riskless profits by performing simultaneous transactions on two or more markets. This is possible when prices on two different markets become inconsistent. As an example, consider a stock which is quoted on Wall Street at $172, while the London quote is £100. Assume that the exchange rate is 1.75 $/£. One can therefore make a riskless profit by simultaneously buying N stocks in New York and selling the same amount, or go short in N stocks, in London. The profit is $3N$. Such arbitrage opportunities cannot last for long. The very action of this arbitrageur will make the price move up in New York and down in London, so that the profit from a subsequent transaction will be significantly lower. With today's computerized trading, arbitrage opportunities of this kind only last very briefly, while triangular arbitrage, involving, e.g., the European, American, and Asian markets, may be possible on time scales of 15 minutes, or so.

Arbitrage is also possible on two national markets, involving, e.g., a futures market and the stock market, or options and stocks. Arbitrage therefore makes different markets mutually consistent. It ensures "market efficiency", which means that all available information is accounted for in the current price of a security, up to inconsistencies smaller than applicable transaction costs.

The absence of arbitrage opportunities is also an important theoretical tool which we will use repeatedly in subsequent chapters. It will allow a consistent calculation of prices of derivatives based on the prices of the underlying securities. Notice, however, that while satisfied in practice on liquid markets in standard circumstances, it is, in the first place, an assumption which should be checked when modeling, e.g., illiquid markets or exceptional situations such as crashes.

2.6 Price Formation at Organized Exchanges

Prices at an exchange are determined by supply and demand. The procedures differ slightly according to whether we consider an auction or continuous trading, and whether we consider a computerized exchange, or traders in a pit.

Throughout this book, we assume a single price for assets, except when stated otherwise explicitly. This is a simplification. For assets traded at an exchange, prices are quoted as bid and ask prices. The bid price is the price at which a trader is willing to buy; the ask price in turn is the price at which he is willing to sell. Depending on the liquidity of the market, the bid–ask spread may be negligible or sizable.

2.6.1 Order Types

Besides the volume of a specific stock, buy and sell orders may contain additional restrictions, the most basic of which we now explain. They allow the investor to specify the particular circumstances under which his or her order must be executed.

A *market order* does not carry additional specifications. The asset is bought or sold at the market price, and is executed once a matching order arrives. However, market prices may move in the time between the decision of the investor and the order execution at the exchange. A market order does not contain any protection against price movements, and therefore is also called an unlimited order.

Limit orders are executed only when the market price is above or below a certain threshold set by the investor. For a buy (sell) order to limit S_L, the order is executed only when the market price is such that the order can be excecuted at $S \leq S_L$ ($S \geq S_L$). Otherwise, the order is kept in the order book of the exchange until such an opportunity arises, or until expiry. A sell order with limit S_L guarantees the investor a minimum price S_L in the sale of his assets. A limited buy order, vice versa, guarantees a maximal price for the purchase of the assets.

Stop orders are unlimited orders triggered by the market price reaching a predetermined threshold. A stop-loss (stop-buy) order issues an unlimited sell (buy) order to the exchange once the asset price falls below S_L. Stop orders are used as a protection against unwanted losses (when owning a stock, say), or against unexpected rises (when planning to buy stock). Notice, however, that there is no guarantee that the price at which the order is executed is close to the limit S_L set, a fact to be considered when seeking protection against crashes, cf. Chap. 5.

2.6.2 Price Formation by Auction

In an auction, every trader gives buy and sell orders with a specific volume and limit (market orders are taken to have limit zero for sell and infinity for buy orders). The orders are now ordered in descending (ascending) order of the limits for the buy (sell) orders, i.e., $S_{L,1} > S_{L,2} > \ldots > S_{L,m}$ for buy orders, and $S_{L,1} < S_{L,2} < \ldots < S_{L,n}$ for the sell orders. Let $V_b(S_i)$ and $V_s(S_i)$ be the volumes of the buy and sell orders, respectively, at limit S_i. We now form the cumulative demand and offer functions $D(S_k)$ and $O(S_k)$ as

$$D(S_k) = \sum_{i=1}^{k} V_b(S_i) , \quad k = 1, \ldots, m \tag{2.1}$$

$$O(S_k) = \sum_{i=1}^{k} V_s(S_i) , \quad k = 1, \ldots, n . \tag{2.2}$$

The market price of the asset determined in the auction then is that price which allows one to execute a maximal volume of orders with a minimal residual of unexecuted order volume, consistent with the order limits. If the order volumes do not match precisely, orders may be partly executed.

We illustrate this by an example. Table 2.1 gives part of a hypothetical order book at a stock exchange. One starts executing orders from top to bottom on both sides, until prices or cumulative order volumes become inconsistent. In the first two lines, the buy limit is above the sell limit so

Table 2.1. Order book at a stock exchange containing limit orders only. Orders with volume in boldface are executed at a price of 162. With a total transaction volume of 900, the buy order of 300 shares at 162 is executed only partly

Buy			Sell		
Volume	Limit	Cumulative	Volume	Limit	Cumulative
200	164	200	**400**	160	400
500	163	700	**400**	161	800
300	162	1000	**100**	162	900
200	161	1200	300	163	1200
300	160	1500	300	164	1500
$V_b(S_i)$	S_i	$D(S_i)$	$V_s(S_i)$	S_i	$O(S_i)$

that the orders can be executed at any price $163 \geq S \geq 161$. In the third line, only 900 (cumulated) shares are available up to 162 compared to a cumulative demand of 1000. A transaction is possible at 162, and 162 is fixed as the transaction price for the stock because it generates the maximal volume of executed orders. However, while the sell order of 100 stocks at 162 is executed completely, the buy order of 300 stocks is exectued only partly (volume 200). Depending on possible additional instructions, the remainder of the order (100 stocks) is either cancelled or kept in the order book.

The problem can also be solved graphically. The cumulative offer and demand functions are plotted against the order limits in Fig. 2.3. The solid line is the demand, and the dash-dotted line is the offer function. They intersect at a price of 162.20. The auction price is fixed as that neighboring allowed price (we restricted ourselves to integers) where the order volume on the lower of both curves is maximal. This happens at 162 with a cumulative volume of 900 (compare to a volume of 750 at 163).

The dotted line in Fig. 2.3 shows the cumulative buy functions if an additional market order for 300 stocks is entered into the order book. The demand function of the previous example is shifted upward by 300 stocks, and the new price is 163. All buy orders with limit 163 and above are executed completely, including the market order (total volume 1000). Sell orders with limit below 163 are executed completely (total volume 900), and the order with limit 163 can sell only 100 shares, instead of 300. The corresponding order book is shown in Table 2.2.

2.6.3 Continuous Trading:
The XETRA Computer Trading System

Elaborate rules for price formation and priority of orders are necessary in the computerized trading systems such as the XETRA (**EX**change **E**lectronic

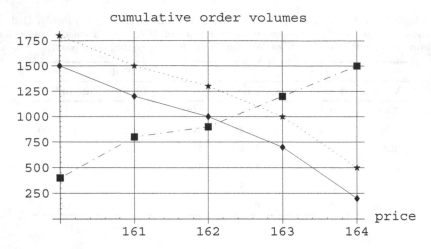

Fig. 2.3. Offer and demand functions in an auction at a stock exchange. The solid line is the demand function with limit orders only, and the dotted line includes a market order of 300 shares. The dash-dotted line is the offer function

Table 2.2. Order book including a market buy order. Orders with volume in boldface are executed at a price of 163. With a total transaction volume of 1000, the sell order of 300 shares at 163 is executed only partly

Buy			Sell		
Volume	Limit	Cumulative	Volume	Limit	Cumulative
300	market	300	**400**	160	400
200	164	500	**400**	161	800
500	163	1000	**100**	162	900
300	162	1300	**300**	163	1200
200	161	1500	300	164	1500
300	160	1800			
$V_b(S_i)$	S_i	$D(S_i)$	$V_s(S_i)$	S_i	$O(S_i)$

Trading) system introduced by the German Stock Exchange in late 1997 [27]. Here, we just describe the basic principles.

Trading takes place in three main phases. In the pretrading phase, the operators can enter, change, or delete orders in the order book. The traders cannot access any information on the order book.

The matching (i.e., continuous trading) phase starts with an opening auction. The purpose is to avoid a crossed order book (e.g., sell orders with limits significantly below those of buy orders). Here, the order book is partly closed, but indicative auction prices or best limits entered, are displayed continuously. Stocks are called to auction randomly with all orders left over from the preceding day, entered in the pretrading phase, or entered during the auction until it is stopped randomly. The price is determined according to the rules of the preceding section. It is clear, especially from Fig. 2.3, that in this way a crossed order book is avoided.

In the matching phase, the order book is open and displays both the limits and the cumulative order volumes. Any newly incoming market or limit order is checked immediately against the opposite side of the order book, for execution. This is done according to a set of at least 21 rules. More complete information is available in the documentation provided by, e.g., Deutsche Börse AG [27]. Here, we just mention a few of them, for illustration. (i) If a market or a limit order comes in and faces a set of limit orders in the order book, the price will be the highest limit for a sell order, resp. the lowest limit for a buy order. (ii) If a market buy order meets a market sell order, the order with the smaller volume is executed completely, while the one with the larger volume is executed partly, at the reference price. The reference price remains unchanged. (iii) If a limit sell order meets a market buy order, and the currently quoted price is higher than the lowest sell limit, the trade is concluded at the currently quoted price. If, on the other hand, the quoted price is below the lowest sell limit, the trade is done at the lowest sell limit. (iv) If trades are possible at several different limits with maximal trading volume and minimal residual, other rules will determine the limit depending on the side of the order book, on which the residuals are located.

If the volatility becomes too high, i.e., stock prices leave a predetermined price corridor, matching is interrupted. At a later time, another auction is held, and continuous trading may resume. Finally, the matching phase is terminated by a closing auction, followed by a post-trading period. As in pretrading, the order book is closed but operators can modify their own orders to prepare next day's trading.

On a trading floor where human traders operate, such complicated rules are not necessary. Orders are announced with price and volume. If no matching order is manifested, traders can change the price until they can conclude a trade, or until their limit is reached.

3. Random Walks in Finance and Physics

The Introduction, Chap. 1, suggested that there is a resemblance of financial price histories to a random walk. It is therefore more than a simple curiosity that the first successful theory of the random walk was motivated by the description of financial time series. The present chapter will therefore describe the random walk hypothesis [28], as formulated by Bachelier for financial time series, in Sect. 3.2 and the physics of random walks [29], in Sect. 3.3. The mathematical description of random walks can be found in many books [30]. A classical account of the random walk hypothesis in finance has been published by Cootner [7].

3.1 Important Questions

We will discuss many questions of basic importance, for finance and for physics, in this chapter. Not all of them will be answered, some only tentatively. These problems will be taken up again in later chapters, with more elaborate methods and more complete data, in order to provide more definite answers. Here is a list:

- How can we describe the dynamics of the prices of financial assets?
- Can we formulate a model of an "ideal market" which is helpful to predict price movements? What hypotheses are necessary to obtain a tractable theoretical model?
- Can the analysis of historical data improve the prediction, even if only in statistical terms, of future developments?
- How must the long-term drifts be treated in the statistical analysis?
- How was the random walk introduced in physics?
- Are there qualitative differences between solutions and suspensions? Is there osmotic pressure in both?
- Have random walks been observed in physics? Can one observe the *one-dimensional* random walk?
- Is a random walk assumption for stock prices consistent with data of real markets?
- Are the assumptions used in the formulation of the theory realistic? To what extent are they satisfied by real markets?

- Can one make predictions for price movements of securities and derivatives?
- How do derivative prices relate to those of the underlying securities?

The correct understanding of the relation of real capital markets to the ideal markets assumed in theoretical models is a prerequisite for successful trading and/or risk control. Theorists therefore have a skeptical attitude towards real markets and therein differ from practitioners. In ideal markets, there is generally no easy, or riskless, profit ("no free lunch") while in real markets, there may be such occasions, in principle. Currently, there is still controversy about whether such profitable occasions exist [3, 31].

We now attempt a preliminary answer at those questions above touching financial markets, by reviewing Bachelier's work on the properties of financial time series.

3.2 Bachelier's "Théorie de la Spéculation"

Bachelier's 1900 thesis entitled "Théorie de la Spéculation" contains both theoretical work on stochastic processes, in particular the first formulation of a theory of random walks, and empirical analysis of actual market data. Due to its importance for finance, for physics, and for the statistical mechanics of capital markets, and due to its difficult accessibility, we will describe this work in some detail.

Bachelier's aim was to derive an expression for the probability of a market or price fluctuation of a financial instrument, some time in the future, given its current spot price. In particular, he was interested in deriving these probabilities for instruments close to present day futures and options, cf. Sect. 2.3, with a FF 100 French government bond as the underlying security. He also tested his expressions for the probability distributions on the daily quotes for these bonds.

3.2.1 Preliminaries

This section will explain the principal assumptions made in Bachelier's work.

Bachelier's Futures

Bachelier considers a variety of financial instruments: futures, standard (plain vanilla) options, exotic options, and combinations of options. However, his basic ideas are formulated on a futures-like instrument which we first characterize.

- The underlying security is a French government bond with a nominal value of FF 100, and 3% interest rate. A coupon worth $Z = 75c$ is detached every three months (at the times t_i below).

- Unlike modern bond futures, Bachelier's futures do not include an obligation of delivery of the bond at maturity. Only the price difference of the underlying bond is settled in cash, as would be done today with, e.g., index futures. The advantage of buying the future, compared to an investment into the bond, then is that only price changes must be settled. The important investment in the bond upfront can thus be avoided, and the leverage of the returns is much higher.
- The expiry date is the last trading day of the month. The price of the futures is fixed on entering the contract, cf. Sect. 2.3.2, and the long position acquires all rights deriving from the underlying bond, including interest.
- The long position receives the interest payments (coupons) from the futures.
- Bachelier's futures can be extended beyond their maturity (expiry) date, to the end of the following month, by paying a prolongation fee K. This is not possible on present day futures. It conveys some option character to Bachelier's futures because its holder can decide to honor the contract at a later stage where the market may be more favorable to him.

Market Hypotheses

Bachelier makes a series of assumptions on his markets which have become standard in the theory of financial markets. He postulates that, at any given instant of time, the market (i.e., the *ensemble* of traders) is neither *bullish* nor *bearish,* i.e., does not believe either in rising or in falling prices, a *hausse* or *baisse* of the market. (Notice that the individual traders may well have their opinion on the direction of a market movement.) This is, in essence, what has become the hypothesis of "efficient and complete" markets. In particular:

- Successive price movements are statistically independent.
- In a perfect market, all information available from the past to the present time, is completely accounted for by the present price.
- In an efficient market, the same hypothesis is made, but small irregularities are allowed, so long as they are smaller than applicable transaction costs.
- In a complete market, there are both buyers and sellers at any quoted price. They necessarily have opposite opinions about future price movements, and therefore on the average, the market does not believe in a net movement.

The Regular Part of the Price of Bachelier's Futures

Let us assume that there are no fluctuations in the market. The price of the futures F is then completely governed by the price movements of the underlying security S which is shown in Fig. 3.1. Due to the accumulation of interest, the value, and therefore the price, of the bond increases linearly in time by $Z = 75c$ over three months. When the coupon is detached from the bond at times t_i $(t_{i+1} - t_i = 3m)$, the value of the bond decreases instantaneously

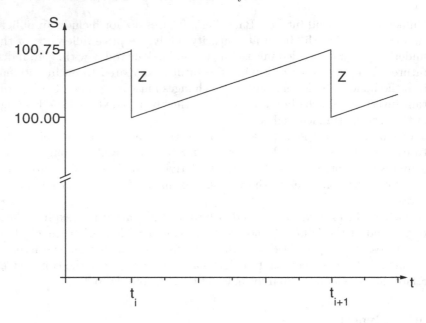

Fig. 3.1. Deterministic part of the spot price evolution of the underlying French government bond. t_i denotes the time where a 75c coupon is detached from the bond

by Z. The movement of the futures price is more dramatic, reflecting only price *changes,* but reproduces the basic pattern of Fig. 3.1. In the absence of prolongation fees ($K = 0$), immediately after the payment of interest at some t_i, the value of the futures contract is zero. Due to the accumulation of interest on the underlying bond, the futures price then increases linearly in time to 75c, immediately before t_{i+1}

$$F(t) = S(t) - S(t_i) = \left(\frac{Z}{t_{i+1} - t_i} \right) (t - t_i) \quad \text{for} \quad t_i \le t \le t_{i+1} . \quad (3.1)$$

This is because at maturity, the price difference accumulated on the underlying bond is settled between the long and short positions. Immediately after the maturity date, the value of the futures falls to zero again as shown as the solid line in Fig. 3.2. The holder of the futures receives the interest payment of the underlying bond. Notice the leverage on the price variations of the futures. The bond price varies by 0.75% each time a coupon is detached while the futures varies by 100% because the interest payment is 0.75% of the bond value, but makes up the entire value of the futures. With a finite prolongation fee K, price movement will be less pronounced. In the extreme case where $K = Z/3$, the value of the futures contract at maturity, after one month, will be equal to the initial investment for carrying it on, i.e., K. It will then jump up by K due to the cost of prolongation, etc. This is the

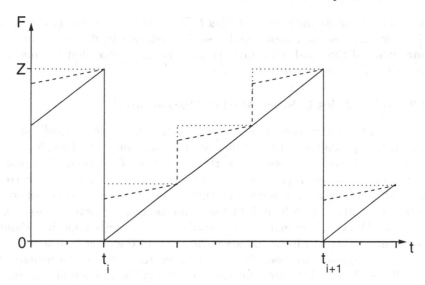

Fig. 3.2. Deterministic part of the evolution of the futures price F for three different prolongation fees: $K = 0$ (*solid line*), $K = Z$ (*dotted line*), and $0 < K < Z$ (*dashed line*)

dotted line in Fig. 3.2. For intermediate $Z > K > 0$, the futures price will vary as represented by the dashed lines in Fig. 3.2: the value is $K < Z/3$ immediately after interest payment, from where it increases linearly to $Z/3$ at the first maturity date, jumps by another K and increases to $2Z/3$ at the second maturity date, etc., to t_{i+1} where interest is paid, and the value falls back to K.

The important observation of Bachelier now is that *all prices on any given line F(t) [or S(t)] are equivalent.* As long as the price evolution is deterministic, the return an investor gets from buying the futures (or bond) at any given time is the same, provided the price is on the applicable curve $F(t)$ [or $S(t)$]. The returns are the same because the slope is independent of time. For a given K, all prices on one given curve represent the true, or fundamental (in modern terms), value of the asset. For a given prolongation cost K the drift of the true futures price is

$$\frac{\mathrm{d}F(t)}{\mathrm{d}t} = \frac{\mathrm{d}S(t)}{\mathrm{d}t} - \frac{3K}{(t_{i+1} - t_i)} \,, \tag{3.2}$$

between two maturity dates.

If now fluctuations are added, and the current spot price of the futures is $F(t)$, the true, or fundamental value of the futures a time $t + T$ from now, is

$$\tilde{F}(t + T) = F(t) + \frac{\mathrm{d}F}{\mathrm{d}t}T \,, \tag{3.3}$$

provided no maturity date occurs during t. The effect of a maturity date can be included as described above, and a similar relation holds for the fundamental price of the bond. Of course, there is no guarantee that the quoted price at $t + T$ will be equal to $\tilde{F}(t + T)$.

3.2.2 Probabilities in Stock Market Operations

Bachelier distinguishes two kinds of probabilities, a "mathematical" and a "speculative" probability. The mathematical probability can be calculated and refers to a game of chance, like throwing dice. The speculative probability may not be appropriately termed "probability", but perhaps better "expectation", because it depends on future events. It is a subjective opinion, and the two partners in a financial transaction necessarily have opposite expectations (in a complete market: necessarily always exactly opposite) about those future events which can influence the value of the asset transacted.

The probabilities discussed here, of course, refer to the mathematical probabilities. Notice, however, that the (grand-) public opinion about stock markets, where the idea of a random walk does not seem to be deeply rooted, sticks more to the speculative probability. Also for speculators and active traders, the future expectations may be more important than the mathematical probability of a certain price movement happening. The mathematical probabilities refer to idealized markets where no easy profit is possible. On the other hand, fortunes are made and lost on the correctness of the speculative expectations. It is important to keep these distinctions in mind.

Martingales

In Sect. 3.2.1, we considered the deterministic part of the price movements both of the French government bond, and of its futures. There is a net return from these assets because the bond generates interest. Between the cash flow dates, there is a constant drift in the (regular part of) the asset prices and most likely, there will also be a finite drift if fluctuations are included. Such drifts are present in most real markets, cf. Figs. 1.1 and 1.2. Consequently, Bachelier's basic hypothesis on complete markets, viz. that *on the average,* the agents in a complete market are neither bullish nor bearish, i.e., neither believe in rising nor in falling prices, Sect. 3.2.1, must be modified to account for these drifts which, of course, generate net positive expectations for the future movements.

The modified statement then is that, *up to the drift* dF/dt, *resp.* dS/dt, the market does not expect a net change of the true, or fundamental, prices. (Bachelier takes the artificial case $K = Z/3$, i.e., the dotted lines in Fig. 3.2, to formalize this idea.) However, deviations of a certain amplitude y, where $y = S(t) - S(0)$ or $F(t) - F(0)$, occur with probabilities $p(y)$, which satisfy

$$\int_{-\infty}^{\infty} p(y)\mathrm{d}y = 1 \tag{3.4}$$

for all t. The expected profit from an investment is then

$$E(y) \equiv \langle y \rangle = \int_{-\infty}^{\infty} y\,p(y)\mathrm{d}y > 0 \quad \text{so long as} \quad \begin{cases} \frac{\mathrm{d}S}{\mathrm{d}t}, \frac{\mathrm{d}F}{\mathrm{d}t} > 0 \\ \\ Z > 3K\,. \end{cases} \tag{3.5}$$

[The notation $E(y)$ for an expectation value is more common in mathematics and econometrics, while physicists often prefer $\langle y \rangle$.] Such an investment is not a fair game of chance because it has a positive expectation. However, for a

$$\text{fair game of chance}: \quad E(y) = 0\,. \tag{3.6}$$

This condition, the vanishing of the expected profit of a speculator, is fulfilled in Bachelier's problem only if $Z = 3K$, or if $\mathrm{d}S/\mathrm{d}t$ or $\mathrm{d}F/\mathrm{d}t$ is either zero or subtracted out. Then a modified price law between the maturity dates

$$x(t) = y(t) - \frac{\mathrm{d}S}{\mathrm{d}t}t \quad \text{or} \quad x(t) = y(t) - \frac{\mathrm{d}F}{\mathrm{d}t}t\,, \tag{3.7}$$

where t is set to zero at the maturity times (nt_i for the bond and $nt_i/3$ for the futures), must be used. This law fulfills the fair game condition

$$E(x) \equiv \langle x \rangle = 0\,. \tag{3.8}$$

With these prices corrected for the deterministic changes in fundamental value, the expected *excess* profit of a speculator now vanishes. A clear separation of the regular, or deterministic price movement, contained in the drift term, and of the fluctuations, has been achieved. Equation (3.8) emphasizes that there is no easy profit possible due to the fair game condition (3.6). Now it is possible to attempt a statistical description of the fluctuation process.

$x(t)$ describes a drift-free time series. This is what is called, in the modern theory of stochastic processes [32], a *martingale*, or a *martingale stochastic process*, i.e., $E(x) = 0$, or more precisely (in discrete time)

$$E(x_{t+1} - x_t | x_t, x_{t-1}, x_{t-2}, \ldots, x_0) = 0\,, \tag{3.9}$$

where $E(x_{t+1} - x_t | x_t, \ldots)$ is the expectation value formed with the *conditional probability* $p(x_{t+1} - x_t | x_t, \ldots)$ of $x_{t+1} - x_t$, conditioned on the observations $x_t, x_{t-1}, x_{t-2}, \ldots, x_0$. One may also say that $y(t)$, the stochastic process (time series) followed by the bond price or any other financial data, is an *equivalent martingale process*. An equivalent martingale process is a stochastic process which is obtained from a martingale stochastic process by a simple change of the drift term, cf. (3.7).

The equivalent martingale hypothesis is equivalent to that of a perfect and complete market, and approximately equivalent to that of an efficient and complete market.

Distribution of Probabilities of Prices

What can we say about the probability density $p(x,t)$ of a price change of a certain amplitude x, at some time t in the future? In attempting to answer this question, Bachelier gave a rather complete, though sometimes slightly inaccurate, formulation of a theory of the random walk, five years before Einstein's seminal paper [4].

From now on, we will assume that the price $S(t)$ itself follows a martingale process, or that all effects of nonzero drifts have been incorporated correctly. The general shape of the probability distribution at some time t in the future is shown in Fig. 3.3. Here, $p(x_1,t)\mathrm{d}x_1$ is the probability of a price change $x_1 \leq x \leq x_1 + \mathrm{d}x_1$ at time t. In a first appoximation, the complete market hypothesis requires the distribution to be symmetric with respect to $x = 0$, and the fair game condition, i.e., the assumption of a martingale process, requires the maximum to be at $x = 0$ at any t, and to have a quadratic variation for sufficiently small x. Also, it must decrease sufficiently quickly for $x \to \pm\infty$ to make $p(x,t)$ normalizable. Strictly speaking, since the price of a bond cannot become negative, $p(x,t) = 0$ for $x < -S(0)$, but this effect is negligible in practice so long as fluctuations are small compared to the bond price.

The Chapman–Kolmogorov–Smoluchowski Equation Bachelier then tries to derive $p(x,t)$ from the law of multiplication of probabilities. If $p(x_1,t_1)\mathrm{d}x_1$ is the probability of a price change $x_1 \leq x \leq x_1 + \mathrm{d}x_1$ at time t_1, and $p(x_2 - x_1,t_2)\mathrm{d}x_2$ is the probability of a change $x_2 - x_1$ in t_2, the joint probability for having a change to x_1 at t_1 *and* to x_2 at $t_1 + t_2$ is $p(x_1,t_1)p(x_2-x_1,t_2)\mathrm{d}x_1\mathrm{d}x_2$. These paths are shown as solid lines in Fig. 3.4. Then, the probability to have a change of x_2 at $t_1 + t_2$, independent of the intermediate values, is

$$p(x_2, t_1 + t_2)\mathrm{d}x_2 = \left[\int_{-\infty}^{+\infty} p(x_1, t_1)p(x_2 - x_1, t_2)\mathrm{d}x_1 \right] \mathrm{d}x_2 . \qquad (3.10)$$

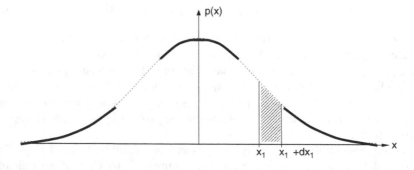

Fig. 3.3. General shape of the probability density function $p(x,t)$ of a price change x at some time t in the future

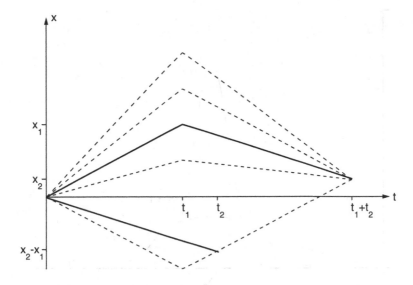

Fig. 3.4. Multiplication of probabilities in the (x, t)-plane. Strictly speaking, only the probabilities at t_1, t_2, and $t_1 + t_2$ are used. For clarity, they have been connected by straight "paths". To derive the Chapman–Kolmogorov–Smoluchowski equation, one must integrate over all values of x at t_1. A few such paths a shown as dashed lines

This equation is known in physics and mathematics as the Chapman–Kolmogorov–Smoluchowski (CKS) equation, and was rederived there some decades after Bachelier. It is a convolution equation for the probabilities of statistically independent random processes (resp. Markov processes more generally).

Bachelier solves this equation by the Gaussian normal distribution

$$p(x, t) = p_0(t) \exp\left[-\pi p_0^2(t) x^2\right] .$$ (3.11)

Inserting this into CKS (3.10) gives the condition

$$p_0^2(t_1 + t_2) = \frac{p_0^2(t_1) p_0^2(t_2)}{p_0^2(t_1) + p_0^2(t_2)}$$ (3.12)

which in turn determines the time evolution of $p(t)$ as

$$p_0(t) = H/\sqrt{t}$$ (3.13)

with a constant H. The substitution $\sigma^2 = t/2\pi H^2$ then gives the normal form of the Gaussian

$$p(x, t) = \frac{1}{\sqrt{2\pi}\sigma(t)} \exp\left(-\frac{x^2}{2\sigma^2(t)}\right) .$$ (3.14)

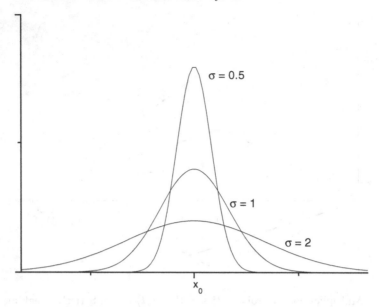

Fig. 3.5. The Gaussian distribution for three different values of the standard deviation σ, i.e., three different times $t \propto \sigma^2$

Its shape, for three different values of σ, i.e., time, is shown in Fig. 3.5. The following facts are important [set $x_0 = 0$ in Fig. 3.5 if you are interested in changes or $x_0 = S(0)$ if you are interested in absolute prices]: (i) for $t = 0$, we have $\sigma = 0$, and this corresponds to $p(x) = \delta(x)$, i.e., certain knowledge of the price at present (not shown in Fig. 3.5); (ii) the peak of the distribution, and its mean do not change with time, reflecting the martingale property; (iii) the distribution function broadens *slowly,* only with $\sigma \propto \sqrt{t}$. This fact (and eventual deviations thereof, of real markets) is of practical importance since it excludes big price movements over moderately long time intervals.

An important problem is, however, that Bachelier did not recognize that his "solution" to (3.10) is not the only solution, and in fact a rather special one. Fortunately enough, Bachelier approached his problem along several different routes. He obtained the same solution (3.14) for two more special problems, where it was both correct and unique. One was the solution of the random walk, the other the formulation of a "diffusion law" for price changes.

The Random Walk A discrete model for asset price changes would consider two mutually exclusive events A (happening with a probability p), and B (with a probability $q = 1 - p$). These events can be thought to represent price changes by $\pm x_0$, in one time step. Then the probability of observing, in m events, α realizations of A and $m - \alpha$ realizations of B is given by the binomial distribution

$$p_{A,B}(\alpha, m - \alpha) = \frac{m!}{\alpha!(m-\alpha)!}p^\alpha (1-p)^{m-\alpha} . \tag{3.15}$$

One may now ask:

1. Which α maximizes $p(\alpha, m - \alpha)$ at fixed m and p? The answer is $\alpha = mp$, and thus $m - \alpha = mq$. In a financial interpretation, this gives the most likely price change after m time steps, e.g., trading days $x_{max} = m(p - q)x_0$. A finite difference $p - q$ would represent a drift in the market (in this argument, one is not restricted to martingale processes).
2. What is the distribution function of price changes? The complete expression for general α, p, m has been derived by Bachelier [6]. It simplifies, however, in the limit $m \to \infty$, $\alpha \to \infty$ with $h = \alpha - mp$ finite, to

$$p(h) = \frac{1}{\sqrt{2\pi mpq}} \exp\left(-\frac{h^2}{2mpq}\right) . \tag{3.16}$$

3. For $p = q = 1/2$, finally, and setting $h \to x$, $m = t/\Delta t$ with Δt the unit time step, and $H = \sqrt{2\Delta t/\pi}$, the Gaussian distribution

$$p(x) = \frac{H}{\sqrt{t}} \exp\left(-\frac{\pi H^2 x^2}{t}\right) \tag{3.17}$$

of (3.11)–(3.14) is recovered. In this limit of large m, one has passed from discrete time and discrete price movements to continuous variables.

This is the first formulation of the random walk, or equivalently of the theory of Brownian motion, or of the "Einstein–Wiener stochastic process".

Other quantities of interest, such as the probability for a price change contained in a window, $P(0 \le x(t) \le X)$, the expected width X of the distribution of price changes, $P(-X \le x(t) \le X) = 1/2$, or of the expected profit associated with a financial instrument whose payoff is x if $x > 0$, and zero if $x < 0$ (i.e., an investment in options), have been derived by simple integration [6].

The Diffusion Law Yet another derivation can be done via the diffusion equation. For this purpose, assume that prices are discretized $\ldots, S_{n-2}, S_{n-1}, S_n, S_{n+1}, S_{n+2}, \ldots$, and that at some time t in the future, these prices are realized with probabilities $\ldots, p_{n-2}, p_{n-1}, p_n, p_{n+1}, p_{n+2}, \ldots$. Then, one may ask for the evolution of these probabilities with time. Specifically, what is the probability p'_n of having S_n at a time step Δt after t? If we assume that a price change $S_n \to S_{n\pm 1}$ must take place during Δt, we find $p'_n = (p_{n-1} + p_{n+1})/2$ because the price S_n can either be reached by a downward move from S_{n+1}, occurring with a probability $p_{n+1}/2$, or by an upward move from S_{n-1} with a probability $p_{n-1}/2$. The change in probability of a price S_n during the time step Δt is then

$$\Delta p_n = p'_n - p_n = \frac{p_{n+1} - 2p_n + p_{n-1}}{2} \to \frac{1}{2}\frac{\partial^2 p(S,t)}{\partial S^2}(\Delta S)^2 \tag{3.18}$$

if the limit of continuous prices and time is taken. On the other hand,

$$\Delta p_n \to \frac{\partial p(S,t)}{\partial t}\Delta t \tag{3.19}$$

in the same limit, and therefore

$$D\frac{\partial^2 p}{\partial S^2} - \frac{\partial p}{\partial t} = 0 . \tag{3.20}$$

$p(S,t)$ therefore satisfies a diffusion equation, and the Gaussian distribution is obtained for special initial conditions. These conditions $p(S,0) = \delta[S - S(0)]$, i.e., knowledge of the price at time $t = 0$, apply here.

Bachelier realized that (3.20) is Fourier's equation, and that consequently, one may think of a diffusion process, or of radiation of probability through a price level.

These considerations are equally valid for Bachelier's bonds and for his futures. As has been discussed above, both prices differ by their drift coefficients, and by an offset corresponding to the nominal value of the bond, but not in their fluctuations. Therefore, the equivalent martingale processes for both assets are the same, and the description of their fluctuations achieved here is valid for both of them. We will see later that the same model, with only minor modifications, became *the* standard model for financial markets.

Bachelier solved many other important problems in the theory of random walks, always motivated by financial questions. He calculated prices for simple and exotic options, and solved the first passage problem (the probability that a certain price S, or price change x, is reached for the first time at time t in the future). He also solved the problem of diffusion with an absorbing barrier – this corresponds to hedging of options with futures, and vice versa, and the corresponding probability distribution is shown in Fig. 3.6. Here, one requires that losses larger than a threshold, $-x_0$, have zero probability. In

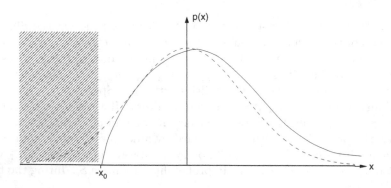

Fig. 3.6. Probability distribution for hedging of options with futures, equivalent to diffusion with an absorbing barrier

modern terms, when one is long in a futures, this hedge can be achieved by going long in a put with a strike price $X = S(0) - x_0$, up to the price of the put.

Bachelier's idea to consider discrete prices at discrete times, and to associate a certain probability with the transition from one price S_t at time t to another price S_{t+1} at the next time is also important in option pricing. When starting from a given price S_0 at the present time $t = 0$, a "binomial tree" for future asset prices is generated by allowing, at each time t, either an upward or a downward move of the asset price with probabilities $p_t(\text{up})$ and $p_t(\text{down}) = 1 - p_t(\text{up})$. Cox, Ross, and Rubinstein show how one can calculate option prices backwards, starting at the maturity date of an option. From there, one iteratively works back to the present date. This method will not be discussed in this book, and the reader is referred to the literature for further details [10].

3.2.3 Empirical Data on Successful Operations in Stock Markets

Bachelier performed a variety of empirical tests of this theory by evaluating five years of quotes (1894-1898) of the French government bond and its associated futures.

The two parameters of the theory, which must be determined from empirical data, are the drift and the volatility (standard deviation). From his empirical data, Bachelier obtains for the drifts of the bond and the futures

$$\frac{\mathrm{d}S}{\mathrm{d}t} = 0.83 \frac{\text{centimes}}{\text{day}} \quad \text{and} \quad \frac{\mathrm{d}F}{\mathrm{d}t} = 0.264 \frac{\text{centimes}}{\text{day}} \tag{3.21}$$

and for the volatility coefficient

$$\frac{1}{2\pi H} = \lim_{t \to 0} \frac{\sigma}{\sqrt{2\pi t}} = 5 \frac{\text{centimes}}{\sqrt{\text{day}}} \tag{3.22}$$

He later corrects these numbers for the difference between calendar days and trading days.

The interval where price changes are contained with 50% probability (50% confidence interval)

$$\int_{-\alpha_t}^{\alpha_t} p(x, t)\mathrm{d}x = \frac{1}{2} \tag{3.23}$$

is then $\alpha_1 = 9c$ for $t = 1d$, and $\alpha_{30} = 46c$ for $t = 30d$. For $t = 30d$, there are 60 data points available, with 33 changes smaller than α_{30}, and 27 larger. For $t = 1d$, Bachelier has 1452 data points, with 815 changes smaller than α_1 and 637 larger than α_1.

One should become suspicious here because the number of changes larger than α_1 deviates from the expected value (776) by more than $38 = \sqrt{1452}$. This may be due, at least in part, to the drift of the prices. Including the drift terms, Bachelier finds that the 50% interval for price changes for $t = 30d$ is $-38c \leq x \leq +54c$, but does not give the corresponding numbers for the 1-day intervals where the disagreement is most serious, nor does he indicate how well the observed price changes fall into this modified interval. In fact, he does not comment even on the unexpectedly small number of large price changes in his observations, compared with his theory. Modern empirical studies find a mean-reversion in the stochastic processes followed by interest rates and bond prices [15], i.e., extreme price changes are *less likely* than in Bachelier's random walk. This trend apparently is present in Bachelier's price history already. Stock, currency or commodity markets, on the other hand, have significantly more big price changes than predicted by a simple random-walk hypothesis.

By integration of the probability distributions, one can calculate the probability of getting a profit from an investment into a bond or a futures. For the bond, the probability for profit after a month, $P(1\text{m}) = 0.64$, and after a year, $P(1\text{y}) = 0.89$. For the futures, on the other hand, $P(1\text{m}) = 0.55$, and $P(1\text{y}) = 0.65$. The difference is due to the different drift rates: that of the futures is lower because there is a finite prolongation fee K, for carrying it on to the next maturity date. (On the other hand, the return on the invested capital is expected to be bigger for the futures.) In Bachelier's times, options were labeled by the premium one had to pay for the *right* to buy or sell (call or put) the underlying at maturity. Bachelier calculated the 50% intervals for the price variations of a variety of such options, with different maturities and premiums, and found rather good agreement with the intervals he derived from his observations. (Needless to say the payoff profiles for calls and puts shown in Figs. 2.1 and 2.2 can already be found in Bachelier's thesis, as well as those of combinations thereof.)

3.2.4 Biographical Information on Louis Bachelier (1870–1946)

Apparently, not much biographical information on Louis Bachelier is available. My source of information is essentially Mandelbrot's book on fractals [33]. Bachelier defended his thesis "Théorie de la spéculation" on March 29, 1900, at the Ecole Normale Supérieure in Paris. Apparently, the examining committee was not overly impressed because they attributed the rating "honorable" where the standard apparently was (and still is in France today) "très honorable". On the other hand, his thesis was translated and annotated into English in 1964 [7], a rather rare event.

Bachelier's work had no influence on any of his contemporaries, but he remained active throughout his scientific life, and published in the best journals. Only very late, did he become a professor of mathematics at the University of Besançon. There is a sharp contrast between the difficulties he experienced in his scientific career, and the posthumous fame he earned for his thesis.

There may be two main reasons for this. One is related to an error in taking limits of a function describing a stochastic process in a publication, which was uncovered by the selection committee of a university where he had applied for a position, and confirmed by the famous French mathematician Paul Lévy. However, it was also Lévy who later realized that Bachelier had derived, long before Einstein and Wiener, the main properties of the stochastic "Einstein–Wiener" process, and of the diffusion equation. The second reason certainly is related to the subject of his dissertation: speculation on financial markets was not considered to be a subject for "pure science" (and perhaps still is not universally recognized so today, as witnessed by a few comments of colleagues on the course underlying this book). There was no community in economics which could have taken up his ideas and achievements: discrete and continuous stochastic processes, martingales, efficient markets and fair games, random walks, etc., and, for mathematicians, he was linked to the error mentioned above. The final part in his tragedy was played by Poincaré who wrote the official report on Bachelier's thesis. While he complained that the subject was rather far from what the other students used to treat, he also realized how far Bachelier had advanced in the theory of diffusion, and of stochastic processes. However, Poincaré also suffered from lapses of memory. A few years later, when he took an active part in discussions on Brownian motion, he had completely forgotten Bachelier's seminal work.

3.3 Einstein's Theory of Brownian Motion

The starting point of Einstein's work on Brownian motion is rather surprising from a present day perspective: the implication of classical thermodynamics that there would not be an osmotic pressure in suspensions [4]. The aim of Einstein's work was *not* to explain Brownian motion (the small irregular motions of particles resulting from the decay of plant pollen in an aqueous solution which the Scottish botanist R. Brown had observed under the microscope – Einstein did not have accurate information on this phenomenon) but to show that the statistical theory of heat required the motion of particles in suspensions, and thereby both diffusion and an osmotic pressure. Such a phenomenon would not be allowed by classical thermodynamics. For the physics concepts discussed here, we refer to any textbook on statistical mechanics or thermodynamics [29].

3.3.1 Osmotic Pressure and Diffusion in Suspensions

The phenomenon of osmotic pressure is commonly discussed for solutions [29]. One considers a solution where the solute is dissolved, in a concentration c, in the solvent in a volume V^* enclosed by a membrane. This membrane is assumed to be permeable only to the solvent, and not to the solute, and immersed in a surrounding volume of solvent. The solvent therefore can freely flow in and out. One then finds that the solute exercises a pressure p on the membrane

$$pV^* = cRT , \qquad\qquad (3.24)$$

the osmotic pressure. Here R is the gas constant, and T the temperature. The idea behind (3.24) is that the solute acts as an ideal gas enclosed in the volume V^* while the solvent does not sense the membrane and can be ignored, an interpretation that goes back to van't Hoff. In a solution, the solute is of microscopic, i.e., atomic or molecular size, the same situation as for a true ideal gas.

In a suspension, on the other hand, the particles immersed in a fluid are macroscopic, though small. (There is some confusion about the notion of "microscopic size" in Einstein's paper which should be interpreted as a "size visible under the microscope".) One may now consider a setup similar to the preceding paragraph, i.e., enclose the suspension in a semipermeable membrane, surrounded by a volume of "solvent" fluid. The statement of classical thermodynamics, according to Einstein, is that there is no osmotic pressure in such a suspension $p_{susp} \equiv 0$.

I have not seen this statement documented in any textbook on thermodynamics that I have consulted, and an informal poll among colleagues demonstrated that this fact is not appreciated today (a consequence of the influence of Einstein's work). One explanation goes as follows: When macroscopic particles are suspended in a liquid, the chemical potential of the liquid is not changed according to thermodynamics. The chemical potentials of both constituents are different but cannot change because there is no exchange of particles, by definition of the suspension. The suspension is a heterogeneous phase whereas the analogous situation for a solution is considered to be homogeneous, though with a different chemical potential. This chemical potential difference is at the origin of the osmotic pressure. It is finite for a solution enclosed by a semipermeable membrane, and zero for a heterogeneous phase of a solvent plus suspended particles in contact with a pure solvent phase. Another argmument is that, in equilibrium, the free energy does not depend on the positions of the suspended particles, assumed to be at rest, and that of the membrane, and therefore $P = -(\partial F/\partial V)_T = 0$. As a corollary, there would be no diffusion of particles in a suspension.

Contrary to thermodynamics which works only with macroscopic state variables, the statistical theory of heat developed by Einstein and others, inquires on the origin of heat, and the connection to the microscopic constituents of matter. The question is what microscopic changes are originated by addition or removal of heat. Heat is related to an irregular state of motion of the microscopic building blocks of matter, such as atoms, molecules or electrons: the addition (removal) of heat simply increases (decreases) this motion. As a consequence, both microscopically small particles (the solute) and macroscopic particles (in the suspension) must follow the same laws of motion, and of statistical mechanics. From this, Einstein finds that osmotic pressure is built up both in solutions and suspensions enclosed in a semipermeable membrane, and that there is a unique expression for the diffusion constant of particles in a liquid

$$D = \frac{RT}{N}\frac{1}{6\pi\kappa r} \ , \tag{3.25}$$

where κ is the viscosity coefficient of the liquid and r is the radius of the particles assumed to be spherical. Due to the different size of particles in solutions and suspensions, there is a quantitative difference in the diffusion constant, but there is no qualitative difference between solutions and suspensions in statistical mechanics.

3.3.2 Brownian Motion

The idea which Einstein puts forward is that the particles of the solvent will hit the suspended particles in shocks of random strength and direction, and thereby impart momentum to them. He assumes that

1. the motion of the individual suspended particles is independent of each other;
2. the motion is completely randomized by the shocks;
3. a one-dimensional approximation is sufficient;
4. within a time interval τ, particle j moves from $x_j \to x_i + \Delta_j$ with some random Δ_j.

The Δ_j are taken from a probability distribution $p(\Delta)$ such that

$$\frac{dn}{n} = p(\Delta)d\Delta \tag{3.26}$$

is the fraction of particles which are shifted by distances between Δ and $\Delta + d\Delta$ in one time step. p is normalized and symmetric

$$\int_{-\infty}^{\infty} p(\Delta)d\Delta = 1 \ , \quad p(\Delta) = p(-\Delta) \ . \tag{3.27}$$

The shape of $p(\Delta)$ can now be found by an argument quite similar to Bachelier's third derivation of the Gaussian distribution.

Consider a long, narrow (ideally 1D) cylinder oriented along the x-axis, and let $f(x,t)dx$ be the number of particles contained between x and $x + dx$ at time t. A time step τ later, this number is

$$f(x,t+\tau)dx = dx \int_{-\infty}^{\infty} d\Delta \, p(\Delta)f(x - \Delta, t) \tag{3.28}$$

which contains nothing more than the statement that all particles at $(x,t+\tau)$ must have been *somewhere* at the previous time step. Expanding in τ on the left and in Δ on the right-hand side gives

$$f(x,t) + \tau \frac{\partial f(x,t)}{\partial t} + \cdots \qquad (3.29)$$

$$= \int_{-\infty}^{\infty} d\Delta p(\Delta) \left[f(x,t) - \Delta \frac{\partial f(x,t)}{\partial x} + \frac{\Delta^2}{2} \frac{\partial^2 f(x,t)}{\partial x^2} + \cdots \right].$$

Using (3.27), this reduces to the diffusion equation

$$\frac{\partial f(x,t)}{\partial t} = D \frac{\partial^2 f(x,t)}{\partial x^2} \quad \text{with} \quad D = \frac{1}{2\tau} \int_{-\infty}^{\infty} d\Delta \, \Delta^2 p(\Delta) . \qquad (3.30)$$

For the initial condition $f(x, t = 0) = \delta(x)$, this is solved by the Gaussian distribution

$$f(x,t) = \frac{n}{\sqrt{4\pi Dt}} \exp\left(-\frac{x^2}{4Dt} \right) , \qquad (3.31)$$

where $n = cN$ is the number of suspended particles.

3.4 Experimental Situation

We now discuss the first empirical evidence for random walks in finance and in physics. We will be quite superficial here. An in-depth discussion of the statistical properties of financial time series is the subject of Chap. 5. For physics, we briefly discuss Jean Perrin's seminal observation of Brownian motion under the microscope, which refers to two- or three-dimensional Brownian motion. Truly one-dimensional Brownian motion is rather difficult to observe, and I will discuss the only example I am aware of: the diffusion of electronic spins in the organic conductor TTF-TCNQ.

3.4.1 Financial Data

Figure 1.3 showed a price chart generated from random numbers. The similarity to the behavior of the DAX, Fig. 1.1, is striking! To expand on this similarity, Fig. 3.7 shows another simulation of a random walk (upper panel), and compares it to the DAX quotes from January 1975 to May 1977 (lower panel), i.e., the left end of Fig. 1.2. In the perspective of an (informed) investor, the central problem therefore is to distinguish pure randomness from correlations, or even components of deterministic evolution! In passing, our simulations also contain a warning: we know that even the best random number generators never produce completely random numbers. This is known and under control to a large extent. What is often forgotten is that many practical random number generators are substandard, and sometimes even have drifts! Using them in a computer simulation may produce completely spurious results!

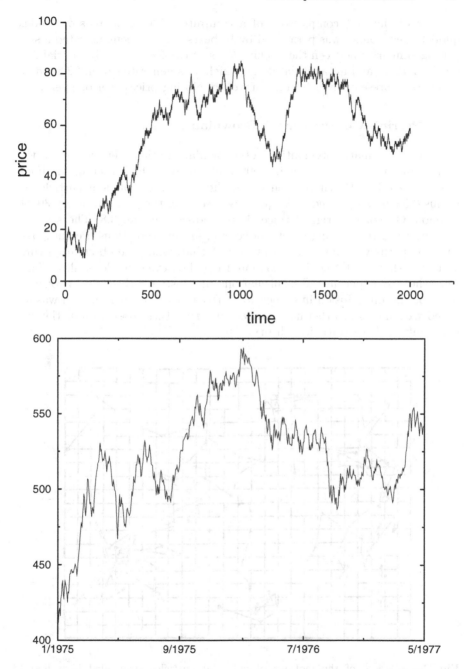

Fig. 3.7. Computer simulation of price charts as a random walk (*upper panel*) and comparison to the evolution of the DAX share index from January 1975 to May 1977 (*lower panel*). DAX data provided by Deutsche Bank Research

One of the first comparisons of a computer simulation to stock index quotes in economics was performed by Roberts [34]. He demonstrated a surprising similarity between the weekly closes of the Dow Jones Industrial Average in 1956, and an artificial index which was generated from 52 random numbers, representing the change of weekly closing prices over one year.

3.4.2 Perrin's Observations of Brownian Motion

The first systematic observations of Brownian motion were made by the French physicist Jean Perrin in 1909 and are described later in his book *Les Atomes* [35]. He charted on paper the motion of colloidal particles of radius $0.53\,\mu$m suspended in a liquid, by recording the positions every 30–50 seconds. One of his original traces is reproduced in Fig. 3.8. The straight lines between the turning points, of course, are interpolations. Perrin noted that the paths were not straight at all but that, when the observation time scale was shortened, they became more ragged on even smaller scales. These were the first experimental confirmations of Einstein's theory on Brownian motion, and on diffusion in suspensions. Recall that no such motion was allowed within classical thermodynamics, and that these observations thereby also confirmed the statistical theory of heat.

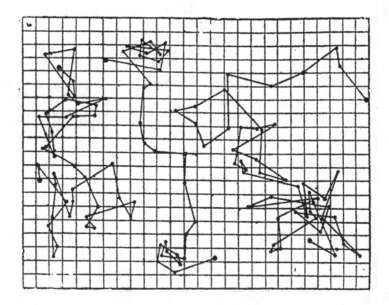

Fig. 3.8. Traces of the motions of colloidal particles suspended in a liquid, by J. Perrin. The grid size is 3.2 μm, and positions have been recorded every 30–50 seconds. Reprinted by permission from J. Perrin: *Les Atomes*, ©1948 Presses Universitaires de France

3.4.3 One-Dimensional Motion of Electronic Spins

Perrin's observations concern three-dimensional random walks. The one-dimensional random walk actually treated by Bachelier and Einstein is not so "easy" to observe. One way to generate a one-dimensional random walk is to simply project the trajectories of a higher-dimensional random walk such as Perrin's, on a line. The first actual measurement of one-dimensional Brownian motion probably is the work of Kappler [36] in 1931, who set out to determine Avogadro's constant from the Brownian motion of a torsion balance. He attached a tiny mirror to the quartz wire of a torsion balance. The wire was a few centimeters long and a few tenths of a micron thick. Molecules in the surrounding air, performing Brownian motion hit the mirror with random velocities in random direction, and thereby impart random momenta to the mirror. The mirror will then perform a one-dimensional rotational Brownian motion in an external mean-reverting potential provided by the restoring force of the quartz fiber, i.e., execute a stochastic Ornstein-Uhlenbeck process [37]. The motion of the mirror is recorded by using it to deflect a narrow light ray onto a photographic film.

A more recent example is provided by the one-dimensional trajectories of a colloidal particle of 2.5 μm diameter, performing Brownian motion in a suspension of deionized water. They have been measured in a study searching for microscopic chaos [38]. However, the paper does not reveal if they have been one-dimensionalized by projection or if the particle motion was one-dimensional.

The main problem in the observation of truly one-dimensional Brownian motion is to fabricate structures which are narrow enough so that the microscopic diffusion process becomes one-dimensional, i.e., that are of the size of the diffusing particles. Organic chemistry was the first to achieve this goal. From the mid-1960s on, there has been a big interest in low-dimensional materials conducting electric current because a theory predicted that superconductivity would be possible at room temperature (or above) in quasi-1D structures [39]. This aim has not been achieved although superconductivity has been found in organic materials at low temperatures. On the way, much interesting physics has been discovered in the many families of 1D organic conductors synthesized so far [40].

One important organic metal is tetrathiafulvalene-tetracyanoquinodimethane. The molecules constituting this material, and the basic crystal structure of TTF-TCNQ are shown in Fig. 3.9. The large planar molecules preferentially stack on top of each other, and the one-dimensionality of the electronic band structure is enhanced by the directional nature of the highest occupied molecular orbitals. Nuclear magnetic resonance now allows one to monitor the diffusive motion of the electronic spin in these one-dimensional bands [41].

In the absence of perturbations, one would observe a sharp δ-function resonance line at the nuclear Larmor frequency $\hbar\omega_N = 2\mu_N H_0$ where H_0

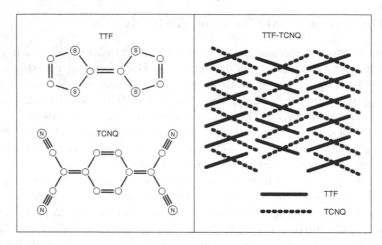

Fig. 3.9. Molecular constituents of TTF-TCNQ, and its schematic crystal structure. TTF = tetrathiafulvalene, TCNQ = tetracyanoquinodimethane

is the external magnetic field and μ_N the nuclear magneton. Perturbations, however, generate random magnetic fields at the site of the nucleus, and broaden the resonance line. Its width is usually measured by the relaxation rate $1/T_1$ (in the case we shall discuss, the spin–lattice relaxation rate $1/T_1$ is appropriate). One prominent source of perturbation is the electronic spins which couple to the nuclear spins through the hyperfine interaction. They create a fluctuating magnetic field at the site of the nucleus which faithfully reflects the dynamics of the electronic spin motion. The influence on the width of the resonance line is given by Moriya's formula (simplified here for our purposes)

$$\frac{1}{T_1 T} \propto \sum_{q} \left.\frac{\mathrm{Im}\chi_\perp(\boldsymbol{q},\omega)}{\omega}\right|_{\omega=\omega_N}, \tag{3.32}$$

where T is the temperature and $\chi_\perp$ is the transverse spin susceptibility *of the electrons*. Its microscopic definition is

$$\mathrm{Im}\chi_\perp(\boldsymbol{q},\omega) = \frac{\pi}{2}(g\mu_B)^2 \sum_{\boldsymbol{k}} \left(f[E_{\boldsymbol{k}\downarrow}] - f[E_{\boldsymbol{k}+\boldsymbol{q}\uparrow}]\right)\delta(\hbar\omega - E_{\boldsymbol{k}+\boldsymbol{q}\uparrow} + E_{\boldsymbol{k}\downarrow}). \tag{3.33}$$

g and μ_B are the electronic g-factor and Bohr magneton, respectively, and $f(E)$ is the Fermi–Dirac distribution function. Equation (3.33) states that resonance absorption is possible when occupied and unoccupied states of different spin direction at the relative wavevector $\boldsymbol{q}$ probed by the measurement, differ by precisely the energy of the external electromagnetic field. Now notice that in the presence of a magnetic field H_0, the electronic spins are shifted from their zero-field dispersions $E_{\boldsymbol{k},s}(H_0) = E_{\boldsymbol{k},s}(0) + s\hbar\omega_E/2$ by the *electronic Larmor frequency* $\hbar\omega_E = 2\mu_B H_0$. This implies that all freqencies in

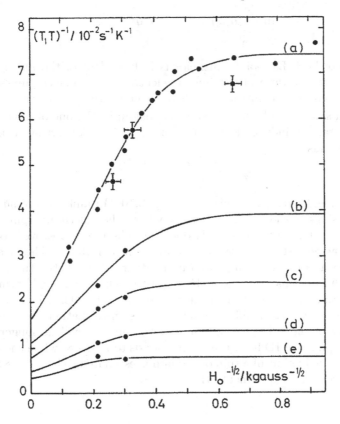

Fig. 3.10. Nuclear magnetic spin–lattice relaxation rate $1/T_1$ for the organic conductor TTF-TCNQ, plotted versus $1/\sqrt{H_0}$. At ambient pressure, curve (**a**), there is a wide range with $1/T_1 \propto 1/\sqrt{H_0}$ indicating 1D diffusion of electronic spins. Only at small fields is a crossover to H_0-independence typical for 3D diffusion observed. Curves (**b**)–(**e**) are for higher pressures where the spin dynamics is less 1D. The temperature was 296 K. By courtesy of D. Jérome. Reprinted by permission from G. Soda, et al., J. Phys. (Paris) **38**, 931 (1977) ©1977 EDP Sciences

the transverse susceptibility are shifted by $\omega_E \gg \omega_N$

$$\frac{1}{T_1 T} \to \sum_q \frac{\mathrm{Im}\chi_\perp(q,\omega_E)}{\omega_N} . \tag{3.34}$$

Let us now we assume that the electrons perform a random walk. Their susceptibility, which is the spin–spin correlation function, is given as

$$\chi(q,t) = \chi_s \exp(-Dq^2|t|) , \quad \chi(q,\omega) = \frac{Dq^2}{Dq^2 - i\omega} , \tag{3.35}$$

where D is the diffusion constant. Then

$$\frac{\mathrm{Im}\chi_\perp(\boldsymbol{q},\omega_E)}{\omega_N} = \frac{Dq^2}{(Dq^2)^2 + \omega_E^2}\frac{\omega_E}{\omega_N} . \tag{3.36}$$

The ratio of both Larmor frequencies is independent of the magnetic field (and equal to the ratio of the inverse electronic and nuclear masses), and will not be considered further. The sum over $\boldsymbol{q}$ in the first fraction on the right-hand side crucially depends on dimension: in 1D, one obtains $\sim \omega_E^{-1/2}$ and in 3D, an ω_E-independent result for small ω_E. Converting to magnetic fields, one finds

$$\frac{1}{T_1 T} \sim \begin{cases} \text{const.} & (3D) \\ 1/\sqrt{H_0} & (1D) . \end{cases} \tag{3.37}$$

The experimental results are shown in Fig. 3.10. For ambient pressure, curve (a), they show a wide range of fields where the electronic spin diffusion is indeed 1D. Only at small fields does one observe a crossover to a field-independent relaxation rate typical for 3D diffusion, (3.37). The idea behind this crossover is the following. Even in a rather 1D band structure, the electrons will have a small but finite chance of tunneling to a neighboring chain. They will thus have a finite lifetime $\tau_\perp$ on one chain. This lifetime will cut off the influence of their diffusive motion on spin-relaxation because, due to the locality of the hyperfine interaction, the nucleus will no longer see the electronic spin. The 1D limit then corresponds to $\tau_\perp \to \infty$ while the 3D limit is $\tau_\perp \to 0$. The lifetime of a spin on a chain is estimated to be $\tau_\perp \sim 8 \times 10^{-12}$s at 300 K from this experiment [41].

4. The Black–Scholes Theory of Option Prices

We now turn to the determination of the prices of derivative securities such as forwards, futures, or options in the presence of fluctuations in the price of the underlying. Such investments for speculative purposes are risky. Bachelier's work on futures already shows that for *relative* prices, even the deterministic movements of the derivative are much stronger than those of the bond, and it seems clear that an investment into a derivative is then associated with a much higher risk (see also Bachelier's evaluation of success rates) than in the underlying security, although the opportunities for profit would also be higher.

Derivative prices depend on certain properties of the stochastic process followed by the price of the underlying security. Remember from Chap. 2 that options are some kind of insurance: the price of an insurance certainly depends on the frequency of occurrence of the event to be insured. We therefore introduce the standard model of stock prices, as used in textbooks of quantitative finance [10], [12]–[16] and place this model in a more general context of stochastic processes.

4.1 Important Questions

Based on these models, we will discuss some of the important questions which are listed below.

- What determines the price of a derivative security?
- What is the role of the return of the underlying security, i.e., the drift in its price?
- What are the appropriate stochastic processes to model financial time series? Are they independent of the assets considered?
- How can we classify stochastic processes?
- How can we calculate with stochastic variables?
- What is geometric Brownian motion? Is it different from Bachelier's model?
- What is the risk of an investment in a derivative?
- What is the price of risk?
- Can risk in financial markets be eliminated? At what cost?

- Can option pricing be related to diffusion? What would be different from standard diffusion problems?
- How can we calculate option prices in ideal markets? What is different in real markets?
- What are "The Greeks"?
- How do traders represent the deviations of traded option prices from those calculated in idealized models?
- How are derivative prices related to the expected payoff of the derivative?
- What is the difference in pricing European and American-style options?
- Can options be created synthetically?
- What is a volatility index, and how is it constructed?

The important achievement of Black and Scholes [42] and Merton [43] was to answer almost all of these questions, at least for a certain idealized market. While of course one can take a speculative position in a derivative involving a big risk, Black, Merton, and Scholes show that the risk can be eliminated in principle by a hedging strategy, i.e., by an investment in another security correlated with the derivative, so as to offset all or part of the price variations. For options, there is a dynamic hedging strategy by which *the risk can be eliminated completely.* At the same time, the possibility of hedging the risk allows to one fix a fair price of an option: it is determined by the expected payoff for the holder and the cost of the hegde, and no *additional risk premium* is necessary on options in idealized markets. Although their assumptions are not necessarily realistic, this is a benchmark result which earned Merton and Scholes the 1997 Nobel Prize in Economics, Black having died meanwhile. For forwards and futures, a static hedge, implemented at the time of writing, is sufficient.

Here, we only present the theoretical framework established in finance [10]. Of course, this heavily draws on the assumption of a random walk followed by financial time series. While we have discussed random walks in finance and physics in the previous chapter quite generally, we will specify in detail the model used by economists. More advanced and more speculative proposals for derivative pricing and hedging will be discussed later in Chap. 7. Also, we will limit our discussion to the most basic derivatives (forwards, futures, and European options): they are sufficient to illustrate the main principles. The methods developed here can then be applied, with only minor extensions, to more complicated instruments [10]. Hull's book [10] also contains much more information on practical aspects, and is highly recommended for reading.

4.2 Assumptions and Notation

4.2.1 Assumptions

Here, we summarize the main economic assumptions underlying the work of Black, Merton, and Scholes, as well as much related work on derivative pricing and financial engineering. More specific assumptions on the stochastic process followed by the underlying security will be developed in Sect. 4.4. We assume:

- a complete and efficient market;
- zero transaction costs;
- that all profits are taxed in a similar way, and that consequently, tax considerations are irrelevant;
- that all market participants can lend and borrow money at the same risk-free interest rate r;
- that all market participants use all arbitrage possibilities;
- continuous compounding of interest, i.e., an amount of cash y accumulates interest as $y(T) = y(t) \exp[r(T - t)]$;
- that short selling with full profits is allowed;
- that there are no payoffs such as dividends, from the underlying securities (*we* shall make this assumption here to simplify matters; it is not realistic, and payoffs can be incorporated into derivative pricing schemes [10]).

4.2.2 Notation

Here we list the most important symbols used in the following chapters:

- T ... time of maturity of a derivative
- t ... present time
- S ... price of the underlying security
- K ... delivery price in a forward or futures contract
- f ... value of a long position in a forward or futures contract
- F ... price of forward contract
- r ... risk-free interest rate
- C ... price of a call option
- P ... price of a put option
- X ... strike price of the option.

4.3 Prices for Derivatives

Some price considerations are independent of the fluctuations of the price of the underlying securities. These are the forward prices and futures prices because they are binding contracts to both parties, and can be perfectly, and statically, hedged. (There are some restrictions to this statement for futures because they can be traded on exchanges.) We shall treat them first. Also some price limits for options can be derived without knowing the stochastic process of the underlying securities. An accurate calculation, however, requires this knowledge and will be deferred to Sect. 4.5.

4.3.1 Forward Price

We claim that the price of a forward contract on an underlying without payoff, such as dividends, is

$$F(t) = S(t) \exp[r(T - t)] \,. \tag{4.1}$$

Notice that this is the price *today* of the contract with maturity T. It is just the spot price with accumulated risk-free interest, and is *independent* of any historical or future drift in the price S of the underlying! We prove this equation in two different ways, in order to illustrate the methods of proofs often used in finance.

First Proof

We prove (4.1) by contradiction, relying on a "no arbitrage" argument. Assume first that $F(t) > S(t) \exp[r(T - t)]$. Then, at time t, an investor can borrow an amount of cash S and use it to buy the underlying at the spot price $S(t)$. At the same time, he goes short in the forward. This involves no cost because the forward is just a contract carrying the obligation to deliver the underlying at maturity. At maturity T, the credit must be reimbursed with interest accrued, i.e., there is a cash flow $-S(t) \exp[r(T-t)]$. The underlying is now sold under the terms of the forward contract, which results in a cash flow $F(T)$, the (yet) undetermined forward price. However, $F(T) = F(t)$, because the price of the forward has been fixed at the time of writing of the contract, and there are no trading opportunities. The total cash flow is therefore $F(t) - S(t) \exp[r(T - t)] > 0$, and a riskless profit can be made. This is contrary to the assumption of no arbitrage opportunities.

For the opposite assumption, $F(t) < S(t) \exp[r(T - t)]$, an investor can generate a riskless profit $S(t) \exp[r(T - t)] - F(t)$ by (i) taking the long position in the forward at t, (ii) short-selling the underlying asset at t, giving a cash flow $+S(t)$, (iii) investinging this money at the risk-free rate r at t, (iv) buying back the underlying asset at T under the terms of the forward contract, resulting in a cash flow $-F(T) = -F(t)$, and (v) getting back $S(t) \exp[r(T - t)]$ from his risk-free cash investment. Consequently, the only price compatible with the absence of arbitrage possibilities is (4.1).

Second Proof

The idea here is to construct two portfolios out of the three assets: forward, underlying and cash. These two portfolios carry the same risk, and their value at some instant of time can be shown to be equal.

Portfolio A contains a long position in the forward with a value $f(t)$, and an amount of cash $K \exp[-r(T - t)]$. At time T, this will be worth K. Portfolio B contains one underlying asset. At maturity T, the long position of the forward is used to acquire the asset, and both portfolios are worth the same because the delivery price K must be spent and both portfolios contain one asset. Moreover, both portfolios carry the same risk for all times because the long position in the forward *necessarily* receives the asset at maturity. Hence both portfolios have the same value for all times, i.e.,

$$f(t) + K \exp[-r(T - t)] = S(t) . \tag{4.2}$$

Now, the forward price can be fixed to the delivery price $F(t) = K$ by requiring that the net value of the long position at the time of writing is zero, i.e., that a fair contract for both parties is written. $f(t) = 0$ in (4.2) directly leads back to (4.1).

While these results may look trivial, they are indeed noteworthy:

- The prices of forwards and (to some extent, to be specified below) futures can be fixed at the time of writing the contract. They do not depend on the future evolution of the price of the underlying, up to maturity. Of course, a forward contract entered at a time $t' > t$, when the price of the underlying has changed to $S(t')$, will have a different price $F(t')$, determined again by (4.1). As the second proof makes clear, the "forward price" F actually is the delivery price of the underlying asset at maturity. It is not a price reflecting the intrinsic value of the contract. Unlike for the options to be discussed later, this intrinsic value is zero. The reason is that the outcome is certain: the underlying asset is delivered at maturity.
- In the above proofs, this fact was used to calculate the forward price in terms of the price of the underlying. A position in the forward, or in the underlying asset, carries a risk, connected to the price variations of the underlying asset. However, this risk can be hedged away statically (i.e., once and for all): for a long position in the forward, one can go short in the underlying, and for a short position in the forward, a long position in the underlying asset will eliminate the risk completely. This allows another interpretation of the forward price (4.1): in such a portfolio with a perfect hedge, there is no longer any risk. In the absence of arbitrage opportunities, it only can earn the risk-free interest rate r. This is precisely what (4.1) states.

4.3.2 Futures Price

Futures are distinguished from forwards mainly by being standardized, tradable instruments. If the interest rates do not vary during the period of the contract, the futures price equals the forward price. The prices are different, however, when interest rates vary. These differences are introduced by details of the trading procedures. For a forward, there is no cash flow for either party until maturity, where it will be settled. For futures, margin accounts (where a fixed fraction of the liabilities of a derivative portfolio is deposited for security) must be opened with the broker, and balanced daily. The money flowing in and out of these margin accounts in the case of a futures contract can then be invested, resp. must have been liquidated, at current market conditions, i.e., based on interest rates that may be different from those at the time the contract was entered. This gives different prices for forwards and futures. Empirically, however, the differences seem to be rather small [10].

4.3.3 Limits on Option Prices

The forward and future prices for contracts written today are independent of the details of the price history of the underlying, such as the drift or variance of the price. This is not so for options, and for accurate price calculations a knowledge of the important parameters of the price variations of the underlying is necessary. This will be developed in Sect. 4.5.1 below. On the other hand, it is fairly simple to obtain certain limits to be obeyed by option prices without knowing the price fluctuations of the underlying. If not stated otherwise, we will always consider European type options.

Upper Limits

A call option, by construction, can never be worth more than the underlying security. Therefore

$$C(t) \leq S(t) . \tag{4.3}$$

The value of a put option can never exceed the strike price

$$P(t) \leq X . \tag{4.4}$$

If one of these inequalities is violated, an arbitrageur can make riskless profit by buying the stock and selling the option (call), or simply selling the option (put). For a European put, a more stringent condition can be given because the strike price is also fixed in the future, and can be discounted from maturity to the present date

$$P(t) \leq X \exp[-r(T-t)] . \tag{4.5}$$

Lower Limits

To determine the lower limits of a call price, we construct two portfolios: A contains one call at price C and $X \exp[-r(T-t)]$ in cash; B contains one stock. At maturity, B is worth $S(T)$. If $S(T) > X$, the call in A is exercised, and A is worth $S(T)$ (X is used to buy the stock). If $S(T) < X$, the call option expires worthless, and portfolio A is worth X. The value of A is therefore $\max[S(T), X] \geq S(T)$, the value of B. This is valid for all times because the value of both portfolios depends only on the same source of uncertainty, the evolution of the stock price S. Consequently,

$$C(t) \geq \max\{S(t) - X \exp[-r(T-t)], 0\} . \tag{4.6}$$

The equivalent relation for a put,

$$P(t) \geq \max\{X \exp[-r(T-t)] - S(t), 0\},\tag{4.7}$$

can be derived in a similar way, using one portfolio (C) containing the put option and the stock, and another (D) with $X \exp[-r(T-t)]$ in cash.

These limits, together with a sketch of the dependence of option prices on those of the underlying, are shown in Figs. 4.1 (call) and 4.2 (put). The arrows in Figs. 4.1 and 4.2 indicate how the curve is displaced, resp. distorted,

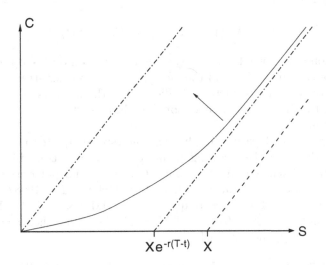

Fig. 4.1. Price limits for call options. The *curved* line sketches a realistic price curve. The *arrow* marks the direction of displacement of the curve when r, $T - t$, or the volatility (standard deviation) σ of the stock price increase

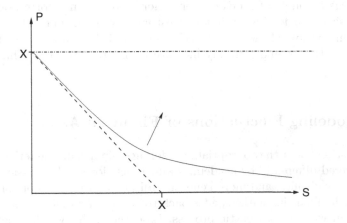

Fig. 4.2. Price limits for put options

when the interest rate r, the time to maturity $T - t$, or the volatility of the underlying stock (measured by the standard deviation σ of the stock price) change. An empirical investigation on 58 US stocks from August 1976 to June 1977, discussed by Hull [10], finds that the lower limits for calls, (4.6) and Fig. 4.1, were violated in 1.3% of the quotations. Out of these, 29% were corrected on the next quote while 71% were smaller than applicable transaction costs. Therefore no arbitrage was possible despite these limit violations.

Another important relation, put–call parity, can be derived by comparing the portfolios A and C:

$$C(t) + X \exp[-r(T - t)] = P(t) + S(t) . \qquad (4.8)$$

This equation does not rely on any specific assumption on the options or on the prices of the underlying and therefore provides a rather stringent test on the correct operation (complete and efficient) of the markets. The empirical study cited by Hull [10] finds occasional violations of put–call parity on a 15-minute time scale.

Checking put–call parity simply from newspaper quotes may be more involved, as shown by the following example with options traded in late 1998 on the EUREX exchange. At $t = 1998/10/21$, call and put options on the Bayer stock with nominal maturity December 1998, i.e., $T = 1998/12/18$ and a strike price of $X = DM\ 65$, were quoted $C = DM\ 2.38$ and $P = DM\ 5.50$. Bayer was quoted $S(1998/10/21) = DM\ 61.25$. Assuming then $r = 3\%$ p.a., and $T - t = (1/6)y$, one has (in DM)

$$2.38 + 65 \exp(-0.005) = 67.05 \neq 66.75 = 5.50 + 61.25 . \qquad (4.9)$$

It is not clear, however, that this is an actual violation of put–call parity. In particular, the assumption on r has been made ad hoc with rates relevant for savings accounts of a private consumer, and may not correspond to the market situation for institutional investors. Assuming put–call parity and calculating backwards, would give $r(T - t) = 0.01$, i.e., twice as much as used above, and then would certainly indicate an interest rate much higher than 3% p.a.

4.4 Modeling Fluctuations of Financial Assets

The question about the appropriate modeling of financial time series may well be answered differently by academics and practitioners. The basic approach taken by academics, and more generally all people with a skeptical attitude towards the financial markets, goes back to Bachelier and assumes some kind of random walk, or stochastic process. Essentially, this will be the attitude adopted in this book. Some aspects of random walks have been discussed in Chap. 3. Others will be introduced below, together with a more general summary of important facts on stochastic processes.

Among the practitioners, traders and analysts classified as "chartists", practicing "technical analysis", would not share this opinion. This group of operators attempts to distinguish recurrent patterns in financial time series and tries to make profit out of their observation. The citation from Malkiel's book *A Random Walk Down Wall Street* reproduced in Chap. 1 testifies to this, as well as numerous books on technical analysis at different levels. However, the issue of correlations in financial time series is nontrivial. We shall discuss simple aspects in Sect. 5.3.2, but subtle aspects are still the subject of ongoing research. It has to be taken seriously because technical analysis is alive and well on the markets, and one therefore must conclude that some money can be earned this way, and that certain correlations indeed exist in financial data, perhaps even introduced by a sufficient number of traders following technical analysis even on purely random samples. Systematic studies of the profitability of technical analysis reach controversial conclusions, however [31].

4.4.1 Stochastic Processes

Classic references on stochastic processes are Cox and Miller, and Lévy [32]. There are two excellent books by J. Honerkamp, concerned with, or touching upon, stochastic processes [44], and presenting a more physics-oriented perspective.

We say that a variable with an unpredictable time evolution follows a *stochastic process*. The changes of this variable are drawn from a probability distribution according to some specified rules. *One* distinction of stochastic processes is made according to whether time is treated as a continuous or a discrete variable, and whether the stochastic variable is continuous or discrete. We will be rather sloppy on this distinction here.

Stochastic processes are described by the specification of their dynamics and of the probability distribution functions from which the random variables are taken. The dynamics is usually given by a stochastic difference equation such as, e.g.,

$$x(t+1) = x(t) + \varepsilon(t) \tag{4.10}$$

where x is the stochastic variable and ε is a random variable whose probability distribution must be specified, or by differential equations such as

$$\dot{x}(t) = ax(t) + b\varepsilon(t) , \tag{4.11}$$

$$\dot{x}(t) = ax(t) + bx(t)\varepsilon(t) . \tag{4.12}$$

Equation (4.11) describes "additive noise" because the random variable is added to the stochastic variable, and (4.12) describes "multiplicative noise".

Next, we must specify the probability distribution function of $\varepsilon(t)$, e.g.,

$$p(\varepsilon, t) = \frac{1}{\sqrt{2\pi\sigma^2 t}} \exp\left(-\frac{\varepsilon^2}{2\sigma^2 t}\right) . \tag{4.13}$$

Correlations in a stochastic process can be described either in its defining equation, e.g., by a dependence on earlier times [cf., e.g., the various autoregressive processes (4.44), (4.46) and (4.48) below], or by the *conditional probability*

$$p\left[x(t_1) = x_1 | x(t_0) = x_0, x(t_{-1}) = x_{-1}, \ldots\right] , \tag{4.14}$$

which measures the probability that the variable x takes the value x_1 at t_1 *provided that* x_0 has been observed at t_0 *and* x_{-1} at t_{-1}, etc. For a continuous variable, the conditional probability density $p[\ldots]\mathrm{d}x_1$ measures the probability that at t_1, $x_1 \leq x \leq x_1 + \mathrm{d}x_1$, provided x_0 has been observed at t_0, etc. The *unconditional probability* (or marginal probability) of observing x_1 at t_1, independently of earlier realizations of x, is then

$$p\left[x(t_1) = x_1\right]$$
$$= \int \mathrm{d}x_0 \mathrm{d}x_{-1} \ldots p\left[x(t_1) = x_1, x(t_0) = x_0, x(t_{-1}) = x_{-1}, \ldots\right] \tag{4.15a}$$
$$= \int \mathrm{d}x_0 \mathrm{d}x_{-1} \ldots p\left[x(t_1) = x_1 | x(t_0) = x_0, x(t_{-1}) = x_{-1}, \ldots\right]$$
$$\times \ p(x_0, t_0)p(x_{-1}, t_{-1}) \ldots \tag{4.15b}$$

where $p[\ldots]$ on the right-hand side of (4.15a) is the *joint probability* which measures the probability of observing x_1 at t_1 and x_0 at t_0, etc. It is related to the conditional probability (4.14) by the second equality (4.15b).

A stochastic process is stationary if

$$p(x, t) = p(x) , \tag{4.16}$$

and it is a martingale stochastic process if

$$E(x_1 | x_0, x_{-1}, \ldots) = \int \mathrm{d}x_1 x_1 p\left[x(t_1) = x_1 | x(t_0) = x_0, x(t_{-1}) = x_{-1}, \ldots\right] = x_0 , \tag{4.17}$$

where E is the expectation value conditioned on earlier observations x_0, x_{-1}, etc.

We now discuss a few important stochastic processes.

Markov Processes

For a Markov process, the next realization only depends on the present value of the random variable. There is no longer-time memory. For $\ldots t_{-2} \leq t_{-1} \leq t_0 \leq t_1 \leq \ldots$, a Markov process satisfies

$$p\left[x(t_1) = x_1 | x(t_0) = x_0, x(t_{-1}) = x_{-1}, \ldots\right] = p\left[x(t_1) = x_1 | x(t_0) = x_0\right] .$$
$$(4.18)$$

Markov processes obey the Chapman–Kolmogorov–Smoluchowski equation (3.10), derived by Bachelier [6].

For Markov processes in continuous time, one can take a short-time limit of the conditional probability distributions

$$p(x, t | x', t') \to \delta(x - x') \quad \text{for} \quad t \to t' , \tag{4.19}$$

and expand it around this limit in first order in $t - t'$:

$$p(x, t | x', t') \approx [1 - a(x, t)(t - t')] \delta(x - x') + (t - t')w(x, x', t) , \tag{4.20}$$

where $a(x, t)$ and $w(x, x', t)$ are expansion coefficients. $a(x, t)$ is the reduction, in first order in the time difference, of the initial "certainty", i.e., the weight of $\delta(x - x')$ due to the widening of the conditional probability distribution, and $w(x, x', t)$ quantifies precisely this effect in first order in $t - t'$. Inserting this expansion into the Chapman–Kolmogorov–Smoluchowski equation (3.10), one obtains the master equation

$$\frac{\partial p(x, t)}{\partial t} = \int dx' w(x, x', t) p(x', t) - \int dx' w(x, x', t) p(x, t) . \tag{4.21}$$

The first term on the right-hand side describes transitions $x' \to x$ at t, and the second term transitions $x \to x'$. We have made an integro-differential equation from the original convolution equation. In special situations, the master equation may reduce to a partial differential equation, the Fokker–Planck equation [37], which will be discussed in later in Chap. 6.

In finance, Markov processes are consistent with an efficient market. If this were not so, technical analysis would allow one to produce above-average profits. Conversely, to the extent that technical analysis generates consistent profits above the market return, the assumption of a Markov process for financial time series must be questioned.

The Wiener Process

The Wiener process, often also called the Einstein–Wiener process, or Brownian motion, is a particular Markov process with continuous variable and continuous time. It was formulated for the first time by Bachelier [6], and discussed on an elementary level in Sect. 3.2.2. If the stochastic variable is called z, its two important properties are:

1. Consecutive Δz are statistically independent.
2. Δz is given, for a small but finite time interval Δt, and for an infinitesimal interval dt, by

$$\Delta z = \varepsilon \sqrt{\Delta t} \tag{4.22}$$
$$dz = \varepsilon \sqrt{dt} . \tag{4.23}$$

ε is drawn from a normal distribution

$$p(\varepsilon) = \frac{1}{\sqrt{2\pi}} \exp\left(-\frac{\varepsilon^2}{2}\right) \qquad (4.24)$$

with zero mean and unit variance.

The passage from a Wiener process in discrete time to one in continuous time is illustrated in Fig. 4.3.

The conditions for a Wiener process are stronger than for a general Markov process, in that it uses independent, identically distributed (abbreviated: IID) random variables. Being independent, the correlations of the random numbers ε are

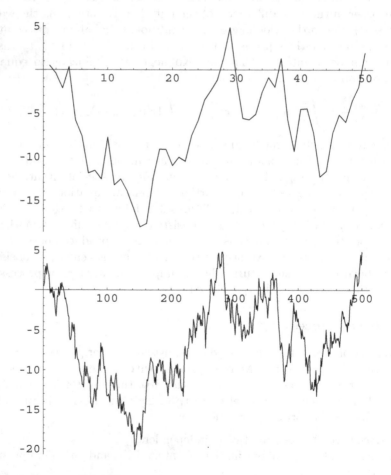

Fig. 4.3. Passage from discrete time to continuous time for a Wiener process. The increments were drawn from a normal distribution with zero mean and unit variance

$$\langle \varepsilon(t)\varepsilon(t')\rangle = \sigma^2 \begin{cases} \delta_{t,t'} \\ \delta(t-t') \end{cases} \qquad (4.25)$$

where σ^2 is the variance of the underlying normal distribution. Its noise spectrum is

$$F(\omega) = \int_{-\infty}^{\infty} d\tau \langle \varepsilon(t)\varepsilon(t+\tau)\rangle e^{i\omega\tau} = \sigma^2 . \qquad (4.26)$$

It is independent of frequency, and therefore "white noise". Often, this is also written as

$$\varepsilon(t) \in \mathrm{WN}(0, \sigma^2) . \qquad (4.27)$$

W characterizes the random variables as "white noise", N denotes "normally distributed", and the arguments are the mean and variance. A stochastic process with an additive white noise term describes algebraic Brownian motion. Notice that some authors, e.g., Hull [10], prefer to take the standard deviation instead of the variance, as the second argument of WN in (4.27).

Equation (4.23) may seem very surprising for those who are not familiar with stochastic processes. It is to be interpreted in the sense of mean square fluctuations, resp. expectation values. A more detailed argument goes as follows. Let a stochastic process be defined by the differential equation

$$\frac{dz(t)}{dt} = \varepsilon(t) \qquad (4.28)$$

where $\varepsilon(t)$ is the random variable. Then, the change $dz(t)$ of the random variable z in an infinitesimal time interval dt is given by integration

$$dz(t) = \int_{t}^{t+dt} dt' \varepsilon(t') . \qquad (4.29)$$

For a nonstochastic variable, this integral would be trivial and given by $dz = \varepsilon(t)dt$. That this can't hold for a stochastic variable is clear from taking the expectation values of (4.29)

$$\langle dz(t)\rangle = \int_{t}^{t+dt} dt' \langle \varepsilon(t')\rangle = 0 . \qquad (4.30)$$

On the other hand, the expectation value of $(dz)^2$ becomes

$$\langle dz(t)dz(t)\rangle = \int_{t}^{t+dt} dt_1 dt_2 \langle \varepsilon(t_1)\varepsilon(t_2)\rangle = \sigma^2 \int_{t}^{t+dt} dt_1 = \sigma^2 dt . \qquad (4.31)$$

For the second equality, we have used (4.25), and the third equality obtains in the usual way because σ^2 is a nonstochastic quantity. These expectation values are consistent with $dz = \varepsilon\sqrt{dt}$, (4.23).

For a Wiener process, the *expectation value* of the stochastic variable in a small time interval vanishes

$$E(\Delta z) \equiv \langle \Delta z \rangle = \int_{-\infty}^{\infty} \mathrm{d}(\Delta z)\, \Delta z\, p(\Delta z) = 0 \ . \tag{4.32}$$

Its *variance* is linear in Δt,

$$\mathrm{var}(\Delta z) = \int_{-\infty}^{\infty} \mathrm{d}(\Delta z)(\Delta z)^2 p(\Delta z) = \Delta t \ , \tag{4.33}$$

and its *standard deviation* behaves as

$$\sqrt{\mathrm{var}(\Delta z)} = \sqrt{\Delta t} \ . \tag{4.34}$$

Finite time intervals T may be considered as being composed of many small intervals ($T = N\Delta t$ fixed, as $N \to \infty$ and $\Delta t \to 0$), each of which corresponds to one time step of a Wiener process. For sums of normally distributed quantities, the mean values and variances are additive:

$$\langle z(T) - z(0) \rangle \quad = 0 \ , \tag{4.35}$$
$$\mathrm{var}[z(T) - z(0)] = T \ , \tag{4.36}$$

and the standard deviation is $\sqrt{T}$.

The Wiener process may be generalized by superposing a drift $a\,\mathrm{d}t$ onto the stochastic process $\mathrm{d}z$

$$\mathrm{d}x = a\,\mathrm{d}t + b\,\mathrm{d}z \ . \tag{4.37}$$

For this *generalized Wiener process,* we have

$$\langle x(T) - x(0) \rangle \quad = aT \ , \tag{4.38}$$
$$\mathrm{var}[x(T) - x(0)] = b^2 T \ . \tag{4.39}$$

This generalized Wiener process is shown in Fig. 4.4.

A further generalization is the Itô process where the drift term and prefactor of the stochastic component depend on the random variable [$a \to a(x,t)$, $b \to b(x,t)$], i.e.,

$$\mathrm{d}x = a(x,t)\mathrm{d}t + b(x,t)\mathrm{d}z \ , \tag{4.40}$$

and $\mathrm{d}z = \varepsilon\sqrt{\mathrm{d}t}$ describes a Wiener process. The Itô process will play an important role in the standard model for stock prices.

Other Important Processes

For completeness, we discuss some more important stochastic processes or classification criteria.

1. **Self-similar stochastic processes** with index, or Hurst exponent, H are defined by

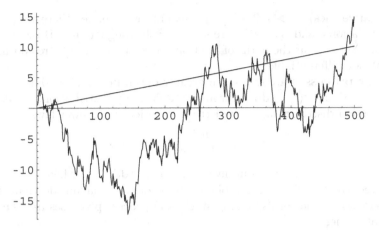

Fig. 4.4. The generalized Wiener process. The straight line shows the drift superposed on the data in the bottom panel of Fig. 4.3

$$p\left[x(at)\right] = p\left[a^{H}x(t)\right] \quad \text{with } a > 0 .\tag{4.41}$$

A rescaling of time leads to a change in length scale, and there is no intrinsic scale associated with this process. Such a process violates (4.16) and therefore cannot be stationary. Brownian motion (cf. above) is self-similar with $H = 1/2$. However, the converse is not true: There are non-Gaussian stochastic processes with independent increments but $H = /2$ [45, 46].

2. In **fractional Brownian motion**, introduced by Mandelbrot [47], the random variables are not uncorrelated, and therefore describe "colored noise". The construction is done starting from ordinary Brownian motion $dz \equiv dz(t)$, (4.23), and a parameter H satisfying $0 < H < 1$. Then fractional Brownian motion of exponent H is essentially a moving average over $dz(t)$ in which past increments of $z(t)$ are weighted by a power-law kernel $(t - s)^{H-1/2}$. Mandelbrot and van Ness define fractional Brownian motion of exponent H, $B_{H}(t)$, as [47]

$$B_H(t) = B_H(0) + \frac{1}{\Gamma(H + \frac{1}{2})} \left\{ \int_{-\infty}^{0} \left[(t - s)^{H-1/2} \right. \right.$$
$$\left. \left. - (-s)^{H-1/2} \right] dz(s) + \int_{0}^{t} (t - s)^{H-1/2} dz(s) \right\} \tag{4.42}$$

$B_H(0)$ is an arbitrary initial starting position. For $H = 1/2$, fractional Brownian motion reduces to ordinary Brownian motion.

The ranges $H < 1/2$ and $H > 1/2$ are very different. For $H < 1/2$, the paths look less ragged than ordinary Brownian motion, and the variations are "antipersistent" (positive variations preferentially followed by negative ones). $H > 1/2$ is the "persistent" regime, i.e., there are positive correlations, and the paths are significantly rougher than Brownian motion. Notice that the paths of fractional Brownian motion are continuous but not differentiable.

3. **Lévy processes** are treated in greater detail in Sect. 5.4. The IID random variable $\varepsilon(t)$ is drawn from a stable Lévy distribution. Unlike the Gaussian distribution, Lévy distributions decay as power laws

$$L_\mu(x) \sim \frac{\mu A^\mu}{|x|^{1+\mu}} , \quad |x| \to \infty . \tag{4.43}$$

They are stable, i.e., form-invariant under addition, when $0 < \mu < 2$. Large events being more probably by orders of magnitude than under a Gaussian, the corresponding stochastic process possesses frequent discontinuities.

4. **Autoregressive processes** are non-Markovian. The equation of motion contains memory terms which depend on past values of variables. The equation

$$x(t) = \sum_{k=1}^{p} \alpha_k x(t - k) + \varepsilon(t) + \sum_{k=1}^{q} \beta_k \varepsilon(t - k) \tag{4.44}$$

describes an autoregressive, moving average, ARMA(p, q), process. It depends on the past p realizations of the stochastic variable x, and on the past q values of the random number ε. ARMA(p, q) processes can be interpreted as stochastically driven oscillators and relaxators [44].

Variants thereof, the ARCH and GARCH processes, are important in econometrics and finance [13]. The acronyms stand for autoregressive [process with] conditional heteroscedasticity, and generalized autoregressive [process with] conditional heteroscedasticity. Heteroscedasticity means that the variance of the process is not constant but depends on random variables. To be specific, an ARCH(q) process [48] is defined by (4.22) with

$$\varepsilon(t) \in \text{WN}\left[0, \sigma^2(t)\right] \tag{4.45}$$

$$\sigma^2(t) = \alpha_0 + \sum_{i=1}^{q} \alpha_i \varepsilon^2(t - i) , \tag{4.46}$$

and a GARCH(p, q) process [49] by (4.22) with

$$\varepsilon(t) \in \text{WN}\left[0, \sigma^2(t)\right] \tag{4.47}$$

$$\sigma^2(t) = \alpha_0 + \sum_{i=1}^{q} \alpha_i \varepsilon^2(t - i) + \sum_{i=1}^{p} \beta_i \sigma^2(t - i) . \tag{4.48}$$

In both cases, the random variable is drawn from a normal distribution with zero mean and a *time-dependent* variance $\sigma^2(t)$ which depends on the last q realizations of the random variable ε and, for the GARCH(p, q) process, in addition on the last p values of the variance σ^2.

4.4.2 The Standard Model of Stock Prices

Bachelier modeled stock or bond prices by a random walk superimposed on a constant drift (with the exception of the liquidation days where coupons were detached from the bonds, or the maturity dates of the futures where a prolongation fee had to be paid eventually). The drift was further eliminated from the problem by considering the equivalent martingale process as the fundamental variable, i.e., a Wiener process with zero mean and a variance increasing linearly in time.

There are two problems with this proposal:

1. The stock or bond prices in the model may become negative, in principle, when the changes $\Delta S(T)$ accumulated over a time interval T exceed the starting price $S(0)$. While this is not likely in practical situations, it should be a point of concern, in principle.

2. In Bachelier's model, the profit of an investment into a stock with price S over a time interval T is

$$\langle S(T) - S(0) \rangle = \frac{\mathrm{d}S}{\mathrm{d}t} T , \tag{4.49}$$

where $\mathrm{d}S/\mathrm{d}t$ is the drift which was assumed fixed and independent of S. More important than the *profit*, for an investor, will be the *return* on his capital invested. An investor will require that the return of an investment will be independent of the price of the asset (in other words, if a return of 15% p.a. is required when a stock is at \$40, it will also be required at \$65). This can be written as

$$\mathrm{d}S = \mu S \mathrm{d}t , \tag{4.50}$$

giving $S(t) = S_0 e^{\mu t}$ where μ is the return rate, and $\mu \Delta t$ the return over a time interval Δt. This has consequences for the risk of an investment, measured by the standard deviation or – in financial contexts – volatility of asset prices. (Being careful, one should distinguish between variances accumulated over certain time intervals, or variance rates, entering the stochastic differential equations, resp. the corresponding quantities for the standard deviations.) A reasonable requirement is that the variance of the returns $\mu = \mathrm{d}S/\mathrm{d}t$ should be independent of S, i.e., that the uncertainty on reaching the 15% return discussed above, is the same regardless of whether the stock price is at \$40 or 80\$. This implies that, over a time interval Δt

$$\sigma^2 \Delta t = \mathrm{var}\left(\frac{\Delta S}{S}\right) \tag{4.51}$$

is independent of the stock price, or that

$$\mathrm{var}(S) = \sigma^2 S^2 \Delta t . \tag{4.52}$$

These requirements suggest that the asset price can be represented as an Itô process

$$dS = \mu S dt + \sigma S dz \ , \ \text{resp.} \quad \frac{dS}{S} = \mu dt + \sigma dz = \mu dt + \sigma \varepsilon \sqrt{dt} \qquad (4.53)$$

with instantaneous drift and standard deviation rates μ and σ. In other words,

$$\frac{dS}{S} \in \text{WN}(\mu dt, \sigma^2 dt) \ , \qquad (4.54)$$

i.e., dS/S is drawn from a normal distribution with mean μdt and standard deviation $\sigma \sqrt{dt}$. Concerning (4.53) and (4.54), notice that

$$\frac{dS}{S} \neq d \ln S \quad \text{for stochastic variables.} \qquad (4.55)$$

The process (4.53) is referred to as *geometric Brownian motion*. S follows a stochastic process subject to multiplicative noise. It avoids the problem of negative stock prices, and apparently is in better agreement with observations.

Notice that the model of stock prices following geometric Brownian motion (4.53) must be considered as a hypothesis which has to be checked critically, and not as an established and universal theory. A critical comparison to empirical market data will be given in Chap. 5. For a superficial comparison, Fig. 4.5 shows the chart of the Commerzbank through the year 1997. This chart is not primarily shown for supportive purposes. More intended to inspire caution, it demonstrates the enormous variety of behavior encountered even for a single blue chip stock, which contrasts with the simplicity of the postulated standard model (4.53). While a priori the parameters μ and σ of the standard model are taken as constants, Fig. 4.5 suggests that this may be a valid approximation – if ever – only over limited time spans. The annualized volatility is 33.66%, and the drift during this year is $\mu = 82\%$. As is apparent from the figure, μ and σ in practice depend on time, and on shorter time scales in the course of the year they may be rather far from the values cited. Analyses taking μ and σ constant will only have a finite horizon of application. This observation has been an important motivation for the study of the ARCH and GARCH processes discussed in Sect. 4.4.1. Due to its simplicity, and the fundamental insights it allows, we shall use the model of geometric Brownian motion in the remainder of this chapter to develop a theory of option pricing. To do so, we must know, however, some properties of functions of stochastic variables.

4.4.3 The Itô Lemma

If we assume that the price process of a financial asset follows a stochastic process, the process followed by a derivative security, such as an option, will

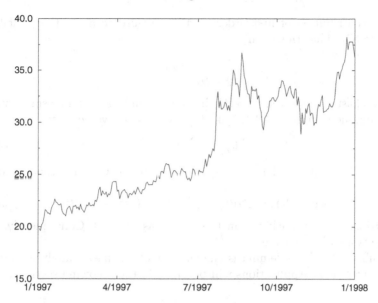

Fig. 4.5. Chart of the Commerzbank share from 1/1/1997 to 31/12/1997. The price has been converted to Euros. The volatility is $\sigma = 33.66\%$

(i) again be stochastic, and (ii) be a function of the price of the underlying. We therefore must know the properties of functions of stochastic variables.

An important result here, and the only one we need for future development, is a lemma due to Itô. Let $x(t)$ follow an Itô process, (4.40),

$$\mathrm{d}x = \dot{a}(x,t)\mathrm{d}t + b(x,t)\mathrm{d}z = a(x,t)\mathrm{d}t + b(x,t)\varepsilon\sqrt{\mathrm{d}t} \ . \tag{4.56}$$

Then, a function $G(x,t)$ of the stochastic variable x and time t also follows an Itô process, given by

$$\mathrm{d}G = \left(\frac{\partial G}{\partial x}a + \frac{\partial G}{\partial t} + \frac{1}{2}b^2\frac{\partial^2 G}{\partial x^2}\right)\mathrm{d}t + b\frac{\partial G}{\partial x}\mathrm{d}z \ . \tag{4.57}$$

The drift of the Itô process followed by G is given by the first term on the right-hand side in parentheses, and the standard deviation rate is given by the prefactor of $\mathrm{d}z$ in the second term.

There is a handwaving way to motivate the different terms in (4.57). We attempt a Taylor expansion of $G(x + \mathrm{d}x, t + \mathrm{d}t)$ about $G(x,t)$ to first order in $\mathrm{d}t$. The first order expansion in $\mathrm{d}x$ produces the first and the last terms on the right-hand side of (4.57), and the first order expansion in $\mathrm{d}t$ produces the second term. Stopping the expansion at this stage would not be consistent, however, because $\mathrm{d}x$ contains a terms proportional to $\sqrt{\mathrm{d}t}$, shown explicitly in (4.56). The second-order expansion in $\mathrm{d}x$ therefore produces another contribution of first order in dt, the third term on the right-hand side of (4.57). That this term

$$\frac{1}{2}b^2\frac{\partial^2 G}{\partial x^2}\varepsilon^2 dt$$

is nonstochastic, and given correctly in (4.57), can be shown in a spirit similar to the argument in Sect. 4.4.1. Take the expectation value of $\varepsilon^2 dt$

$$\langle\varepsilon^2 dt\rangle = \langle\varepsilon^2\rangle dt = dt \ , \tag{4.58}$$

where the last equality follows from $\varepsilon \in \mathrm{WN}(0,1)$. On the other hand, its variance,

$$\mathrm{var}(\varepsilon^2 dt) = \langle\varepsilon^4 dt^2\rangle - \langle\varepsilon^2 dt\rangle^2 = \left(\langle\varepsilon^4\rangle - 1\right)dt^2 \tag{4.59}$$

tends to zero more quickly than the mean, as $dt \to 0$. Consequently, $\varepsilon^2 dt$ represents a sharp variable.

A full proof of this lemma is the subject of stochastic analysis and will not be given here. Applications will be given in the following sections.

4.4.4 Log-normal Distributions for Stock Prices

We now derive the probability distribution for the stock prices, based on the assumption of geometric Brownian motion. To do that, we start from the stochastic differential equation (4.53) for the price changes

$$dS = \mu S dt + \sigma S dz \ , \tag{4.60}$$

and apply the Itô lemma with $G(S, t) = \ln S(t)$ [remember (4.55)!]

$$\frac{\partial G}{\partial S} = \frac{1}{S} \ , \quad \frac{\partial^2 G}{\partial S^2} = -\frac{1}{S^2} \ , \quad \frac{\partial G}{\partial t} = 0 \quad \Rightarrow \tag{4.61}$$

$$dG = \left(\mu - \frac{\sigma^2}{2}\right)dt + \sigma dz \ . \tag{4.62}$$

With $\mu = $ const. and $\sigma = $ const., $\ln S$ follows a generalized Wiener process with an effective drift $\mu - \sigma^2/2$ and standard deviation rate σ. Notice that both S and G are affected by the same source of uncertainty: the stochastic process dz. This will become important in the next section, where S and G will represent the prices of the underlying and the derivative securities, respectively. [As is clear from (4.53), dS/S also follows, under the same assumptions, a generalized Wiener process, however with an unrenormalized drift μ. This illustrates (4.55). The consequences will be discussed below.]

If t denotes the present time, and T some future time, the probability distribution of $\ln S$ will be a normal distribution with mean and variance

$$\langle\ln S\rangle \quad = \left(\mu - \frac{\sigma^2}{2}\right)(T - t) \tag{4.63}$$

$$\mathrm{var}(\ln S) = \sigma^2(T - t) \ , \tag{4.64}$$

i.e.,

$$p(\ln S_T/S_t) = \frac{1}{\sqrt{2\pi\sigma^2(T-t)}} \exp\left(-\frac{\left[\ln\left(\frac{S_T}{S_t}\right) - \left(\mu - \frac{\sigma^2}{2}\right)(T-t)\right]^2}{2\sigma^2(T-t)}\right).$$

(4.65)

The stock price changes themselves are then distributed according to a log-normal distribution [use $p(\ln S_T/S_t)\mathrm{d}\ln S_T/S_t = \tilde{p}(S_T)\mathrm{d}S_T$]

$$\tilde{p}(S_T) = \frac{1}{S_T}p\left(\ln\frac{S_T}{S_t}\right)$$

$$= \frac{1}{\sqrt{2\pi\sigma^2(T-t)}}\frac{1}{S_T}\exp\left(-\frac{\left[\ln\left(\frac{S_T}{S_t}\right) - \left(\mu - \frac{\sigma^2}{2}\right)(T-t)\right]^2}{2\sigma^2(T-t)}\right).$$

(4.66)

This distribution is shown in Fig. 4.6.

Using this distribution and the substitution $S_T/S_t = \exp(\omega)$, we find that the expectation value of S_T evolves as

$$\langle S_T\rangle = \int_0^\infty \mathrm{d}S_T\, S_T\, \tilde{p}(S_T) = S_t\exp[\mu(T-t)],$$

(4.67)

and its variance as

$$\mathrm{var}(S_T) = S_t^2\exp[2\mu(T-t)]\{\exp[\sigma^2(T-t)] - 1\}.$$

(4.68)

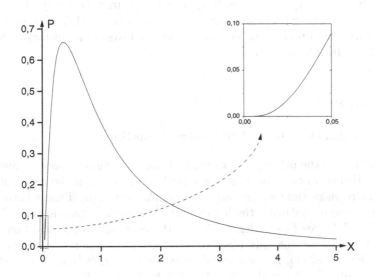

Fig. 4.6. The log-normal distribution $\tilde{p}(S)$

Observe that the expectation value of S_T grows with a rate μ, $\ln\langle S_T \rangle \propto \mu(T-t)$, in line with the definition of μ as the expectation value of the rate of return. Notice, however, that from (4.63), the expectation value of $\ln S$ grows with a different rate $\mu - \sigma^2/2$. The two different results correspond to two different situations where return rates are measured. Equation (4.53) shows that μ is the average of the return rate over a short time interval. The expectation value of the stock price grows with the average return rate over short time intervals. On the other hand, if one takes an actual investment with a specific return rate history with the same average, and calculates its, say yearly, return, this will be less than the average of the yearly returns determined on the way. For a specific example, assume an average growth rate of 10% p.a. over four years. Then, the expected price of the stock after four years is $S_T = S_t(1.1)^4 = 1.464 S_t$. Now assume that the actual growth rates in the four years are $\mu_1 = 5\%$, $\mu_2 = 12\%$, $\mu_3 = 13\%$, $\mu_4 = 10\%$. Then $S_T = 1.05 \times 1.12 \times 1.13 \times 1.1 S_t = 1.462 S_t$, and the actual rate of return over the four years is only 9.5% p.a. If many such investments at a given average return rate μ are considered and their returns are averaged over, the average rate of return will converge to $\mu - \sigma^2/2$. Moreover, the binomial theorem $(1+x)(1-x) = 1 - x^2 \leq 1$ shows that the average short-term growth rate can only be reached in the absence of randomness ($x = 0$), and that the general conclusion is independent of the particular realization assumed in the example. Of course, this is common experience of any investor who determines the return of his investments.

Another way of looking at the different return rates is to notice that, due to the skewness of the log-normal distribution, the rather frequent small prices from negative returns are less weighted in the expectation value than the less frequent very high prices from positive returns. Few very high profits count more in the expectation value than the same number of almost total losses, while the opposite is true for an actual investment history with the same short-time return rate.

4.5 Option Pricing

4.5.1 The Black–Scholes Differential Equation

We now turn to the pricing of options, and the hedging of positions involving options. Investments in options are usually considered to be risky, significantly more risky than investments into stocks or bonds. This is because of the finite time to maturity, the high volatility of options (significantly higher than the volatility of its underlying), and the possibility of a total loss of the invested capital for the long position, and losses even potentially unlimited for the short position, in the case of unfavorable market movements (cf. the discussion in Sect. 2.4, and Figs. 2.1 and 2.2). With f the price of an option ($f = C, P$, for call and put options, respectively), we have

$$\sqrt{\text{var}\,(\Delta f)} = \frac{\partial f}{\partial S}\sqrt{\text{var}\,(\Delta S)}\,, \quad \sqrt{\text{var}\left(\frac{\Delta f}{f}\right)} = \frac{S}{f}\frac{\partial f}{\partial S}\sqrt{\text{var}\left(\frac{\Delta S}{S}\right)} \quad (4.69)$$

for the volatility of the option in terms of the volatility of the underlying. Figs. (4.1) and (4.2) show that $\partial f/\partial S < 1$ in general. While the volatility of the option *prices* is smaller than that of the prices of the underlyings, the volatility of the option *returns* described by the second equation in (4.69) is much higher than that of the returns of their underlyings because the option prices usually are much lower than the prices of the underlyings, $S/f \gg 1$.

Moreover, the writer of an option engages a liability when entering the contract, while the holder has a freedom of action depending on market movement, i.e., an insurance: buy or not buy (sell or not sell) the underlying at a fixed price, in the case of a call (put) option. The question then is: What is the risk premium for the writer of the option, associated with the liability taken over? Or what is the price of the insurance, the additional freedom of choice for the holder? What is the value of the asymmetry of the contract?

These questions were answered by Black and Scholes [42] and Merton [43], and the answer they came up with, under the assumptions specified in Sect. 4.2.1 and developed thereafter, i.e., geometric Brownian motion, is surprising: *There is no risk premium required for the option writer!* The writer can entirely eliminate his risk by a dynamic and self-financing hedging strategy using the underlying security only. The price of the option contract, the value for the long position, is then determined completely by some properties of the stock price movements (volatility) and the terms of the option contract (time to maturity, strike price). For simplicity, and because we are interested only in the important qualitative aspects, we shall limit our discussion to European options, mostly calls, and ignore dividend payments and other complications. For other derivatives or more complex situations, the reader should refer to the literature [10, 12]–[15].

The main idea underlying the work of Black, Merton, and Scholes [42, 43] is that it is possible to form a *riskless* portfolio composed of the option to be priced and/or hedged, and the underlying security. Being riskless, it must earn the risk-free interest rate r, in the absence of arbitrage opportunities. The formation of such a riskless portfolio is possible because, and only because, at any instant of time the option price f is correlated with that of the underlying security. This is shown by the solid lines in Figs. 4.1 and 4.2, which sketch the possible dependences of option prices on the prices of the underlying. The dependence of the option price on that of the underlying is given by $\Delta = \partial f/\partial S$ which, of course, is a function of time. In other words, both the stock and the option price depend on the same source of uncertainty, resp. the same stochastic process: the one followed by the the stock price. Therefore the stochastic process can be eliminated by a suitable linear combination of both assets.

To make this more precise, we take the position of the writer of a European call. We therefore form a portfolio composed of

1. a short position in one call option,
2. a long position in $\Delta = \partial f/\partial S$ units of the underlying stock. Notice that Δ fluctuates with the stock price, and a continuous adjustment of this position is required.

The stochastic process followed by the stock is assumed to be geometric Brownian motion, (4.53),

$$dS = \mu S dt + \sigma S dz .\tag{4.70}$$

A priori, we do not know the stochastic process followed by the option price. We know, however, that it depends on the stock price, and therefore, we can use Itô's lemma, (4.57),

$$df = \left(\frac{\partial f}{\partial S}\mu S + \frac{\partial f}{\partial t} + \frac{1}{2}\sigma^2 S^2 \frac{\partial^2 f}{\partial S^2}\right) dt + \frac{\partial f}{\partial S}\sigma S dz .\tag{4.71}$$

The value of our portfolio is

$$\Pi = -f + \frac{\partial f}{\partial S}S ,\tag{4.72}$$

and it follows the stochastic process

$$d\Pi = -df + \frac{\partial f}{\partial S}dS = \left(-\frac{\partial f}{\partial t} - \frac{1}{2}\sigma^2 S^2 \frac{\partial^2 f}{\partial S^2}\right) dt .\tag{4.73}$$

Notice that the stochastic process dz, the source of uncertainty in the evolution of both the stock and the option prices, no longer appears in (4.73). Moreover, the drift μ of the stock price has disappeared, too. Eliminating the risk from the portfolio also eliminates the possibilities for profit, i.e., the risk premium $\mu > r$ associated with an investment into the underlying security alone (an investor will accept putting his money in a risky asset only if the return is higher than for a riskless asset). The portfolio being riskless, it must earn the risk-free interest rate r,

$$d\Pi = r\Pi dt = r\left(-f + \frac{\partial f}{\partial S}S\right) dt .\tag{4.74}$$

Equating (4.73) and (4.74), we obtain

$$\frac{\partial f}{\partial t} + rS\frac{\partial f}{\partial S} + \frac{1}{2}\sigma^2 S^2 \frac{\partial^2 f}{\partial S^2} = rf ,\tag{4.75}$$

the Black–Scholes (differential) equation. This is a linear second-order partial differential equation of parabolic type. Its operator structure is very similar to the Fokker–Planck equation in physics or the Kolmogorov equation in mathematics (two different names for the same equation) [37]. There are two differences, however: (i) the sign of the term corresponding to the diffusion constant is negative, and (ii) this is a differential equation for a (at present rather arbitrary) function f while the Fokker–Planck equation usually refers to a differential equation for a normalized distribution function $p(x, t)$ whose norm is conserved in the time evolution. (For the use of Fokker–Planck equations in the statistical mechanics of capital markets, see Chap. 6).

For a complete solution to the Black–Scholes equation, we still have to specify the boundary or initial conditions. Unlike physics, here we deal with a final value problem. At maturity $t = T$, we know the prices of the call and put options, (4.6) and (4.7),

$$\left. \begin{array}{l} \text{Call} . \ f - C = \max(S - X, 0) \\ \text{Put} : \ f = P = \max(X - S, 0) \end{array} \right\} \ t = T . \tag{4.76}$$

The solution of this final value problem, (4.75) and (4.76) will be given in the next section. Notice that for second-order partial differential equations, the number and type of conditions (initial, final, boundary) required for a complete specification of the solution depends on the type of problem considered. For diffusion problems such as (3.20), (3.30), or (4.75), a single initial or final condition is sufficient.

Stock prices change with time. Keeping the portfolio riskless in time therefore requires a *continuous* adjustment of the stock position $\Delta = \partial f / \partial S$, as it varies with the stock price. It is clear that this can only be done in the idealized markets considered here, and subject to the assumptions specified earlier. Transaction costs, e.g., would prevent a continuous adjustment of the portfolio, and immediately make it risky. The same applies to credit costs incurred by the adjustments. In practice, therefore, a riskless portfolio will usually not exist, and there will be a finite risk premium on options (often determined empirically by the writing institutions).

The important achievement of Black, Merton, and Scholes was to show that, in idealized markets, the risk associated with an option can be hedged away completely by an offsetting position in a suitable quantity Δ of the underlying security (this hedging strategy is therefore called Δ-hedging), and that no risk premium need be asked by the writer of an option. The hedge can be maintained dynamically, and is self-financing, i.e., does not generate costs for the writer. Of course, this is an approximation in practice because none of the assumptions on which the Black–Scholes equation is based, are fulfilled. This will be discussed in Chap. 5. Despite this limitation, it allows fundamental insights into the price processes for derivatives, and we now proceed to solve the equation.

4.5.2 Solution of the Black–Scholes Equation

The following solution of (4.75) essentially follows the original Black–Scholes article [42], and consists in a reduction to a 1D diffusion equation with special boundary conditions. (This may not be too surprising: Fisher Black held a degree in physics.)

We substitute

$$f(S,t) = e^{-r(T-t)} y(u,v) , \tag{4.77}$$

$$u \quad = \frac{2\rho}{\sigma^2} \left(\ln \frac{S}{X} + \rho[T-t] \right) ,$$

$$v \quad = \frac{2}{\sigma^2} \rho^2 (T-t) , \quad \rho = r - \frac{\sigma^2}{2} . \tag{4.78}$$

Then, the derivatives $\partial f/\partial S$, $\partial^2 f/\partial S^2$, and $\partial f/\partial t$ are expressed through $\partial y/\partial u$, $\partial y/\partial v$, etc., and $y(u,v)$ satisfies the 1D diffusion equation

$$\frac{\partial y(u,v)}{\partial v} = \frac{\partial^2 y(u,v)}{\partial u^2} . \tag{4.79}$$

The boundary conditions (4.76) for a call option translate into

$$y(u,0) = \begin{cases} 0 & u < 0 \\ X \left(e^{u\sigma^2/2\rho} - 1 \right) & u \geq 0 . \end{cases} \tag{4.80}$$

Diffusion equations are solved by Fourier transform in the spatial variable(s)

$$y(u,v) = \int_{-\infty}^{\infty} dq e^{iqu} y(q,v) , \tag{4.81}$$

reducing (4.79) to an ordinary differential equation in v with the solution

$$y(q,v) = y(q,0) \exp\left(-q^2 v\right) . \tag{4.82}$$

$y(q,0)$, formally, is given by the Fourier transform of the boundary conditions (4.80) which, however, should NOT be performed explicitly. The trick, instead, is to transform the solution (4.82) back to u-variables, giving a convolution integral

$$y(u,v) = \frac{1}{2\pi} \int_{-\infty}^{\infty} dw \, y(w,0) f(u-w) \quad \text{with} \quad f(x) = \sqrt{\frac{\pi}{v}} \exp\left(-\frac{x^2}{4v}\right) . \tag{4.83}$$

Another substitution $z = (w-u)/\sqrt{2v}$ almost gives the final result

$$y(u,v) = \frac{X}{\sqrt{2\pi}} \int_{-u/\sqrt{2v}}^{\infty} dz e^{-z^2/2} \left[\exp\left(\frac{\sigma^2}{2\rho} \left\{ \sqrt{2vz} + u \right\} \right) - 1 \right] . \tag{4.84}$$

The only task remaining is to complete the square in the exponent, and insert all substituted quantities. This gives the Black–Scholes equation for a European call option (remember that the boundary conditions for a call have been used in the derivation)

$$C(S,t) \equiv f(S,t) = SN(d_1) - Xe^{-r(T-t)} N(d_2) . \tag{4.85}$$

The equivalent solution for a European put option is

$$P(S,t) = Xe^{-r(T-t)}N(-d_2) - SN(-d_1) .\tag{4.86}$$

$N(d)$ is the cumulative normal distribution

$$N(d) = \frac{1}{\sqrt{2\pi}} \int_{-d}^{\infty} dx\, e^{-x^2/2} ,\tag{4.87}$$

and its two arguments in (4.85) are given by

$$d_1 = \frac{\ln\frac{S}{X} + \left(r + \frac{\sigma^2}{2}\right)(T - t)}{\sigma\sqrt{T - t}} ,\tag{4.88}$$

$$d_2 = \frac{\ln\frac{S}{X} + \left(r - \frac{\sigma^2}{2}\right)(T - t)}{\sigma\sqrt{T - t}} .\tag{4.89}$$

Clearly, $S \equiv S(t)$. The behavior of $C(S)$ is sketched in Fig. 4.1 as the solid line, and the equivalent put price is sketched in Fig. 4.2. The time evolution of a call price, as given by the Black–Scholes equation (4.85), is displayed in Fig. 4.7. In that figure, all parameters have been kept fixed, and only time elapses. We therefore monitor the *time value* of the options. The intrinsic value is given by $S(t) - X$, i.e., the payoff if the option was exercised today. While the intrinsic value fluctuates with the evolution of the stock price,

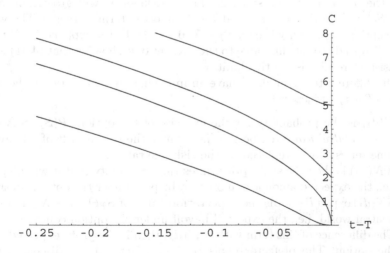

Fig. 4.7. Time evolution of the price of a European call option as a function of time *before* maturity in years. Fixed stock price $S = 100$, interest rate $r = 6\%/y$, and volatility $\sigma = 30\%/\sqrt{y}$ have been assumed. The curves represent different strike prices $X = 95, 98, 100, 105$ from top to bottom, i.e., the options are in the money (top two lines), at the money, and out of the money, respectively

the time value always decreases. It measures the probability left at time t for a favorable stock price movement to occur before maturity T. It varies strongest for options at the money, and less for options far in or out of the money.

There are a few interesting limiting cases of (4.85). If $S \gg X$, the option is exercised almost certainly. In this case, it will become equivalent to a forward contract with a delivery price X. If $S \gg X$, $d_1, d_2 \to \infty$, and $N(d_{1,2}) \to 1$. The Black–Scholes equation then reduces to

$$f(S,t) = S - Xe^{-r(T-t)} . \tag{4.90}$$

This was precisely the expression for the value of the long position in a forward contract derived earlier, (4.2). In that problem, the delivery price was to be fixed so that the value of the contracts for both parties came out to $f = 0$. Here, the strike price of the option is fixed from the outset, and f therefore represents the intrinsic value of the long position in the option, which has become equivalent to a forward by the assumption $S \gg X$. Notice that S must be exponentially large compared to X for our derivation to hold.

If $\sigma \to 0$, the stock becomes almost riskless. In (4.85), two different cases must be considered. If $\ln(S/X) + r(T-t) > 0$, $d_{1,2} \to \infty$, $N(d_i) \to 1$, and (4.90) continues to hold. If, on the other hand, $\ln(S/X) + r(T-t) < 0$, $d_{1,2} \to -\infty$, $N(d_i) \to 0$, and $f(S,t) \to 0$. Putting both cases together,

$$C(S,t) \equiv f(S,t) = \max(S - Xe^{-r(T-t)}, 0) . \tag{4.91}$$

If on the other hand, the stock is almost riskless, it will grow from S to $S_T = Se^{r(T-t)}$ in the time interval $T - t$ almost deterministically. The value of the option at maturity is $\max(S_T - X, 0)$, and a factor $\exp[-r(T-t)]$ must be applied to discount this value to the present day, showing that (4.91) gives a consistent result also in this limit.

The different terms in (4.85) have an immediate interpretation if the term $\exp[-r(T-t)]$ is factored out:

1. $N(d_2)$ is the probability for the exercise of the option, $P(S_T > X)$, *in a risk-neutral world* (cf. below), i.e., where the actual drift of a financial time series can be replaced by the risk-free rate r.
2. $XN(d_2)$ is then the strike price times the probability that it will be paid, i.e., the expected amount of money to be paid under the option contract.
3. $SN(d_1)\exp[r(T-t)]$ is the expectation value of $S_T\Theta(S_T - X)$ in a risk-neutral world, i.e., the expected payoff under the option contract.
4. The difference of this term with $XN(d_2)$ then is the profit expected from the option. The prefactor $\exp[-r(T-t)]$ factored out discounts that profit, realized at maturity T, down to the present day t. The option price is precisely this discounted difference.

This interpretation is consistent with the capital asset pricing model which deals with the relation of risk and return in market equilibrium. It states that the expected return on an investment is the discounting rate which one must apply to the profit expected at maturity, in order to obtain the present price. In our interpretation of (4.85), one would just read this sentence from the backwards.

For an option, no specific risk premium is necessary. The entire risk is contained in the price of the underlying security, and can be hedged away.

Because of their importance, we reiterate some statements made in earlier sections, or implicitly contained therein:

1. The construction of a risk-free portfolio is possible only for Itô–Wiener processes.
2. Because of the nonlinearity of $f(S)$, $\partial f/\partial S$ is time-dependent.
3. The portfolio is risk-free only instantaneously. In order to keep it risk-free over finite times, a continuous adjustment is required.
4. Beware of calculating the option price by a naïve expectation value of the profit, and discounting such as

$$
\begin{aligned}
& e^{-r(T-t)} \int_0^\infty p_{\text{hist}}(S_T)(S_T - X)\Theta(S_T - X) \\
& = \langle \max(S_T - X, 0)\rangle_{\text{hist}} \neq C(S,t) \,,
\end{aligned}
\tag{4.92}
$$

using the historical (recorded) distribution of prices $p_{\text{hist}}(S)$. This will give the wrong result! Such a calculation will give too high a price for the option because p_{hist} is based on a stochastic process with the historic drift μ which ignores the possibility of hedging and overestimates the risk involved in the option position. This will be discussed further in the next section.

We have just discussed the simplest option contract possible, a European call option. The equivalent pricing formulae for a put option can be derived straightforwardly by the reader: they only differ in the boundary condition (4.76) used in the solution of the Black–Scholes differential equation. Many generalizations are possible, such as for options on dividend paying stocks, currencies, interest rates, indices or futures, combi or exotic options, etc. The interested reader is referred to the finance literature [10, 12]–[15] for discussions using similar assumptions as made here (geometric Brownian motion, etc.).

Also path integral methods familiar from physics may be useful [50]. In fact, one can solve the Black–Scholes equation (4.75) by noting the similarity to a time-dependent Schrödinger equation. Time, however, is imaginary, $\tau = it$, identifying the problem as one of quantum statistical mechanics rather than one of zero-temperature quantum mechanics corresponding to real times. The "Black–Scholes Hamiltonian" entering the Schrödinger equation then becomes

$$
H_{BS} = -\frac{\sigma^2}{2}\frac{\partial^2}{\partial x^2} + \left(\frac{\sigma^2}{2} - r\right)\frac{\partial}{\partial x} = \frac{p^2}{2m} + \frac{i}{\hbar}\left(\frac{\sigma^2}{2} - r\right)p
\tag{4.93}
$$

$$\text{with} \quad x = \ln S \ , \quad p = -\mathrm{i}\hbar \frac{\partial}{\partial x} \ , \quad \text{and} \quad m = \frac{\hbar^2}{\sigma^2} \ .$$

The Black–Scholes equation (4.85) is then obtained by evaluating the path integral using the appropriate boundary conditions (4.76). This method can also be generalized to more complicated problems such as option pricing with a stochastically varying volatility $\sigma(t)$ [51]. That such a method works is hardly surprising from the similarity between the Black–Scholes and Fokker–Planck equations. For the latter, both path-integral solutions, and the reduction to quantum mechanics, are well established [37]. We will use the path integral method in Chap. 7 to price and hedge options in market situations where some of the assumptions underlying the Black–Merton–Scholes analysis are relaxed.

4.5.3 Risk-Neutral Valuation

As mentioned in Sect. 4.5.1, eliminating the stochastic process in the Black–Scholes portfolio as a necessary consequence also eliminates the drift μ of the underlying security. μ, however, is the only variable in the problem which depends on the risk aversion of the investor. The other variables, $S, T - t, \sigma$ are independent of the investor's choice. (Given values for these variables, an operator will only invest his money, e.g., in the stock if the return μ satisfies his requirements.) Consequently, the solution of the Black–Scholes differential equation does not contain any variable depending on the investor's attitude towards risk such as μ, cf. (4.85).

One can therefore assume any risk preference of the agents, i.e., any μ. In particular, the assumption of a risk-neutral (risk-free) world is both possible and practical. In such a world, all assets earn the risk-free interest rate r. The solution of the Black–Scholes found in a risk-neutral world is also valid in a risky environment (our solution of the problem above takes the argument in reverse). The reason is the following: in a risky world, the growth rate of the stock price will be higher than the risk-free rate. On the other hand, the discounting rate applied to all future payoffs of the derivative, to discount them to the present day value, then changes in the same way. Both effects offset each other.

Risk-neutral valuation is equivalent to assuming martingale stochastic processes for the assets involved (up to the risk-free rate r). Equation (4.92) shows that simple expectation value pricing of options, using the historical probability densities for stock prices $p_{\text{hist}}(S)$, does not give the correct option price. In other words, if an option price was calculated according to (4.92), arbitrage opportunities would arise. On the other hand, intuition would suggest that some form of expectation value pricing of a derivative should be possible: the present price of an asset should depend on the expected future cash flow it generates.

Indeed, even in the absence of arbitrage, expectation value pricing is possible, but at a price: a price density $q(S)$ different from the historical density

$p_{\text{hist}}(S)$ must be used [52]. This is the consequence of a theorem which states that under certain conditions (which we assume to be fulfilled), for a stochastic process with a probability density $p_{t,T}(S_T)$ for S_T, and conditional densities including the information available up to t, $p_{t,T}(S_T|S_t, S_{t-1}, S_{t-2}, \ldots)$, there is an equivalent martingale stochastic process described by a different probability $q_{t,T}(S_T)$, such that in the absence of arbitrage opportunities, the price of an asset with a payoff function $h(S_T)$ is given by a discounted expectation value using $q_{t,T}$

$$f(t) = e^{-r(T-t)} \int_{-\infty}^{\infty} dS_T \, h(S_T) q_{t,T}(S_T) \,. \tag{4.94}$$

As an example, for a call option, the payoff function is $h(S_T) = \max(S_T - X, 0)$ and, with the correct probability density for the equivalent martingale process, involving the risk-free rate r instead of the drift μ of the underlying, the price

$$C(T) = e^{-r(T-t)} \int_{-\infty}^{\infty} dS_T \, \max(S_T - X, 0) q_{t,T}(S_T) \tag{4.95}$$

will produce the Black–Scholes solution (4.85). Also, the discounted stock price is an equivalent martingale

$$S_t = e^{-r(T-t)} \int_{-\infty}^{\infty} dS_T S_T q_{t,T}(S_T) \,. \tag{4.96}$$

Using equivalent martingales, expectation value pricing for financial assets is possible. Martingales are tied to the notion of risk-neutral valuation.

4.5.4 American Options

The valuation of American options employs the same general risk-neutral framework as for European options. In principle, a riskless hedge of the option position is possible by holding a suitable quantity of the underlying asset. A short position in one American call option still is hedged by a long position in Δ shares of the underlying – the difference to European options is in the numerical value of Δ. The valuation therefore can be based on equivalent martingale processes, with the risk-free rate r as the drift. However, the possibility of early exercise introduces significant complexity and prevents an exact analytic solution.

The basic principle for the valuation of an American option can be illustrated easily. Assume first that time is a discrete variable, $t_i = i\Delta t$, $i = 0, \ldots, N$, $\Delta t = T/N$, where T is the maturity of the option. An American option then can be exercised at any t_i. For geometric Brownian motion, the probability distributions (4.65) and (4.66) are obtained with the trivial replacements $t \to t_i$ and $S_t \to S_i$. The transition probability (conditional probability density) for an elementary time step of the equivalent martingale process in the risk-neutral world, for geometric Brownian motion becomes

$$q_{t_{i-1},t_i}(S_i) \equiv q(S_i, t_i \mid S_{i-1}, t_{i-1})$$

$$= \frac{1}{\sqrt{2\pi\sigma^2\Delta t}} \exp\left(-\frac{\left[\ln\left(\frac{S_i}{S_{i-1}}\right) - \left(r - \frac{\sigma^2}{2}\right)\Delta t\right]^2}{2\sigma^2\Delta t}\right).$$

$$(4.97)$$

One time step before expiry, at t_{N-1}, it is advantageous to exercise the option if its immediate payoff exceeds its value on the assumption of holding it to maturity,

$$h\left(S_{N-1}\right) > f(t_{N-1}),$$ (4.98)

where $h(S_i)$ is the payoff function, and $f(t_i)$ is the value of the option, cf. (4.94). To be specific, an American call with payoff $h(S_i) = \max(S_i - X, 0)$ should be exercised at t_{N-1} when

$$S_{N-1} - X > C(t_{N-1})$$ (4.99)

with $C(t_{N-1})$ given by using the discretized version of (4.95). This argument can be iterated backward in time because for an American option, no particular significance is attached to the time of maturity. Consequently, at time t_{i-1}, early exercise is advantageous when the payoff received immediately exceeds the value of the option derived from holding it until the next possibility to exercise, i.e. t_i. The early exercise condition is

$$h\left(S_{i-1}\right) > e^{-r\Delta t} \int_{-\infty}^{\infty} dS_i \, h(S_i) q_{t_{i-1},t_i}(S_i).$$ (4.100)

The right-hand side has been taken from (4.94) and rewritten for a single time step. For an American call, we get

$$S_{i-1} - X > e^{-r\Delta t} \int_{-\infty}^{\infty} dS_i \, \max(S_i - X, 0) q_{t_{i-1},t_i}(S_i),$$ (4.101)

in analogy to (4.95). The option at $t = t_0$ then is priced, and hedged, by iterating the problem backward from maturity, t_T, to $t = t_0$, and taking the continuum limit of time, $\Delta t \to 0$, $N \to \infty$ with $T = N\Delta t$ fixed. Of course, a closed solution of this problem is impossible because for every possible price S_i, a decision on early exercise must be taken at each step i.

A variety of approximate solutions has been developed, all suffering from drawbacks though. Monte Carlo simulations are an obvious choice. Random price increments are drawn from a normal distribution (in the case of geometric Brownian motion) to simulate the price history of the underlying, and the average over many runs is taken when ensemble properties are required.

While Monte Carlo simulations in principle give the desired answer, they are computationally inefficient because the errors on averages over finitely many realizations decrease rather slowly. For plain vanilla options, the use of binomial trees provides an alternative. In a binomial tree, price increments have fixed modulus ΔS, i.e. only $\pm \Delta S$ are allowed. This restriction gives enough simplification to make calculations for plain vanilla options practical. However, for exotic, path-dependent options, the discretization of the price increments is an undesirable feature.

General arguments suggest that American call options should never be exercised early in the absence of dividend payments. Dividend payments have not been considered for European options, and will not be discussed here for American options. The role of dividend payments in option pricing, hedging, and exercise is discussed in the standard financial literature [10].

4.5.5 The Greeks

The derivatives of option prices with respect to the parameters and variables upon which the option price depends, play important roles in trading and hedging strategies. Most of them are labelled by greek letters. Collectively, they are called "the Greeks".

We already encountered one of the Greeks, Delta, and its application in hedging, when setting up the riskless Black–Scholes portfolio in (4.72). There, a short position in a call option was combined with a long position in

$$\Delta_C = \frac{\partial C}{\partial S} \tag{4.102}$$

units of the underlying resulting in a portfolio which was riskless agains infinitesimal variations of the price of the underlying, all other things remaining constant. Similarly, the Delta for a put option is

$$\Delta_P = \frac{\partial P}{\partial S} \ . \tag{4.103}$$

The definition of Delta, as well as that of the other Greeks is valid for all options. For European options described by the Black–Scholes equations (4.85) and (4.86), we can evaluate Delta explicitly as

$$\Delta_C = N(d_1) \ , \quad \Delta_P = N(d_1) - 1 \ , \tag{4.104}$$

where $N(d_1)$ and d_1 are defined in (4.87) and (4.88). Its dependence on the price of the underlying, for different times to maturity, is shown in Fig. 4.8.

Delta describes the dollar variation of an option when the price of the underlying changes by one dollar. More important to investors is the leverage of an option, defined as the percentage variation of the option when the price of the underlying varies by one percent. This quantity is given by

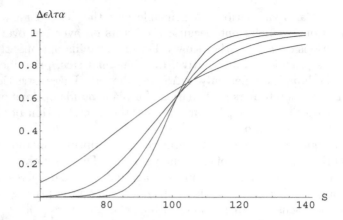

Fig. 4.8. Delta of a European call option described by the Black–Scholes equation as a function of the price of the underlying, for times to maturity of one, two, four and twelve months, from bottom to top at the left margin. The other parameters are $r = 6\%/y$ and $\sigma = 30\%/\sqrt{y}$ as in Fig. 4.7

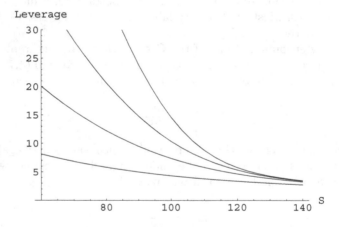

Fig. 4.9. Leverage of a European call option described by the Black–Scholes equation as a function of the price of the underlying, for times to maturity of one, two, four and twelve months, from top to bottom. The other parameters are $r = 6\%/y$ and $\sigma = 30\%/\sqrt{y}$ as in Fig. 4.7

$$\frac{S}{C}\frac{\partial C}{\partial S} \quad \text{and} \quad \frac{S}{P}\frac{\partial P}{\partial S}$$

for call and put options, respectively. The dependence of the leverage on the price of the underlying is displayed in Fig. 4.9 for a European call option. Quite generally, out-of-the money options possess a higher leverage than in-the-money options, and the leverage of a call option decreases when the price of the underlying increases. The downside risk of an option therefore always is superior to its upside chances. Also, all other things remaining constant, the leverage of an option increases when the time to maturity decreases. As a consequence of these two observations, speculative investments in options are advisable only when the investor holds a strong view on the price movement of the underlying, and on the time scale over which this price movement is realized.

The sensitivity of the option price with respect to time to maturity is expressed by Theta,

$$\Theta_C = \frac{\partial C}{\partial t} , \quad \Theta_P = \frac{\partial P}{\partial t} . \tag{4.105}$$

For European call and put options described by the Black–Scholes equation, we have

$$\Theta_{C,P} = -\frac{S\sigma}{2\sqrt{2\pi(T-t)}} e^{-d_1^2/2} \mp rX e^{-r(T-t)} N(\pm d_2) . \tag{4.106}$$

The upper signs apply for a call option, the lower signs for a put. The dependences of Theta on the price of the underlying and on time to maturity is shown in Fig. 4.10. Theta diverges for an at-the-money option when the time to expiration goes to zero. Theta tends towards a finite value when the option is in the money, i.e., in such a case, the loss in value of the call is linear in time shortly before expiration. Theta converges to zero for an out-of-the money call, i.e., such an option has lost all of its value already some time before expiration. Notice that, at least for the European call considered here, the schematic figures in Hull's book [10] seem to indicate an incorrect behavior close to maturity.

Gamma captures the curvature in the derivative prices with respect to the underlying and is defined as

$$\Gamma_C = \frac{\partial^2 C}{\partial S^2} , \quad \Gamma_P = \frac{\partial^2 P}{\partial S^2} . \tag{4.107}$$

In the Black–Scholes framework,

$$\Gamma_C = \Gamma_P \equiv \Gamma = \frac{1}{S\sigma\sqrt{2\pi(T-t)}} e^{-d_1^2/2} . \tag{4.108}$$

The dependence on the price of the underlying has the same functional form as the probability density function of a lognormal distribution. The dependence on time to maturity is more interesting and shown in Fig. 4.11. When an option expires at the money, Gamma diverges. Gamma tends towards zero, on the other hand, both for options in and out of the money. This behavior is easily understood by considering the payoff profiles of call and put options shown in Fig. 2.1. At expiry, there is a discontinuity in slope in the option payoff at $S = X$. In and out of the money, on the other hand, the payoffs are linear in the price of the underlying.

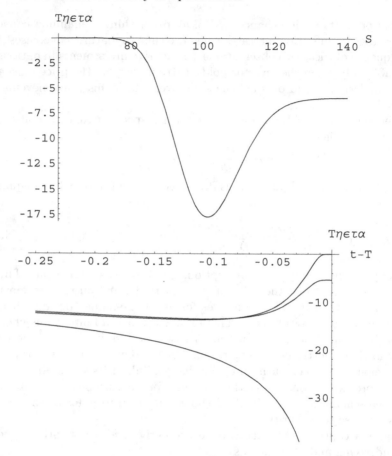

Fig. 4.10. Theta for European call options. The upper panel displays the dependence on the price of the underlying ($X = 100$, $r = 6\%/y$, $\sigma = 30\%/\sqrt{y}$, $T-t = 2m$. The lower panel shows the dependence on time to maturity for $S = 100$ and strike prices $X = 110$ (*top curve*, out of the money), $X = 90$ (*middle curve*, in the money), and $X = 100$ (*bottom curve*, at the money)

The sensitivity of the price of an option with respect to a variation in volatility is important, too. This derivative is called Vega, and is defined as

$$\mathcal{V}_C = \frac{\partial C}{\partial \sigma}, \quad \mathcal{V}_P = \frac{\partial P}{\partial \sigma} \ . \tag{4.109}$$

Vega is the same for call and put options. When the Black–Scholes equation applies, we have

$$\mathcal{V} = \frac{S\sqrt{T-t}}{\sqrt{2\pi}} e^{-d_1^2/2} \ . \tag{4.110}$$

The variation of Vega with the price of the underlying is S^2 times the lognormal probability density function. For an option at the money, the dependence

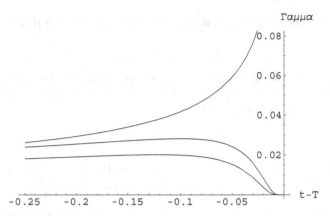

Fig. 4.11. Gamma European call options described by the Black–Scholes equation as a function of time to expiration. The parameters are $S = 100$, $r = 6\%/y$ and $\sigma = 30\%/\sqrt{y}$ and $X = 100$ (*top curve*, at the money), $X = 90$ (*middle curve*, in the money) and $X = 110$ (*bottom curve*, out of the money)

on time to maturity is

$$\mathcal{V} \sim \sqrt{T - t} \quad \text{as} \quad T - t \to 0 \quad (S = X). \tag{4.111}$$

For options in and out of the money,

$$\mathcal{V} \sim \sqrt{T - t}\, e^{-1/(T-t)} \quad \text{as} \quad T - t \to 0 \quad (S \neq X). \tag{4.112}$$

Except for a different power-law prefactor, this behavior is similar to that shown for Gamma in Fig. 4.11.

Finally, a parameter Rho

$$\mathcal{R}_C = \frac{\partial C}{\partial r}, \quad \mathcal{R}_P = \frac{\partial P}{\partial r} \tag{4.113}$$

measures the sensitivity of the prices of call and put options against variations of the risk-free interest rate r. In a Black–Scholes world,

$$\mathcal{R} = \pm X(T - t)e^{-r(T-t)} N(\pm d_2), \tag{4.114}$$

where the upper and lower signs apply to calls and puts, respectively.

We will come back to Vega later in Sect. 4.5.8 on volatility indices. The use of the Greeks in hedging option positions is discussed in Chap. 10 on risk management.

4.5.6 Synthetic Replication of Options

When the risk-free Black–Scholes portfolio was set up for a short position in a European call option with price C in Sect. 4.5.1, a long position in

$\Delta_C = \partial C / \partial S$ units of the underlying S was added to form a riskless portfolio Π_r

$$\Pi_r = -C + \Delta_C S . \tag{4.115}$$

The portfolio consisting of the short option position and the long position in the underlying is exactly equivalent to a long position in a riskless asset of value Π_r. We can transform (4.115) into

$$C = -\Pi_r + \Delta_C S . \tag{4.116}$$

A long call position is equivalent to a short position of value Π_r in a riskless asset and a long position in Δ_C units of the underlying of the call, priced at S.

For a short position in a European put option, the risk-free Black–Scholes portfolio is

$$\Pi_r = -P + \Delta_P S = -P - |\Delta_P| S . \tag{4.117}$$

The short put position is hedged by a short position in $|\Delta_P|$ units of the underlying, as $\Delta_P < 0$. A long position in a put option then is equivalent to

$$P = -\Pi_r - |\Delta_P| S , \tag{4.118}$$

i.e., to a short position of value Π_r in a risk-free asset and another short position in $|\Delta_P|$ units of the underlying.

These equivalences are general and do not assume the validity of the Black–Scholes model. Only the numerical values of Δ_C and Δ_P depend on the price dynamics of the underlying, and on the exercise features of the options. Also, they are not limited to call and put options. The important message is that *any option can be created synthetically by a suitable combination of a position in a riskless asset and another position in the underlying.* This is a result of great practical importance. Whenever an investor wishes to take a position in an option which is not available in the market, he can synthetically replicate the option by taking positions in a risk-free asset and in the underlying. Many portfolio managers and risk managers use this technique to implement their trading and hedging strategies when standard options are not available.

4.5.7 Implied Volatility

Writing the option price in (4.85) symbolically as $C_{\mathrm{BS}}(S, t; r, \sigma; X, T)$, most parameters of the Black–Scholes equation can be observed directly either in the market, or on the option contract under consideration. S and t are independent variables, X and T contract parameters, and r and σ market resp. asset parameters. The volatility σ stands out in that it cannot be observed directly. At best, it can be estimated from historical data on the underlying – a procedure which leaves many questions unanswered.

For a variety of reasons which are the principal motivation of the remainder of this book, the traded prices of options usually differ from their Black–Scholes prices. This is shown in Fig. 4.12 for a series of European calls on the DAX with a lifetime of one month to maturity. The horizontal axis "moneyness", $m = X/S$, represents the dimensionless ratio of strike price over underlying. For comparison, the Black–Scholes solution is also displayed as solid lines. The upper line uses a volatility of $35\% y^{-1/2}$, while the lower one takes $20\% y^{-1/2}$. Under the assumptions of the Black–Scholes theory and geometric Brownian motion, a single value of the volatility should be sufficient to describe the entire series of call options, and the prices should fall on one of the solid lines. Figure 4.12 rejects this hypothesis for real-world option markets.

In the absence of an accurate *ab initio* estimation of the volatility, a rough and pragmatic procedure consists in taking the traded prices for granted and invert the Black–Scholes equation (4.85) for the *implied volatility* σ_{imp} [10]

$$C_{\mathrm{market}}(S, t; r, \sigma; X, T) \equiv C_{\mathrm{BS}}(S, t; r, \sigma_{\mathrm{imp}}; X, T) . \qquad (4.119)$$

The idea is to pack all factors leading to deviations from Black–Scholes theory, independently of their origin, into the single parameter σ_{imp}. Volatility, anyway, is difficult to estimate *a priori*. For the series of options used in Fig. 4.12, the implied volatilites are shown in Fig. 4.13. Apparently, there are deviations of traded option prices from a Black–Scholes equation which depend on the contract to be priced. In this representation, they turn into an implied volatility which explicitly depends on the moneyness of the options. In a purist perspective, implied volatility adds nothing new to the theory of option

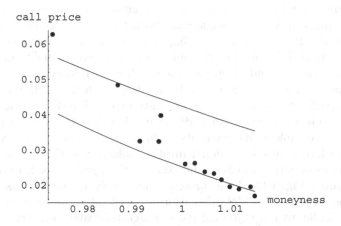

Fig. 4.12. Prices of a series of European call options on the DAX index with one month to maturity, given in units of the index value, against moneyness X/S (*dots*). The two *solid lines* represent the dependence of the Black–Scholes solutions on moneyness with two volatilities $\sigma = 35\% y^{-1/2}$ (*top*) and $\sigma = 20\% y^{-1/2}$ (*bottom*)

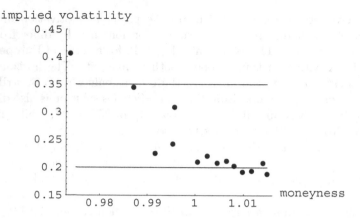

Fig. 4.13. Implied volatilities of a series of European call options on the DAX index with one month to maturity, against moneyness X/S (*dots*) in $\%y^{-1/2}$. Geometric Brownian motion and the Black–Scholes theory take volatility independent of the option contract to be priced. The two *solid lines* mark the contract-independent volatilities used to generate the *solid lines* in Fig. 4.12

pricing, and might even lead to confusion. However, it is a simple transformation of option prices and therefore is an observable on equal footing with the prices. This is similar to physics: When temperature is measured, the basic observable most often is an electric current or voltage drop, or height of a mercury column, etc., which then is transformed into a temperature reading with a suitable calibration. Also, implied volatility is the standard language of derivatives traders and analysts to describe option markets.

The generic shapes of implied volatilities against moneyness are shown in Fig. 4.14. Apparently, a pure smile was characteristic of the US option markets before the 1987 october crash [53]. Ever since, it has become a rather smirky structure. The aim of market models more sophisticated than geometric Brownian motion and of option pricing theories beyond Black–Merton–Scholes, can be restated as to correctly describe implied volatility smiles.

When a series of options with the same strike price but different maturities is analyzed, a term structure (maturity dependence) of the implied volatility is obtained in complete analogy to its moneyness dependence. The volatility smile turns into a two-dimensional implied volatility surface. Figure 4.15 shows a series of cuts through an implied volatility surface of European call options. Unlike Fig. 4.13, these curves do not represent market observations but are the results of a model calculation. Superficially, the one-month curve is not dissimilar to the empirical data, suggesting that theoretical models indeed may be capable of correctly describing option markets. Attempts to fit volatility smiles for a fixed time to maturity usally employ quadratic functions with different parameters for in- and out-of-the-money options, to account for the systematic asymmetry [53].

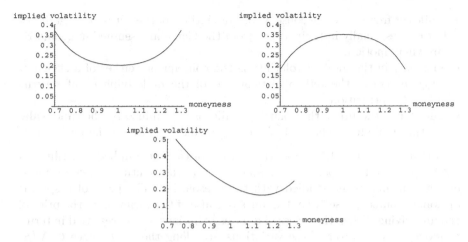

Fig. 4.14. Sketches of implied volatilities against moneyness. Three generic shapes can be observed: a smile (*top left*), a frown (*top right*) and a smirk resp. skewed smile (*bottom*). In equity markets, the smirk is observed most frequently. Often, the term "volatility smile" includes all three shapes

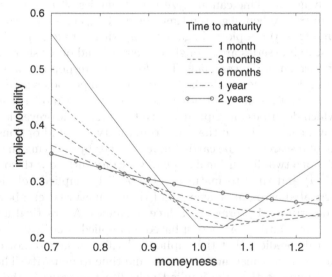

Fig. 4.15. Term structure and moneyness dependence of the implied volatility of a series of European call options based on a model calculation

With reference to the subsequent chapters where we will develop an in-depth description of financial markets, the main handles on the volatility smiles and term structures are:

• The ordinate scale is determined by the average volatility of the market/model.

- Smiles or frowns are the consequence deviations of the actual return distributions, especially in their wings, from the Gaussian assumed in geometric Brownian motion.
- The skew in the implied volatility is the consequence either of a skewness (asymmetry) of the return distribution of the underlying or of return–volatility correlations.
- The term structure of the volatility smiles is determined by the time scales (or time-scale-free behavior) of the important variables in the problem.

Figures 4.12 and 4.13 show the option prices and implied volatilites of DAX options on one particular trading day. Both quantities show an interesting dynamics when studied with time resolution. The price of a specific option, of course, possesses a dynamics because of the variation in the price of the underlying. When the prices of a series of options are represented in terms of moneyness, however, these variations are along the price curve $C(X/S)$ once the effects of changing time to maturity are eliminated, and should not lead to dynamical variations of the price curve itself. Additional dynamics may come, e.g., from the increasing autonomy of option markets which are increasingly driven by demand and supply, in addition to the price movements of the underlying [54]. One can analyze this dynamics of $\sigma_{\mathrm{imp}}(m)$ almost at the money, $m \approx 1$. When, e.g., the time series of $\sigma_{\mathrm{imp}}(1 - \delta) - \sigma_{\mathrm{imp}}(1)$ and $\sigma_{\mathrm{imp}}(1) - \sigma_{\mathrm{imp}}(1 + \delta)$ are plotted against time, there are long periods where both stochastic time series are strongly correlated, and other shorter periods where their correlation is weak [53]. The former correspond to almost rigid shifts of the smile patterns while the latter appear in periods where the smile predominantly changes shape. Both time series can be modeled as AR(1) processes which describes an implied volatility with a mean-reversion time of about 30 days, comparable to the time to maturity of liquid options.

This line of research can be carried much further by studying the dynamical properties of a two-dimensional implied volatility surface with coordinates moneyness (m) and time to maturity $(T - t)$ [54]. Implied volatilities are strongly correlated across moneyness and time to maturity, cf. above, which suggests a description in terms of surface dynamics. A practical aspect are trading rules for volatility prediction based on implied volatility. The "sticky moneyness" rule predicts that the implied volatility surface tomorrow is the same as that today at constant moneyness and time to maturity. The "sticky strike" rule stipulates that the implied volatility tomorrow is the same as today at constant strike and constant maturity (i.e. absolute quantities).

Volatility surfaces can be generated for various series of liquid options such as calls and puts on the S&P500, the FTSE, or the DAX. With a generalization of principal component analysis – a technique widely used in image processing – the implied volatility surfaces can be described as fluctuating random surfaces driven by a small number of dominant eigenmodes. These eigenmodes parameterize the shape fluctuations of the surface. Their fluctuating prefactors describe the amplitude of surface variations. The first eigenmode which accounts for about 80% of the daily variance of the implied volatility surface is a flat sheet in $\sigma_{\text{imp}} - m - (T - t)$ space, almost independent of $T - t$ and with a small positive slope in m. This mode essentially has the same properties as the time series discussed in the second preceding paragraph. It is also negatively correlated with the price of the underlyings, i.e. contributes to a "leverage effect" to be discussed in Sect. 5.6.3. The second eigenmode changes sign at the money and is positive for $m > 1$ and negative for $m < 1$. A positive variation of this mode increases the volatilities of out-of-the money calls and descreases it for out-of-the-money puts. It contributes to the skewness of the risk-neutral distributions (when thinking backwards from implied volatility to risk-neutral measures) and, due to its slope in $T - t$, to the term structure. It also possesses the dynamics of a mean-reverting AR(1)-process. The third mode is a butterfly modes which changes the convexity of the implied volatility surface. It leads to a fattening of the tails of the risk-neutral distributions, cf. the mechanistic rules listed above [54].

This dynamics can be cast in a low-dimensional factor model

$$X(t; m, T-t) \equiv \ln \sigma_{\text{imp}}(t; m, T-t) = X(0; m, T-t) + \sum_{k=1}^{d} x_k(t) f_k(m, T-t) .$$

$$(4.120)$$

f_k is one of the d dominant eigenfunctions of the principal component decomposition. They are time-independent and describe the spatial variation of the fluctuations. The dynamics comes from the randomly fluctuating prefactors $x_k(t)$ which, according to the findings above, can be modeled as Ornstein-Uhlenbeck processes

$$\mathrm{d}x_k(t) = -\lambda_k \left[x_k(t) - \bar{x}_k \right] \mathrm{d}t + \gamma_k \mathrm{d}z_k . \tag{4.121}$$

λ_k is the rate of mean reversion and $\bar{x}_k$ is the average of the kth eigenmode. The stochastic increments $\mathrm{d}z_k$ are uncorrelated and may be drawn from a Gaussian (consistent with the lognormal distribution of implied volatilites, cf. below) or a more general distribution. The ranks of the fluctuating expansion coefficients $x_k(t)$ in (4.120) are ordered according to their variances γ_k^2 which measures the amplitude of the fluctuations they impart on $\sigma_{\text{imp}}(t; m, T - t)$. The dynamics of the implied volatility surfaces analyzed above can be faithfully represented by three factors $x_1(t) \ldots x_3(t)$ [54].

4.5.8 Volatility Indices

Volatility is the most important and least accessible quantity in option theory. Volatility can be inferred either from historical time series [estimate σ in (4.53)] or from implied volatility of options by inverting the Black–Scholes equation as in (4.119). For derivative markets, the second method is preferrable because the information is derived directly from derivative instruments, and implied volatility is more forward-looking than historical volatility.

Derivative trading requires high-frequency information on volatility, resp. implied volatility. The question arises if information on implied volatility can be provided in a standardized manner to assist traders in their decisions. Volatility indices have been constructed by various option exchanges to fill this gap. As volatility often signifies financial turmoil, volatility indices play the role of "investor fear gauges". In the following, we discuss two two indices using different construction principles.

VDAX Index

The VDAX index is provided by Deutsche Börse AG, and measures the implied volatility of at-the-money options on the DAX with 45 days to maturity [55]. Options on the DAX are among the most liquid instruments in the European derivatives markets. Although it is not directly relevant for other options, the knowledge of the VDAX value gives a good indication of the volatilities traded in the broader derivatives market.

Conceptually, the VDAX is based on implied volatilities, i.e. the Black–Scholes equation (4.85), resp. (4.86), is inverted numerically as formally done in (4.119). The practical calculation is more difficult, though, most importantly because, except for accidental circumstances, no option at the money with 45 days to maturity is traded in the markets.

Moreover, in practice, the traded futures price on the DAX are used for the VDAX calculation instead of the DAX itself. For options on futures, the Black–Scholes equation can be rewritten most easily by equating forward and futures prices F and using (4.1) in (4.85) and (4.86) to obtain

$$C_F = e^{-r(T-t)} \left[FN(d_{1F}) - XN(d_{2F}) \right] \tag{4.122}$$

$$P_F = e^{-r(T-t)} \left[XN(-d_{2F}) - FN(-d_{1F}) \right] \tag{4.123}$$

for the prices of call and put options on futures, respectively. d_{1F} and d_{2F} differ from (4.88) and (4.89) and are given by

$$d_{1F} = \frac{\ln \frac{F}{X} + \left(\frac{\sigma^2}{2} \right)(T-t)}{\sigma \sqrt{T-t}}, \tag{4.124}$$

$$d_{2F} = \frac{\ln \frac{F}{X} - \left(\frac{\sigma^2}{2} \right)(T-t)}{\sigma \sqrt{T-t}}. \tag{4.125}$$

The risk-free interest rate no longer appears explicitly in (4.124) and (4.125) and is implicitly accounted for by the use of the futures price F, cf. (4.1). The Black–Scholes problem for options on futures can also be solved *ab initio* following the lines of Sects. 4.5.1 and 4.5.2, with (4.122) and (4.123) as the solutions of the modified differential equations [10, 56]. This solution is known as Black's 1976 model.

The VDAX is based on a set of eight subindices calculated for DAX options with maturities of up to two years. Each subindex is based on four at-the-money options for the given maturity. After data filtering, the best bid and ask prices of each call and put option and of the DAX futures are averaged. Next, the risk-free interest rate is not a universal constant but depends on the maturity of the bonds it is taken from. Under normal conditions, r is lower for short-maturity bonds than for long maturities ("normal interest rate curve"). Only under exceptional circumstances is the interest rate curve inverted, i.e. the long maturities bring less interest than the short maturities. In general, risk-free interest rates are not available for the maturities of the options considered. In practice, they are generated by linear interpolation from two values bracketing the option contract maturity.

When the maturity of futures contracts differs from that of options, put-call parity is used to generate an effective forward price from the option prices. Using (4.1), (4.8) for put-call parity can be rewritten in terms of the forward price as

$$F(t) = [C(t) - P(t)] e^{r(T-t)} + X . \tag{4.126}$$

This equation is used for up to eight pairs of options for four strike prices above and below the "at-the-money point", and averaged at the end. Once the forward price is available, (4.122) and (4.123) are inverted for the implied volatility σ_{imp}.

The implied volatility constituting a volatility subindex for a specific maturity T_i is calculated as a weighted average of the implied volatilities a pair of put and call options with strike prices bracketing the futures price

$$\sigma_{\text{imp}}(T_i) = \frac{[X_h - F(T_i)] \left[\sigma_{\text{imp},l}^{\text{call}} + \sigma_{\text{imp},l}^{\text{put}}\right] + [F(T_i) - X_l] \left[\sigma_{\text{imp},h}^{\text{call}} + \sigma_{\text{imp},h}^{\text{put}}\right]}{2(X_h - X_l)} . \tag{4.127}$$

The subscripts h and l label the options with maturity T_i above and below the futures price. The eight volatility subindices are published by Deutsche Börse AG as additional information.

The VDAX then is the implied volatility generated from the two subindices with maturity closest to 45 days, by interpolation of the variances

$$\sigma_{\text{imp}}^{\text{VDAX}} = \sqrt{\frac{T_{i+1} - T}{T_{i+1} - T_i} \sigma_{\text{imp}}^2(T_i) + \frac{T - T_i}{T_{i+1} - T_i} \sigma_{\text{imp}}^2(T_{i+1})} , \tag{4.128}$$

where the maturities satisfy

$$T_i \leq T = 45\text{d} < T_{i+1} . \tag{4.129}$$

The VDAX is the implied volatility of a *hypothetical* at-the-money option with a 45-day maturity. At the time of writing, the VDAX is quoted every minute from 9 a.m. to 5:30 p.m.

There have been attempts to create derivatives on the VDAX. It is reported that the pricing and hedging of these products encountered many difficulties. Most likely, it was done in a way similar to that described in the subsequent text, supplemented by rules of thumb for the inevitable differences between the VDAX and the quantity effectively priced and hedged.

VIX

The VIX is the volatility index of the Chicago Board Options Exchange (CBOE) [57]. It measures the volatility of options on the S&P500 index with 30 days to expiration.

Since its introduction in 1993 until 2003, it was based on the implied volatility of at-the-money options on the S&P100 index, and calculated in a manner similar to the VDAX described above. In 2003, the method of calculation was changed, and the index now refers to the S&P500 index. The change was made in response to advances in quantitative finance which were driven by the desire to trade volatility derivatives deriving their measure of volatility directly from a series of option prices [57, 58]. Here, we continue to call this volatility measure "implied volatility" although this formally is not justified by (4.119) which we used to define implied volatility. The labeling is justified, however, (i) when implied volatility is understood as the market's expectation of future realized volatility, and (ii) by the strong similarity of the old VIX based on implied volatility, and the new VIX extended backwards in time to cover the period of the old VIX [57].

To understand the general problem behind the creation of volatility derivatives, notice that the hedging of a derivative, say on option, based on σ_{imp} as obtained by (4.119), is a highly nontrivial task. When volatility can be represented, e.g., as a linear combination of traded instruments, hedging is much easier.

How can one create an instrument that allows pure trading of volatility? With a position in an option, an investor is exposed both to the directional movements of the underlying and to its volatility. Can one eliminate the exposure to directional moves?

The simplest derivative instrument on volatility is a volatility, or variance swap. A swap is a contract which exchanges ("swaps") two cash flows. Swaps are most common in the fixed income sector (bonds and credits), and often the parties exchange the cash flows from fixed interest rate payments against variable interest rate payments. The payoff of a *variance swap* at expiration is

$$\text{VS}(T) = (\sigma_R^2 - K_{\text{var}})N \,, \tag{4.130}$$

where σ_R^2 is the variance of the underlying realized over the lifetime of the swap, K_{var} is variance delivery price and N is the notional of the contract. The holder of the swap receives N dollars for every point by which the variance σ_R^2 exceeds the delivery price K_{var} [58]. Alternatively, the variance swap may be understood as a forward contract.

To understand the construction of such a swap, go back to the definition of the Vega of an option. Vega, as defined in (4.109) measures the sensitivity to changes in volatility. The variance exposure of a call option is measured by the "Variance Vega"

$$\mathcal{V}_{var} = \frac{\partial C}{\partial \sigma^2} = \frac{S\sqrt{T-t}}{2\sqrt{2\pi}\sigma} \exp\left(-\frac{d_1^2}{2}\right), \tag{4.131}$$

where the second equality is valid only for Black–Scholes option prices, and d_1 was given in (4.88). Variance Vega is peaked at $S = X$ with a peak height proportional to X due to the explicit prefactor S. When many options with slightly different strikes are superposed with equal weight in a portfolio, the variance exposure of this portfolio is given by the superposition of the Variance Vegas. This leads to a triangular shape (in S) with Gaussian roundings at the edge. When the portfolio weighs the options with a weight factor X^{-2}, on the other hand, the dependence on S drops out, and the portfolio has an exposure to variance only (provided the price of the underlying remains in the range covered by the option strikes). This result becomes exact when the strike price X is treated as a continuous variable, and the portfolio is expressed as an integral over X with a weight factor X^{-2} [58]. In practice, out-of-the-money options are more liquid. For this reason, both out-of-the-money call and put options are used in setting up the portfolio

$$\Pi_\sigma(t) = \int_0^{S_*} \frac{dX}{X^2} P(X,t) + \int_{S_*}^\infty \frac{dX}{X^2} C(X,t). \tag{4.132}$$

S_* is an arbitrary reference price close to the at-the-money point. This portfolio's Delta and Variance Vega are [58]

$$\Delta = \frac{\partial \Pi_\sigma(t)}{\partial S} \equiv 0, \quad \mathcal{V}_{var} = \frac{\partial \Pi_\sigma(t)}{\partial \sigma^2} = \frac{T-t}{2}. \tag{4.133}$$

At expiration, the value of the portfolio $\Pi_\sigma(T)$ is

$$\Pi_\sigma(T) = \int_0^{S_*} \frac{dX}{X^2} \max\left[X - S(T), 0\right] + \int_{S_*}^\infty \frac{dX}{X^2} \max\left[S(T) - X, 0\right]$$

$$= \frac{S(T) - S_*}{S_*} - \ln\left[\frac{S(T)}{S_*}\right]. \tag{4.134}$$

The first term in the second equation essentially is an ordinary forward contract with a payoff linear in the deviation from the reference price S_*. The second term is a log-contract whose payoff equals the logarithm of the price ratio.

As with any other derivative, the fair delivery price of variance K_{var} is fixed by the requirement that the expected present value of the future payoff in a risk-neutral world is zero. The variance realized over the lifetime of the swap is

$$\sigma_R^2 = \frac{1}{T} \int_0^T \mathrm{d}t\, \sigma^2(t) \,, \tag{4.135}$$

where σ – unlike geometric Brownian motion – may be a time-dependent, perhaps even stochastic quantity. The criterion of zero expected value of the payoff then translates in

$$F(t) = \mathrm{e}^{-r(T-t)} \langle \sigma_R^2 - K_{\text{var}} \rangle = 0 \,. \tag{4.136}$$

When $S(t)$ follows an Itô process (4.40) even with a time-dependent volatility $\sigma(t)$, we can combine (4.53) and (4.62) to obtain

$$\frac{\mathrm{d}S(t)}{S(t)} - \mathrm{d}\left[\ln S(t)\right] = \frac{1}{2}\sigma^2 \mathrm{d}t \,. \tag{4.137}$$

Insert this into (4.135) and solve the second equality in (4.136)

$$K_{\text{var}} = \frac{2}{T} \left\langle \int_0^T \frac{\mathrm{d}S(t)}{S(t)} - \ln\left[\frac{S(T)}{S(0)}\right] \right\rangle \,. \tag{4.138}$$

For an Itô process in a risk-neutral world,

$$\left\langle \int_0^T \frac{\mathrm{d}S(t)}{S(t)} \right\rangle = \left\langle \int_0^T r\mathrm{d}t + \sigma(t)\mathrm{d}z \right\rangle = rT \,. \tag{4.139}$$

The last term in (4.138) is related to the log-contract in our portfolio of options with a continuous strike distribution. Combining everything gives the fair delivery price of the variance swap [58]

$$K_{\text{var}} = \frac{2}{T} \left[\mathrm{e}^{rT} \int_0^{S_\star} \frac{\mathrm{d}X}{X^2} P(X,t) + \mathrm{e}^{rT} \int_{S_\star}^\infty \frac{\mathrm{d}X}{X^2} C(X,t) \right.$$
$$\left. + rT - \left(\frac{S(0)\mathrm{e}^{rT}}{S_\star} - 1\right) - \ln\left(\frac{S_\star}{S(0)}\right) \right] \,. \tag{4.140}$$

This derivation does not require geometric Brownian motion, or the validity of the Black–Scholes assumptions. An instrument trading volatility alone thus can be constructed based on a weighted portfolio of options with a continuous strike distribution and weights inversely proportional to X^2.

Clearly, the value of such an instrument is a measure of the market's expected volatility over the lifetime of the contract, and therefore constitutes a valid volatility index. This is precisely what the CBOE's VIX does. As with the instruments acutally traded in the markets, the ideal continuous strike portfolio is approximated by a set of options with a discrete distribution of strikes. The VIX is [57]

$$\text{VIX} = 100 \sqrt{\frac{2e^{rT}}{T} \sum_i \frac{\Delta X_i}{X_i^2} f(X_i) - \frac{1}{T}\left[\frac{F}{X_0} - 1\right]^2}. \tag{4.141}$$

ΔX_i is the interval between the strike prices, $F = S(0)e^{rT}$ is the forward price, X_0 is the first strike price below the forward level and plays the role of the reference price $S_\star$ in (4.140). $f(X_i) = P(X_i)$ is the price of the put option with strike X_i for $X_i < X_0$ and $f(X_i) = C(X_i)$ is the call price for $X_i > X_0$. Of course, option and forward prices are averages of the bid and ask prices quoted in the market, and interpolation procedures similar to those described for the VDAX are necessary to roll along the fixed time to maturity of 30 days.

Both volatility indices, VDAX and VIX, can be used as underlyings for derivative instruments. In particular, a VIX futures has been traded on CBOE since shortly after the reformulation of the VIX based on volatility swap pricing, and options on the VIX are being introduced. In April 2005, Deutsche Börse AG announced that, in order to facilitate the creation of derivatives on the VDAX, it would change the calculation of the VDAX during the year 2005. While the new method has not been disclosed yet, in can be inferred from details of the press release that it will be similar to the method used by CBOE for the VIX.

5. Scaling in Financial Data and in Physics

The Black–Scholes equation for option prices is based on a number of hypotheses and assumptions. Subsequent price changes were assumed to be statistically independent, and their probability distribution was assumed to be the normal distribution. Moreover, the risk-free interest rate r and the volatility σ were assumed constant (in the simplest version of the theory). In this chapter, we will examine financial data in the light of these assumptions, develop more general stochastic processes, and emphasize the parallels between financial data and physics beyond the realm of Brownian motion.

5.1 Important Questions

We will be interested, among others, in answering the following important questions:

- How well does geometric Brownian motion describe financial data? Can the apparent similarities between financial time series and random walks emphasized in Sect. 3.4.1 be supported quantitatively?
- What are the empirical statistics of price changes?
- Are there stochastic processes which do not lead to Gaussian or log-normal probability distributions under aggregation?
- Is there universality in financial time series, i.e., do prices of different assets have the same statistical properties?
- Are financial markets stationary?
- Are real markets complete and efficient, as assumed by Bachelier?
- Why is the Gaussian distribution so frequent in physics?
- What are Lévy flights? Are they observable in nature?
- Are there correlations in financial data?
- How can we quantify temporal correlations in a financial time series?
- How can we quantify cross-correlations between various asset price histories?

Before discussing in detail the stochastic processes underlying real financial time series, we address the stationarity of financial markets.

5.2 Stationarity of Financial Markets

Geometric Brownian motion underlying the Black–Scholes theory of option pricing works with constant parameters: the drift μ and volatility σ of the return process, and the risk-free interest rate r are assumed independent of time. Is this justified? And is the dynamics of a market the same irrespective of time? That is, are the rules of the stochastic process underlying the return process time-independent?

For a practical option-pricing problem with a rather short maturity, say a few months, the estimation of the Black–Scholes parameters should pose no problem. For an answer to the questions posed above, on longer time scales, we will investigate various time series of returns. The following quantities will be of interest:

- The time series of (logarithmic) returns of an asset priced at $S(t)$ over a time scale τ

$$\delta S_\tau(t) = \ln\left(\frac{S(t)}{S(t-\tau)}\right) \approx \frac{S(t) - S(t-\tau)}{S(t-\tau)} . \tag{5.1}$$

- The time series of returns normalized to zero mean and unit variance

$$\delta s_\tau(t) = \frac{\delta S_\tau(t) - \langle \delta S_\tau(t)\rangle}{\sqrt{\langle [\delta S_\tau(t)]^2\rangle - \langle \delta S_\tau(t)\rangle^2}} , \tag{5.2}$$

where the expectation values are taken over the entire time series under consideration.

We first examine the time series of DAX daily closes from 1975 to 2005 shown in Fig. 1.2. The daily returns $\delta S_{1d}(t)$ derived from the data up to 5/2000 are shown in Fig. 5.1. At first sight, the return process looks stochastic with zero mean. The impressive long-term growth of the DAX up to 2000 and sharp decline thereafter, emphasized in Fig. 1.2, here show up in a small, almost invisible positive resp. negative mean of the return, of much smaller amplitude, however, than the typical daily returns. We also clearly distinguish periods with moderate (positive and negative) returns, i.e., low volatility (more frequent in the first half of the time series) from periods with high (positive and negative) returns, i.e., high volatility (more frequent in the second half of the time series). The main question is if data like Fig. 5.1 are consistent with a description, and to what accuracy, in terms of a simple stochastic process with constant drift and constant volatility. Or, to the contrary, do we have to take these parameters as time dependent, such as in the ARCH(p) or GARCH(p,q) models of Sect. 4.4.1? Or, worse even, do the constitutive functional relations of the stochastic process change with time?

As a first, admittedly superficial test of stationarity, we now divide the DAX time series into seven periods of approximately equal length, and evaluate the average return and volatility in each period. The result of this evaluation is shown in Table 5.1. The central column shows the increase resp.

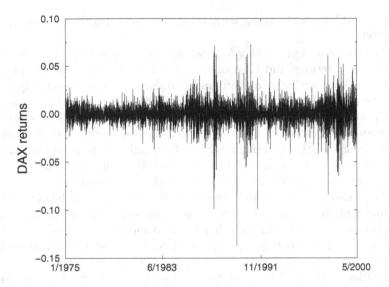

Fig. 5.1. Time series of daily returns of the DAX German blue chip index from 1975 to 2000. Analysis courtesy of Stephan Dresel based on data provided by Deutsche Bank Research

Table 5.1. Average return $\langle \delta S_{1d}(t) \rangle$ and volatility σ of the DAX index in seven approximately equally long periods from January 2, 1975, to December 31, 2004. Analysis courtesy of Stephan Dresel based on data provided by Deutsche Bank Research supplemented by data downloaded from Yahoo, http://de.finance.yahoo.com

Period	Return $[d^{-1}]$	Volatility $[d^{-1/2}]$
02.01.1975–15.03.1979	0.00028	0.0071
16.03.1979–10.06.1983	0.00021	0.0078
13.06.1983–03.09.1987	0.00072	0.0104
04.09.1987–02.12.1991	0.00002	0.0155
03.12.1991–14.02.1996	0.00042	0.0091
16.02.1996–05.05.2000	0.00106	0.0149
08.05.2000–31.12.2004	−0.00049	0.0184

decrease of the average returns with time, which is responsible for the increasing slope of the DAX index in Fig. 1.2. The average return increases by a factor of three to four from 1975 to 2000, and decreases to even become negative in the drawdown period from 2000 to 2005. The rather low value in the fourth period is due to the October crash in 1987 right after the beginning of our period, and another crash in 1991. The last column shows the volatilities which also increase with time. The volatility is particularly big after 2000.

In the six periods up to May 5, 2000, we now subtract the average return from the daily returns and then divide by the standard deviation, in order to obtain a process with mean zero and standard deviation unity. Figure 5.2 shows the probability distributions of the returns normalized in this way, in the six periods. Except for a few points in the wings, the six distributions do not deviate strongly from each other. One therefore would conclude that the rules of the stochastic process underlying financial time series do not change with time significantly, and that most of the long-term evolution of markets can be summarized in the time dependence of its parameters.

Notice, however, that, strictly speaking, this finding invalidates geometric Brownian motion as a model for financial time series because μ and σ were assumed constant there. On the other hand, if such time dependences of parameters only are important on sufficiently long time scales (which we have not checked for the DAX data), one might take a more generous attitude, and consider geometric Brownian motion as a candidate for the description of the DAX on time scales which are short compared to the time scale of variations of the average returns or volatilities. Physicists take a similar attitude,

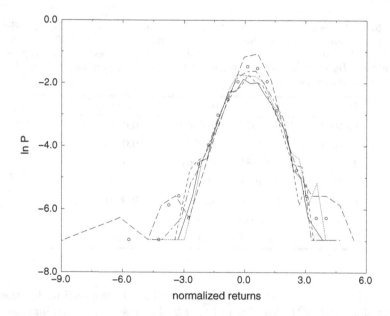

Fig. 5.2. Probability distributions of normalized daily returns of the DAX German blue chip index in the six equally long periods from 1975 to 2000. The normalization procedure is explained in the text and the parameters are summarized in Table 5.1. *Solid line*: period 1, *dotted line*: period 2, *dashed line*: period 3, *long-dashed line*: period 4, *dot-dashed line*: period 5, *circles*: period 6. Analysis courtesy of Stephan Dresel based on data provided by Deutsche Bank Research

e.g., with temperature, in systems slightly perturbed away from equilibrium. While being an equilibrium property in the strict sense, one may introduce local temperatures in an inhomogeneous system on scales that are small with respect to those over which the temperature gradients vary appreciably.

Returning to the probability distributions of the DAX returns, Fig. 5.3 shows the probability distributions of three periods (1, 4, and 6) displaced for clarity. Period 1 is not clearly Gaussian although its tails are not very fat, a fact that we qualitatively reproduce in periods 2 and 3. The distributions of periods 4, 5 (not shown), and 6 do possess rather fat tails whose importance, however, changes with time. In the DAX sample, period 4 including the October crash in 1987, and some more turmoil in 1990 and 1991, clearly has the fattest tails. One therefore should be extremely careful in analyzing market data from very long periods. Markets certainly change with time, and there may be more time dependence in financial time series than just a slow variation of average returns and volatilities. As Fig. 5.3 suggests, even the shape of the probability distribution might change with time.

These complications have not been studied systematically, and are ignored in the following discussion. Depending on the underlying time scales, they may or may not affect the conclusions of the various studies we review. We first proceed to a critical examination of geometric Brownian motion.

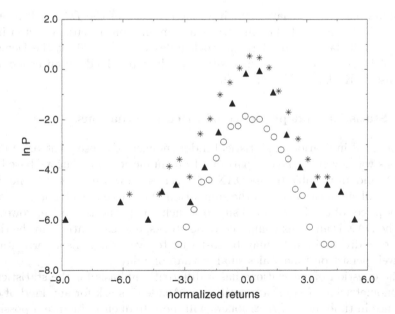

Fig. 5.3. Probability distributions (vertically displaced for clarity) of normalized daily returns of the DAX German blue chip index in the periods 1 (*open circles*), 4 (*filled triangles*), and 6 (*stars*) specified in Table 5.1. Analysis courtesy of Stephan Dresel based on data provided by Deutsche Bank Research

5.3 Geometric Brownian Motion

Geometric Brownian motion makes two fundamental hypotheses on a sto-chastic process:

1. Successive realizations of the stochastic variable are statistically indepen-dent.
2. Returns of financial markets, or relative changes of the stochastic vari-able, are drawn from a normally distributed probability density func-tion, i.e., the probability density function of the stochastic variable, resp. prices, is log-normal.

Here, we examine these properties for financial time series.

5.3.1 Price Histories

Figure 5.4 shows three financial time series which we shall use to discuss correlations: the S&P500 index (top), the DEM/US$ exchange rate (center), and the BUND future (bottom) [17]. The BUND future is a futures contract on long-term German government bonds, and thereby a measure of long-term interest-rate expectations. The data range from November 1991 to February 1995.

Figure 5.5 gives a chart of high-frequency data of the DAX taken on a 15-second time interval. The history is a combination of data collected in a purpose-built database of German stock index data [59, 60] at the Depart-ment of Physics, Bayreuth University, and data provided by an economics database at Karlsruhe University [61].

5.3.2 Statistical Independence of Price Fluctuations

A superficial indication of statistical independence of subsequent price fluc-tuations was given by the comparison of our numerical simulations based on an IID random variable to the DAX time series, shown in Figs. 1.3 and 3.7. The overall similarity between the simulation of a random walk and the daily closing prices of the DAX would support such a hypothesis. Notice, however, that the DAX is an index composed of 30 stocks, and correlations in the time series of individual stocks may be lost due to averaging. Also, correlations may well persist on time scales smaller than one day.

The question of correlations has a different emphasis for the statistician or econometrician, and for a practitioner. Academics ask for any kind of de-pendence in time series. Practitioners will more frequently inquire if possible dependences can be used for generating above-average profits, and if success-ful trading rules can be built on such correlations. Despite what has been said in the preceding paragraph, the apparent importance of technical analy-sis suggests that there may indeed be tradable though subtle correlations.

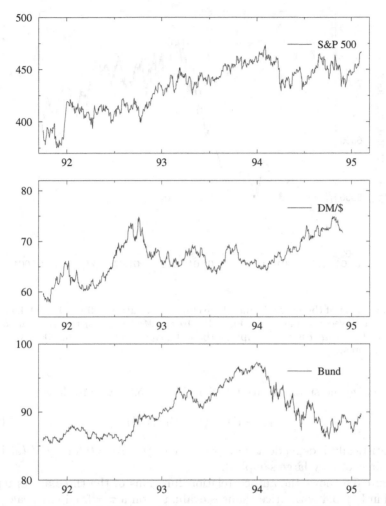

Fig. 5.4. Three financial time series from November 1991 to February 1995: the S&P500 index (*top*), the DEM/US$ exchange rate (*center*), and the BUND futures (*bottom*). From J.-P. Bouchaud and M. Potters: *Théorie des Risques Financiers*, by courtesy of J.-P. Bouchaud. ©1997 Diffusion Eyrolles (Aléa-Saclay)

Correlation Functions

We now analyze correlation functions of returns on a fixed time scale τ, $\delta S_\tau(t)$, (5.1). The autocorrelation function of this quantity is

$$C_\tau(t - t') = \frac{1}{D\tau} \langle [\delta S_\tau(t) - \langle \delta S_\tau(t) \rangle] [\delta S_\tau(t') - \langle \delta S_\tau(t') \rangle] \rangle , \qquad (5.3)$$

where

$$D\tau = \text{var} [\delta S_\tau(t)] , \qquad (5.4)$$

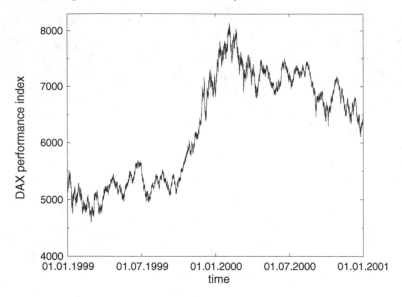

Fig. 5.5. Chart of the DAX German blue chip index during 1999 and 2000. Data are taken on a 15-second time scale. From S. Dresel: *Modellierung von Aktienmärkten durch stochastische Prozesse,* Diplomarbeit, Universität Bayreuth, 2001, by courtesy of S. Dresel

to emphasize the similarity to diffusion. Using (5.2), we also have

$$C_\tau(t - t') = \langle \delta s_\tau(t) \delta s_\tau(t') \rangle \ . \tag{5.5}$$

For statistically independent data, we have $C_\tau(t - t') = 0$ for $t \neq t'$ (at least in the limit of very large samples).

Figure 5.6 shows the autocorrelation functions of the three assets represented in Fig. 5.4 with price changes evaluated on a $\tau = 5$-minute scale [17]. For time lags below 30 minutes, there are weak correlations above the 3σ level. Above 30-minute time lags, correlations are not significant.

When errors are random and normally distributed (a standard assumption), the standard deviation determines the confidence levels as

$$\left.\begin{array}{l} \sigma : 32\% \\ 2\sigma : 5\% \\ 3\sigma : 0.2\% \\ 10\sigma : 2 \times 10^{-23} \end{array}\right\} = 2 \int_\Lambda^\infty P(S)\mathrm{d}S \ . \tag{5.6}$$

Under a null hypothesis of vanishing correlations, 32% of the data may randomly lie outside a 1σ corridor, or 0.2% of the data may be outside a 3σ corridor.

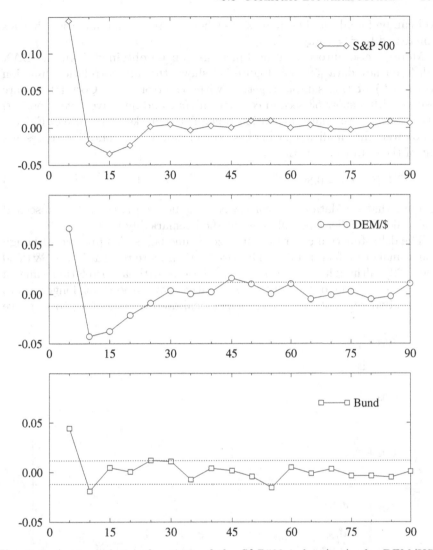

Fig. 5.6. Autocorrelation functions of the S&P500 index (*top*), the DEM/US$ exchange rate (*center*), and the BUND future (*bottom*), over a time scale $\tau = 5$ minutes. The horizontal scale is the time separation $t - t'$ in minutes. The horizontal dotted lines are the 3σ confidence levels. From J.-P. Bouchaud and M. Potters: *Théorie des Risques Financiers*, by courtesy of J.-P. Bouchaud. ©1997 Diffusion Eyrolles (Aléa-Saclay)

In Fig. 5.6 for time lags above 30 minutes, the (null) hypothesis of statistically independent price changes therefore cannot be rejected for the three assets studied. The non-random deviations out of the 3σ corridor for smaller time lags, on the other hand, indicate non-vanishing correlations in this range. Consistent with this is the finding that no correlations significant on the 3σ

level can be found for the same assets when the time scale for price changes is increased to $\tau = 1$ day [17].

More precise autocorrelation functions can be obtained from the DAX high-frequency data [59, 60]. Figure 5.7 shows the autocorrelation function $C_{15''}(t - t')$ of this sample together with 3σ error bars. Correlations are positive with a short 53-second correlation time and negative (overshooting) with a longer 9.4-minute correlation time. The remarkable feature of Fig. 5.7 is, however, the small weight of these correlations! The solid line represents a fit of the data to a function

$$C_{15''}^{\text{fit}}(t - t') = 0.89\delta_{t,t'} + 0.12\mathrm{e}^{-|t-t'|/53''} - 0.01\mathrm{e}^{-|t-t'|/9.4'}, \qquad (5.7)$$

implying that the data are uncorrelated to almost 90%, even at a 15-second time scale. Bachelier's postulate is satisfied remarkably well.

The delta-function contribution at zero time lag is also present, although with a smaller prefactor, in a study based on 1-minute returns in the S&P500 index [62], although only positive correlations with a correlation time of 4 minutes and no overshooting to negative correlations at longer times are found there. A strong zero-time-lag peak and overshooting to negative

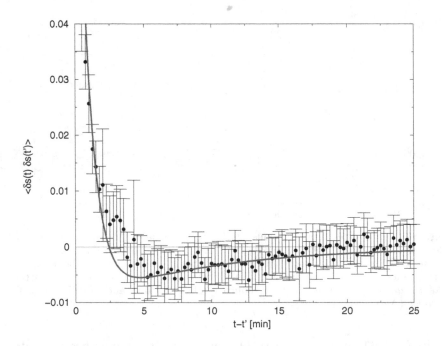

Fig. 5.7. Linear autocorrelation function $C_{15''}(t - t')$ for 15-second DAX returns (*dots*) with 3σ error bars. The solid line is a fit to (5.7) and demonstrates that the data are almost uncorrelated. From S. Dresel: *Modellierung von Aktienmärkten durch stochastische Prozesse,* Diplomarbeit, Universität Bayreuth, 2001, by courtesy of S. Dresel

correlations at about 15 minutes are also visible in 1-minute data from the Hong Kong Hang Seng stock index [63].

That subsequent price changes are essentially statistically independent is not a new finding. It was established, based on time-series analysis, back in 1965 by Fama [64] (and before, perhaps, by others). In the next section, we shall discuss another interesting aspect of Fama's work.

Filters

Fama's work was motivated by Mandelbrot's objections (to be discussed below in Sect. 5.3.3) to the standard geometrical Brownian motion model of price changes of financial assets, Sect. 4.4.2. In the course of his criticism, Mandelbrot also pointed to the "fallacies of filter trading" [33]. Filters were invented by Alexander [65] and were trading rules purported to generate above-average profits in stock market trading.

An $x\%$-filter works like this: if the relative daily price change of an asset $\Delta S/S > x\%$ after a local minimum, then buy the stock and hold until $\Delta S/S < -x\%$ after a local maximum. At this point, sell the stock and simultaneously go short until $\Delta S/S > x\%$ after another local minimum. Close out the short position and go long at the same time, etc. If filters are successful, more successful than, e.g., a naïve buy-and-hold strategy, there must be non-trivial correlations in the stock market.

Fama conducted a systematic investigation of such filters on all Dow Jones stocks from late 1957 to September 1962 [64]. Important results of his study are summarized in Table 5.2. The comparison with simple buy-and-hold is rather negative. Even ignoring transaction costs, only 7 out of the 30 Dow Jones stocks generated higher profits by filter trading than by buy-and-hold. Filter trading, however, involves frequent transactions, and when transaction costs are included, buy-and-hold was the better strategy for all 30 stocks, leading Fama to the conclusion: "From the trader's point of view, the independence assumption of the random-walk model is an adequate description of reality" [64].

Notice that Fama's investigation addresses correlations in the time series of individual stocks, as well as the practical aspects. We now turn to the statistics of price changes.

5.3.3 Statistics of Price Changes of Financial Assets

Early "tests" of the statistics of price changes did not reveal obvious contradictions to a (geometrical) Brownian motion model. Bachelier himself had conducted empirical tests of certain of his calculations, and of the underlying theory of Brownian motion [6]. Within the uncertainties due to the finite (small) sample size, there seemed to be at least consistency between the data and his theory. The problem we remarked on in Sect. 3.2.3, that price changes

Table 5.2. Comparison of profits of filter trading and buy-and-hold on the Dow Jones stocks from late 1957 to September 1962. Transaction costs have been ignored in the first column and have been included in the second column. From *J. Business* **38**, *34 (1965)* courtesy of E. F. Fama. ©The University of Chicago Press 1965

SUMMARY OF FILTER PROFITABILITY IN RELATION TO
NAÏVE BUY-AND-HOLD TECHNIQUE*

STOCK	PROFITS PER FILTER†		
	Without Commissions (1)	With Commissions (2)	Buy-and-Hold (3)
Allied Chemical..............	648.37	−10,289.33	2,205.00
Alcoa......................	3,207.40	− 3,929.42	− 305.00
American Can................	− 844.32	− 5,892.85	1,387.50
A.T.&T......................	16,577.26	4,912.84	20,005.00
American Tobacco...........	8,342.61	− 1,467.71	7,205.00
Anaconda....................	− 28.26	− 7,145.82	862.50
Bethlehem Steel.............	− 837.94	− 6,566.80	652.50
Chrysler....................	− 954.68	−12,258.61	− 1,500.00
Du Pont	6,564.21	− 465.35	9,550.00
Eastman Kodak..............	6,584.95	− 5,926.10	11,860.50
General Electric.............	− 107.06	− 8,601.28	2,100.00
General Foods..............	11,370.33	2,266.89	11,420.00
General Motors.............	− 1,099.40	− 8,440.42	2,025.00
Goodyear...................	− 2,241.28	−17,323.20	2,920.70
International Harvester.......	− 735.95	− 7,444.92	3,045.00
International Nickel..........	5,231.25	− 3,509.97	5,892.50
International Paper..........	2,266.82	− 7,976.68	− 278.10
Johns Manville..............	− 1,090.22	− 8,368.44	1,462.50
Owens Illinois...............	727.27	− 5,960.05	3,437.50
Procter & Gamble...........	12,202.83	4,561.52	8,550.00
Sears......................	4,871.36	408.65	5,195.00
Standard Oil (Calif.).........	− 3,639.79	−21,055.08	5,326.50
Standard Oil (N.J.)..........	− 1,416.48	− 6,208.68	1,380.00
Swift & Co..................	− 923.07	− 8,161.76	552.50
Texaco.....................	2,803.98	− 5,626.11	6,546.50
Union Carbide..............	3,564.02	− 1,612.83	1,592.50
United Aircraft.............	− 1,190.10	− 8,369.88	562.50
U.S. Steel..................	1,068.23	− 5,650.03	475.00
Westinghouse...............	− 338.85	−12,034.56	745.00
Woolworth..................	4,190.78	− 2,403.34	3,225.00

too often fell outside the bounds predicted by Bachelier, was not noticed in his thesis.

A similar rough test is provided by the apparent similarity of the random-walk simulations by Roberts [34] and the variations of the Dow Jones index. One may remark that the actual financial data possess more big changes than his simulation. However, this was not tested for in a systematic manner.

In summary, the model of geometric Brownian motion was pretty well established in the finance community in the early 1960s. It therefore came as a surprise when Mandelbrot postulated in 1963 that the stochastic process describing financial time series would deviate fundamentally and dramatically from geometric Brownian motion [66].

Mandelbrot's Criticism of Geometric Brownian Motion

Mandelbrot examined the prices of a commodity – cotton – on various exchanges in the United States [66]. He used various time series of daily and mid-month closing prices. From them, he calculated the logarithmic price changes, (5.1), for $\tau = 1d$, $1m$. Logarithmic price changes are postulated to be normally distributed by the geometric Brownian-motion model, (4.65). Mandelbrot's results are shown in Fig. 5.8 in a log–log scale where δS_τ is denoted u. In such a scale, a log-normal distribution function would be represented by an inverted parabola

$$\ln p_{\log-\mathrm{nor}}(\delta S_\tau) \propto - \left[\ln \frac{S(t)}{S(t-\tau)} \right]^2 = -[\delta S_\tau(t)]^2 . \tag{5.8}$$

The disagreement between the data and the prediction, (5.8), of the geometric Brownian motion model is striking! The data rather behave approximately as

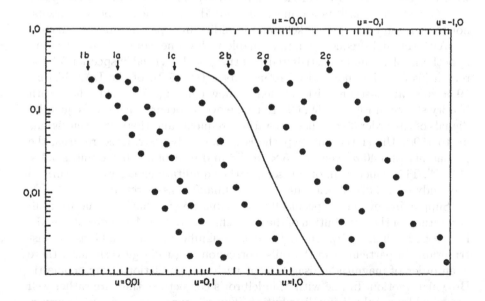

Fig. 5.8. Frequency of positive (*lower left part*, label 1) and negative (*upper right part*, label 2) logarithmic price changes of cotton on various US exchanges. a, b, c represent different time series. u in the legend is δS_τ in the text. Notice the double-logarithmic scale! The solid line is the cumulated density distribution function of a stable Lévy distribution with an index $\mu \approx 1.7$. From *J. Business* **36**, *394 (1963)* and *Fractals and Scaling in Finance* (Springer-Verlag, New York 1997) courtesy of B. B. Mandelbrot. ©The University of Chicago Press 1963

straight lines for large $|\delta S_\tau|$, i.e., are consistent with the asymptotic behavior of a stable Lévy distribution (4.43). A value of $\mu \approx 1.7$ describes the data rather well. Fama, later on, also studied price variations on stock markets, and found evidence further supporting Mandelbrot's claim for Lévy behavior [64].

We shall discuss Lévy distributions is more detail in Sect. 5.4.3. Here, it is sufficient to mention that Lévy distributions asymptotically decay with power laws of their variables, (5.44), and are stable, i.e., form-invariant, under addition if the index $\mu \leq 2$. The Gaussian distribution is a special case of stable Lévy distributions with $\mu = 2$ (cf. below).

It is obvious that, for price changes drawn from Lévy distributions, extreme events are much more frequent than for a Gaussian, i.e., the distribution is "fat-tailed", or "leptokurtic". An immediate consequence of (5.44) is that the variance of the distribution is infinite for $\mu < 2$. Moreover, the underlying stochastic process must be dramatically different from geometric Brownian motion.

One may wonder if Mandelbrot's observation only applies to cotton prices, or perhaps commodities in general, or if stock quotes, exchange rates, or stock indices possess similar price densities. And to what extent does it pass tests with the very large data samples characteristic of trading in the computer age? Commodity markets are much less liquid than stock or bond markets, not to mention currency markets, and liquidity may be an important factor.

With the high-frequency data available today, one can easily reject a null hypothesis of normally distributed returns just by visual inspection of the return history. The normalized returns $\delta s_{15''}(t)$, (5.2), of the DAX history 1999–2000 at 15-second tick frequency shown in Fig. 5.5 yields the return history shown in Fig. 5.9 [59, 60]. Extreme events occur much too frequently! Signals of the order $30\sigma \ldots 60\sigma$ are rather frequent, and there are even signals up to 160σ. Under the null hypothesis of normally distributed returns, the probability of a 40-σ event is 1.5×10^{-348} and that of a 160-σ event is 4.3×10^{-5560}. This conclusion, of course, is rather qualitative, and we now turn to the study of the distribution functions of financial asset returns.

Supporting evidence specifically for stable Lévy behavior came from an early study of the distribution of the daily changes of the MIB index at the Milan Stock Exchange [67]. The data deviate significantly from a Gaussian distribution. In particular, in the tails, corresponding to large variations, there is an order of magnitude disagreement with the predictions from geometric Brownian motion. In line with Mandelbrot's conjecture, they are rather well described by a stable Lévy distribution. The tail exponent $\mu = 1.16$, however, is rather lower than the values found by Mandelbrot.

While this work represents the first determination of the scaling behavior of a stock market index published in a physics journal, ample evidence in favor of stable Lévy scaling behavior had been gathered before in the economics literature. Fama performed an extensive study of the statistical properties

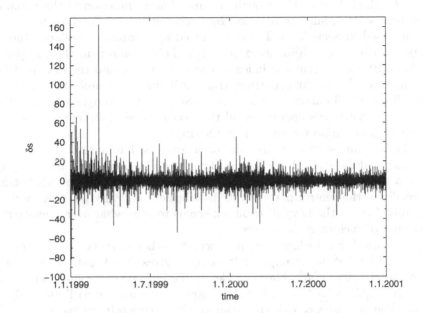

Fig. 5.9. Return history of the DAX German blue chip index during 1999 and 2000, normalized to the sample standard deviation. Data are taken on a 15-second time scale. Notice the event at 160σ and numerous events in the range $30\sigma \ldots 60\sigma$. From S. Dresel: *Modellierung von Aktienmärkten durch stochastische Prozesse*, Diplomarbeit, Universität Bayreuth, 2001, by courtesy of S. Dresel

of US companies listed in the Dow Jones Industrial Average in the 1960s [64]. As suggested in the preceding section, he found that the assumption of statistical independence of subsequent price changes was satisfied to a good approximation. Concerning the statistics of price changes, he found that "Mandelbrot's hypothesis does seem to be supported by the data. This conclusion was reached only after extensive testing had been carried out" [64]. Stable Lévy scaling was also found by economists in other studies of stock returns, foreign exchange markets, and futures markets [68].

Mantegna and Stanley performed a systematic investigation of the scaling behavior of the American S&P500 index [69]. Index changes $Z \equiv \delta S_\tau(t)$ have been determined over different time scales τ (denoted Δt in the figures) ranging from 1 to 1000 minutes (≈ 16 hours). If these data are drawn from a stable Lévy distribution, they should show a characteristic scaling behavior, i.e., one must be able, by a suitable change of scale, to collapse them onto a single master curve. Rescale the variable and probability distribution according to

$$Z_s = \frac{Z}{\tau^{1/\mu}} \quad \text{and} \quad L_\mu(Z_s, 1) = \frac{L_\mu(Z, \tau)}{\tau^{-1/\mu}} . \tag{5.9}$$

Here, $L_\mu(Z, \tau)$ denotes the probability distribution function of the variable Z at time scale τ, and the notation L_μ is chosen to make it consistent with the one used in Sect. 5.4.3. The data indeed approximately collapse onto a single distribution with an index $\mu = 1.4$. This is shown in the top panel of Fig. 5.10. Notice that the index of the distribution and the one used for rescaling must be the same, putting stringent limits on the procedure. Scaling, the collapse of all curves onto a single master curve, strongly suggests that the same mechanisms operate at all time scales, and that there is a single universal distribution function characterizing it.

The bottom panel compares the data for $\tau = 1$ minute with both the Gaussian and the stable Lévy distributions. It is clear that the Gaussian provides a bad description of the data. The Lévy distribution is much better, especially in the central parts of the distribution. For very large index fluctuations $Z \geq 8\sigma$, the Lévy distribution seems to somewhat overestimate the frequency of such extremal events.

Comparable results have been produced for other markets. For the Norwegian stock market, for example, R/S analysis gives an estimate of the Hurst exponent $H \approx 0.614$ [46]. The tail index μ of a Lévy distribution is related to H by $\mu = 1/H \approx 1.63$, in rather good agreement with the S&P500 analysis above. The tail index can also be estimated independently, giving similar values. Using these values, the probability distributions $p(\delta S_\tau)$ for different time scales τ can be collapsed onto a single master curve, as for the S&P500 in Fig. 5.10. Although the data extend out to 15 standard deviations, the truncation for extreme returns is much less pronounced than for the US stock market [46].

Closely related to the stable Lévy distributions are hyperbolic distributions. They also produce very good fits of stock market data [70].

Some kind of truncation is apparently present in the data of Fig. 5.10, and a "truncated Lévy distribution" (to be discussed below) has been invented for the purpose of describing them [71]. Figure 5.11, which displays the probability that a price change $\delta S_{15\mathrm{min}} > \delta x$

$$P_>(\delta x) = \int_{\delta x}^{\infty} \mathrm{d}\left(\delta S_{15\mathrm{min}}\right) p\left(\delta S_{15\mathrm{min}}\right) \qquad (5.10)$$

rather than the probability density function itself, shows that this distribution indeed fits very well the observed variations of the S&P500 index on a 15-minute scale [17]. Similarly good fits are obtained for different time scales, and for different assets, e.g., the BUND future or the DEM/$ exchange rate [17].

Practical Consequences, Interpretation

From the preceding section, it is clear that a Gaussian distribution does not fit the probability distribution of financial time series. Although Mandelbrot's

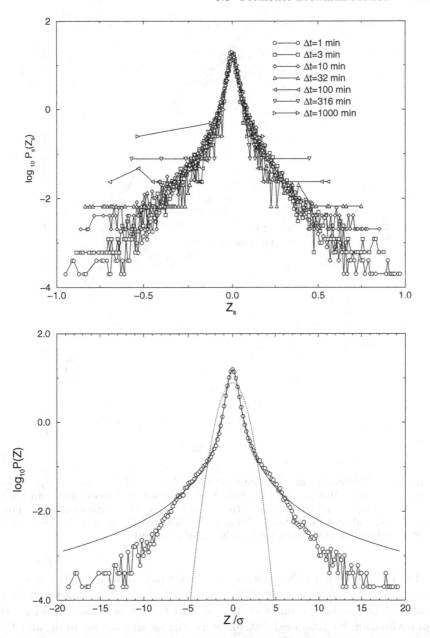

Fig. 5.10. Probability distribution of changes of the S&P500 index. *Top panel*: changes of the S&P500 index rescaled as explained in the text. If the data are drawn from a stable Lévy distribution, they must fall onto a single master curve. Δt in the figure is τ in the text, and $Z \equiv \delta S_\tau$. *Bottom panel*: comparison of the $\tau = 1$-minute data with Gaussian and stable Lévy distributions. By courtesy of R. N. Mantegna. Reprinted by permission from *Nature* **376**, *46 (1995)* ©1995 Macmillan Magazines Ltd.

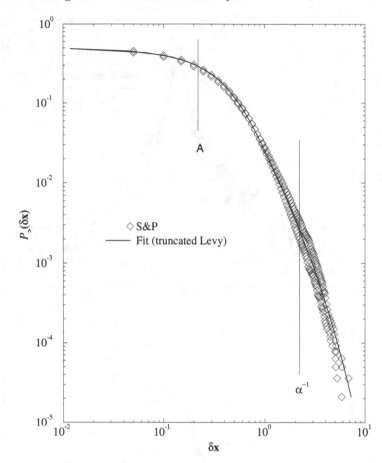

Fig. 5.11. Probability of 15-minute changes of the S&P500 index, $\delta S_{15\text{min}}$, exceeding δx, plotted separately for upward and downward movements, and a fit to a truncated Lévy distribution with $\mu = 3/2$. α is the truncation scale. From J.-P. Bouchaud and M. Potters: *Théorie des Risques Financiers,* by courtesy of J.-P. Bouchaud. ©1997 Diffusion Eyrolles (Aléa-Saclay)

stable Lévy paradigm may not be the last word and although the actual data may decay more quickly than a stable Lévy distribution for very large values of the variables, one certainly should take it seriously (i) as a first approximation for fat-tailed distributions, (ii) as an extreme limit, and (iii) as a worst-case scenario. Here, we summarize important findings, interpret them, and point to some consequences.

1. All empirical data have fat-tailed (leptokurtic) probability distributions.
2. To the extent that they are described by a stable Lévy distribution with index $1 \leq \mu \leq 2$, the variance of an infinite data sample will be infinite.

For finite data samples, the variance of course is finite, but it will not converge smoothly to a limit when the sample size is increased.

3. Quantities derived from the probability distribution, such as the mean, variance, or other moments, will be extremely sample-dependent.

4. Statistical methods based on Gaussian distributions will become questionable.

5. What is wrong the central limit theorem? It apparently predicts a convergence to a Gaussian, which does not take place here.

6. Apparently, special time scales are eliminated by arbitrage.

7. The actual stock price is much less continuous than a random walk.

8. In a Gaussian market, big price changes are very likely the consequence of many small changes. In real markets, they are very likely the consequence of very few big price changes.

9. The trading activity is very non-stationary. There are quiescent periods changing with hectic activity, and sometimes trading is stopped altogether.

10. According to economic wisdom, stock prices reflect both the present situation as well as future expectations. While the actual situation most likely evolves continuously, future expectations may suffer discontinuous changes because they depend on factors such as information flow and human psychology.

11. One consequence, namely that filters cannot work, has been discussed in Sect. 5.3.2. A necessary condition is that the stock price follows a continuous stochastic process. On the contrary, the processes giving rise to Lévy distributions must be rather discontinuous.

12. The assumption of a complete market is not always realistic. With discontinuous price changes, there will be no buyer or no seller at certain prices.

13. Stop-loss orders are not suitable as a protection against big losses. They require a continuous stochastic process to be efficient. Despite this, stop-loss orders may be useful, even necessary, in practice. The point here is that, given the discontinuities in financial time series, the actual price realized in a transaction triggered by a stop loss (or stop buy) order may be quite far from the one targeted when giving the order. Is there an alternative to stop-loss and stop-buy orders in a Lévy-type market?

14. The risk associated with an investment is strongly underestimated by Gaussian distributions or geometric Brownian motion.

15. The standard arguments for risk control by diversification (cf. below) may no longer work (cf. Sect. 10.5.5).

16. The Black–Scholes analysis of option pricing becomes problematic. Geometric Brownian motion is a necessary condition. Risk-free portfolios can no longer be constructed in theory – not to mention the problems encountered in Black and Scholes' continuous adjustment of positions when the stochastic process followed by the underlying security is discontinuous.

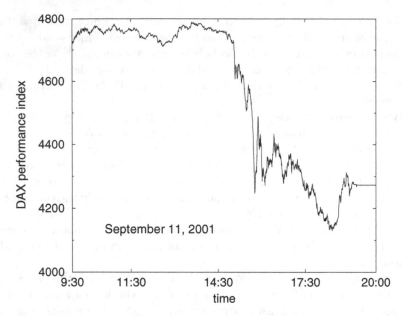

Fig. 5.12. Variation of the DAX German blue chip index during September 11, 2001. Notice the alternation of discontinuous with more continuous index changes. On September 11, 2001, terrorists flew two planes into the World Trade Center in New York

We illustrate these points in the following two figures. Figure 5.12 shows the DAX history (15-second frequency) for the most disastrous day for capital markets during the recent years, September 11, 2001. The first terrorist plane hit the north tower of the World Trade Center in New York at about 14:30 h local time in Germany. The south tower was hit about half an hour later. The reaction of the markets was dramatic. There is a series of crashes followed by strong rebounds, alternating with periods of more continuous price histories. The two biggest losses, 2% and 8% over just a few minutes time scale, clearly stand out. Figure 5.13 shows two hours of DAX history on September 30, 2002. We also see a discontinuous price variation around 16:00 h amidst more continuous changes of the index before and after that time. However, unlike September 11, 2001, no particular catastrophes happened that day – not even exceptionally bad economic news was diffused. Still, the DAX lost about 1% in a 15-second interval, and 3% over a couple of minutes.

5.4 Pareto Laws and Lévy Flights

We now want to discuss various distribution functions which may be appropriate for the description of the statistical properties of economic time series.

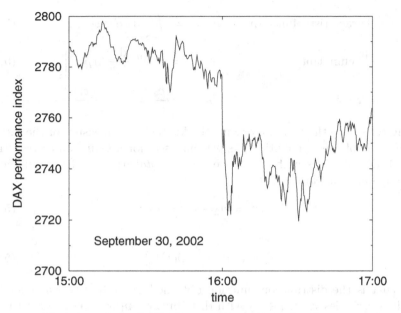

Fig. 5.13. Variation of the DAX German blue chip index during two hours of September 30, 2002. Unlike September 11, 2001, on September 30, 2002, no particular events were reported. Still, a 3% loss over a time scale of about one minute is reported around 16:00 h local time in Germany

Many key words have been mentioned already in the previous section, and are given a precise meaning here.

5.4.1 Definitions

Let $p(x)$ be a normalized probability distribution, resp. density,

$$\int_{-\infty}^{\infty} dx\, p(x) = 1 . \tag{5.11}$$

Then we have the following definitions:

$$\text{expectation value} \quad E(x) \;\equiv\; \langle x \rangle = \int_{-\infty}^{\infty} dx\, x\, p(x) , \tag{5.12}$$

$$\text{mean absolute deviation} \quad E_{\text{abs}}(x) = \int_{-\infty}^{\infty} dx\, |x - \langle x \rangle|\, p(x) , \tag{5.13}$$

$$\text{variance} \quad \sigma^2 \;=\; \int_{-\infty}^{\infty} dx (x - \langle x \rangle)^2 p(x) , \tag{5.14}$$

$$n^{\text{th}} \text{ moment} \quad m_n \;=\; \int_{-\infty}^{\infty} dx\, x^n p(x) , \tag{5.15}$$

$$\text{characteristic function} \quad \hat{p}(z) \;=\; \int_{-\infty}^{\infty} \mathrm{d}x \, \mathrm{e}^{\mathrm{i}zx} p(x) \,, \qquad (5.16)$$

$$n^{\text{th}} \text{ cumulant} \quad c_n \;=\; (-\mathrm{i})^n \left.\frac{\mathrm{d}^n}{\mathrm{d}z^n} \ln \hat{p}(z)\right|_{z=0} , \qquad (5.17)$$

$$\text{kurtosis} \quad \kappa \;=\; \frac{c_4}{\sigma^4} = \frac{\langle (x - \langle x \rangle)^4 \rangle}{\sigma^4} - 3 \,. \qquad (5.18)$$

Being related to the fourth moment, the kurtosis is a measure of the fatness of the tails of the distribution. As we shall see, for a Gaussian distribution, $\kappa = 0$. Distributions with $\kappa > 0$ are called *leptokurtic* and have tails fatter than a Gaussian. Notice that

$$\sigma^2 = m_2 - m_1^2 = c_2 \qquad (5.19)$$

and

$$m_n = (-\mathrm{i})^n \left.\frac{\mathrm{d}^n}{\mathrm{d}z^n} \hat{p}(z)\right|_{z=0} . \qquad (5.20)$$

What is the distribution function obtained by adding two independent random variables $x = x_1 + x_2$ with distributions $p_1(x_1)$ and $p_2(x_2)$ (notice that p_1 and p_2 may be different)? The joint probability of two independent variables is obtained by multiplying the individual probabilities, and we obtain

$$p(x,2) = \int_{-\infty}^{\infty} \mathrm{d}x_1 \, p_1(x_1) p_2(x - x_1) \,, \quad \text{i.e.,} \quad \hat{p}(z,2) = \hat{p}_1(z)\hat{p}_2(z) \,. \quad (5.21)$$

The probability distribution $p(x,2)$ (where the second argument indicates that x is the sum of two random variables) is a convolution of the probability distributions, while the characteristic function $\hat{p}(z,2)$ is simply the product of the characteristic functions of the two variables.

This can be generalized immediately to a sum of N independent random variables, $x = \sum_{i=1}^{N} x_i$. The probability density is an N-fold convolution

$$p(x,N) = \int_{-\infty}^{\infty} \mathrm{d}x_1 \ldots \mathrm{d}x_{N-1} \, p_1(x_1) \ldots p_{N-1}(x_{N-1}) \, p_N \left(x - \sum_{i=1}^{N-1} x_i \right). \qquad (5.22)$$

The characteristic function is an N-fold product,

$$\hat{p}(z,N) = \prod_{i=1}^{N} \hat{p}_i(z) \,, \quad \ln \hat{p}(z,N) = \sum_{i=1}^{N} \ln \hat{p}_i(z) \,, \qquad (5.23)$$

and the cumulants are therefore additive,

$$c_n(N) = \sum_{i=1}^{N} c_n^{(i)} \,. \qquad (5.24)$$

For independent, identically distributed (IID) variables, these relations simplify to

$$\hat{p}(z, N) = [\hat{p}(z)]^N \ , \quad c_n(N) = N c_n \ . \tag{5.25}$$

In general, the probability density for a sum of N IID random variables, $p(x, N)$, can be very different from the density of a single variable, $p_i(x_i)$. A probability distribution is called *stable* if

$$p(x, N)\mathrm{d}x = p_i(x_i)\mathrm{d}x_i \quad \text{with} \quad x = a_N x_i + b_N \ , \tag{5.26}$$

that is, if it is form-invariant up to a rescaling of the variable by a dilation ($a_N \neq 1$) and a translation $b_N \neq 0$. There is only a small number of stable distributions, among them the Gaussian and the stable Lévy distributions. More precisely, we have a

$$\text{stable distribution} \Leftrightarrow \hat{p}(z) = \exp\left(-a|z|^\mu\right) \ , \quad 0 < \mu \leq 2 \ . \tag{5.27}$$

[This statement is slightly oversimplified in that it only covers distributions symmetric around zero. The exact expression is given in (5.41)]. The Gaussian distribution corresponds to $\mu = 2$, and the stable Lévy distributions to $\mu < 2$.

5.4.2 The Gaussian Distribution and the Central Limit Theorem

The Gaussian distribution with variance σ^2 and mean m_1,

$$p_G(x) = \frac{1}{\sqrt{2\pi}\sigma} \exp\left(-\frac{(x - m_1)^2}{2\sigma^2}\right), \tag{5.28}$$

has the characteristic function

$$\hat{p}_G(z) = \exp\left(-\frac{\sigma^2 z^2}{2} + \mathrm{i}m_1 z\right) \ , \tag{5.29}$$

that is, a Gaussian again. It satisfies (5.27) and is therefore a stable distribution, as can be checked explicitly by using the convolution or product formulae (5.22) resp. (5.23). Under addition of N random variables drawn from Gaussians,

$$m = \sum_{i=1}^{N} m_{1,(i)} \ , \quad \text{and} \quad \sigma^2 = \sum_{i=1}^{N} \sigma_i^2 \ . \tag{5.30}$$

$\ln \hat{p}_G(z)$ is a second-order polynomial in z which implies

$$c_n = 0 \text{ for } n > 2 \ , \quad \text{specifically } \kappa = 0 \ . \tag{5.31}$$

Any cumulant beyond the second can therefore be taken as a rough measure for the deviation of a distribution from a Gaussian, in particular in the tails.

Among them, the kurtosis κ is most practical because (i) in general, it is finite even for symmetric distributions and (ii) it gives less weight to the tails of the distribution, where the statistics may be bad, than even higher cumulants would. Distributions with $\kappa > 0$ are called leptokurtic.

Gaussian distributions are ubiquitous in nature, and arise in diffusion problems, the tossing of a coin, and many more situations. However, there are exceptions: turbulence, earthquakes, the rhythm of the heart, drops from a leaking faucet, and also the statistical properties of financial time series, are not described by Gaussian distributions.

Central Limit Theorem

The ubiquity of the Gaussian distribution in nature is linked to the central limit theorem, and to the maximization of entropy in thermal equilibrium. At the same time, it is a consequence of fundamental principles both in mathematics and in physics (statistical mechanics).

Roughly speaking, the central limit theorem states that any random phenomenon, being a consequence of a large number of small, independent causes, is described by a Gaussian distribution. At the same handwaving level, we can see the emergence of a Gaussian by assuming N IID variables (for simplicity – the assumption can be relaxed somewhat)

$$p(x, N) = [p(x)]^N = \exp[N \ln p(x)] . \tag{5.32}$$

Any normalizable distribution $p(x)$ being peaked at some x_0, $p(x, N)$ will have a very sharp peak at x_0 for large N. We can then expand $p(x, N)$ to second order about x_0,

$$p(x, N) \approx \exp\left(-\frac{(x - Nx_0)^2}{2\sigma^2}\right) \quad \text{for } N \gg 1 , \tag{5.33}$$

and obtain a Gaussian. Its variance will scale with N as $\sigma^2 \propto N$.

More precisely, the central limit theorem states that, for N IID variables with mean m_1 and *finite* variance σ, and two finite numbers u_1, u_2,

$$\lim_{N \to \infty} P\left(u_1 \le \frac{x - m_1 N}{\sigma\sqrt{N}} \le u_2\right) = \int_{u_1}^{u_2} \frac{du}{\sqrt{2\pi}} \exp\left(-\frac{u^2}{2}\right) . \tag{5.34}$$

Notice that the theorem only makes a statement on the limit $N \to \infty$, and not on the finite-N case. For finite N, the Gaussian obtains only in the center of the distribution $|x - m_1 N| \le \sigma\sqrt{N}$, but the form of the tails may deviate strongly from the tails of a Gaussian. The weight of the tails, however, is progressively reduced as more and more random variables are added up, and the Gaussian then emerges in the limit $N \to \infty$. The Gaussian distribution is a fixed point, or an attractor, for sums of random variables with distributions of finite variance.

The condition $N \to \infty$, of course, is satisfied in many physical applications. It may not be satisfied, however, in financial markets. Moreover, the central limit theorem requires σ^2 to be finite. This, again, may pose problems for financial time series, as we have seen in Sect. 5.3.3. While, in mathematics, σ^2 finite is just a formal requirement, there is a deep physical reason for finite variance in nature.

Gaussian Distribution and Entropy

Thermodynamics and statistical mechanics tell us that a closed system approaches a state of maximal entropy. For a state characterized by a probability distribution $p(x)$ of some variable x, the probability W of this state will be

$$W[p(x)] \propto \exp\left(\frac{S[p(x)]}{k_B}\right) \tag{5.35}$$

with k_B Boltzmann's constant, and the entropy

$$S[p(x)] = -k_B \int_{-\infty}^{\infty} dx\, p(x) \ln[\sigma p(x)] . \tag{5.36}$$

Here, σ is a positive constant with the same dimension as x, i.e., a characteristic length scale in the problem.

Our aim now is to maximize the entropy subject to two constraints

$$\int_{-\infty}^{\infty} dx\, p(x) = 1 , \quad \int_{-\infty}^{\infty} dx\, x^2 p(x) = \sigma^2 . \tag{5.37}$$

This can be done by functional derivation and the method of Lagrange multipliers

$$\frac{\delta}{\delta p(x)} \left\{ S[p(x)] - \mu_1 \int_{-\infty}^{\infty} dx'\, x'^2 p(x') - \mu_2 \int_{-\infty}^{\infty} dx'\, p(x') \right\} = 0 . \tag{5.38}$$

This is solved by

$$p(x) = \frac{e^{-x^2/2\sigma^2}}{Z} \quad \text{with} \quad Z = \int_{-\infty}^{\infty} dx\, e^{-x^2/2\sigma^2} = \sqrt{2\pi\sigma^2} . \tag{5.39}$$

The identification with temperature

$$2\sigma^2 = k_B T \tag{5.40}$$

is then found by bringing two systems, either with $\sigma = \sigma'$ or $\sigma \neq \sigma'$, into contact and into thermal equilibrium. One will see that σ^2 behaves exactly as we expect from temperature, allowing the identification.

5.4.3 Lévy Distributions

There is a variety of terms related to Lévy distributions. *Lévy distributions* designate a family of probability distributions studied by P. Lévy [32]. The term *Pareto laws,* or *Pareto tails,* is often used synonymously with Lévy distributions. In fact, one of the first occurrences of power-law distributions such as (5.44) is in the work of the Italian economist Vilfredo Pareto [72]. He found that, in certain societies, the number of individuals with an income larger than some value x_0 scaled as $x_0^{-\mu}$, consistent with (5.44). Finally, *Lévy walk,* or better *Lévy flight,* refers to the stochastic processes giving rise to Lévy distributions.

A stable Lévy distribution is defined by its characteristic function

$$\hat{L}_{a,\beta,m,\mu}(z) = \exp\left\{-a|z|^{\mu}\left[1 + i\beta \mathrm{sign}(t)\tan\left(\frac{\pi\mu}{2}\right)\right] + imz\right\} . \qquad (5.41)$$

β is a skewness parameter which characterizes the asymmetry of the distribution. $\beta = 0$ gives a symmetric distribution. μ is the index of the distribution which gives the exponent of the asymptotic power-law tail in (5.44). a is a scale factor characterizing the width of the distribution, and m gives the peak position. For $\mu = 1$, the tan function is replaced by $(2/\pi)\ln|z|$.

For our purposes, symmetric distributions ($\beta = 0$) are sufficient. We further assume a maximum at $x = 0$, leading to $m = 0$, and drop the scale factor a from the list of indices. The characteristic function then becomes

$$\hat{L}_{\mu}(z) = \exp\left(-a|z|^{\mu}\right) . \qquad (5.42)$$

In general, there is no analytic representation of the distributions $L_{\mu}(x)$. The special case $\mu = 2$ gives the Gaussian distribution and has been discussed above. $\mu = 1$ produces

$$L_1(x) = \frac{1}{\pi}\frac{a}{a^2 + x^2} , \qquad (5.43)$$

the Lorentz–Cauchy distribution. Asymptotically, the Lévy distributions behave as ($\mu \neq 2$)

$$L_{\mu}(x) \sim \frac{\mu A^{\mu}}{|x|^{1+\mu}} , \quad |x| \to \infty \qquad (5.44)$$

with $A^{\mu} \propto a$. These power-law tails have been shown in Figs. 5.8 and 5.10.

For $\mu < 2$, the variance is infinite but the mean absolute value is finite so long as $\mu > 1$

$$\mathrm{var}(x) \to \infty , \quad E_{\mathrm{abs}}(x) < \infty \quad \text{for} \quad 1 < \mu < 2 . \qquad (5.45)$$

All higher moments, including the kurtosis, diverge for stable Lévy distributions.

What happens when we use an index $\mu > 2$ in (5.41)? Do we generate a distribution which would decay with higher power laws and possess a finite second moment? The answer is no. Fourier transforming (5.41) with $\mu > 2$, we find a function which is no longer positive semidefinite and which therefore is not suitable as a probability density function of random variables [17, 59].

Lévy distributions with $\mu \leq 2$ are stable. The distribution governing the sum of N IID variables $x = \sum_{i=1}^{N} x_i$ has the characteristic function [cf. (5.25)]

$$\hat{L}_\mu(z, N) = \left[\hat{L}_\mu(z) \right]^N = [\exp(-a|z|^\mu)]^N = \exp(-aN|z|^\mu) , \qquad (5.46)$$

and the probability distribution is its Fourier transform

$$L_\mu(x, N) = \int_{-\infty}^{\infty} dz \, e^{-izx} e^{-aN|z|^\mu} . \qquad (5.47)$$

Now rescale the variables as

$$z' = zN^{1/\mu} , \quad x' = xN^{-1/\mu} \qquad (5.48)$$

and insert into (5.47):

$$L_\mu(x, N) = N^{-1/\mu} \int_{-\infty}^{\infty} dz' \, e^{-iz'x'} e^{-a|z'|^\mu} = N^{-1/\mu} L_\mu(x') , \qquad (5.49)$$

that is, the distribution of the sum of N random variables has the same form as the distribution of one variable, up to rescaling. In other words, the distribution is self-similar. The property (5.49) is at the origin of the rescaling (5.9) used by Mantegna and Stanley in Fig. 5.10. The amplitudes of the tails of the distribution add when variables are added:

$$(A^\mu)^{(N)} = NA^\mu . \qquad (5.50)$$

This relation replaces the additivity of the variances in the Gaussian case. If the Lévy distributions have finite averages, they are additive, too:

$$\langle x \rangle = \sum_{i=1}^{N} \langle x_i \rangle . \qquad (5.51)$$

There is a generalized central limit theorem for Lévy distributions, due to Gnedenko and Kolmogorov [73]. Roughly, it states that, if many independent random variables are added whose probability distributions have power-law tails $p_i(x_i) \sim |x_i|^{-(1+\mu)}$, with an index $0 < \mu < 2$, their sum will be distributed according to a stable Lévy distribution $L_\mu(x)$. More details and more precise formulations are available in the literature [73]. The stable Lévy distributions $L_\mu(x)$ are fixed points for the addition of random variables with infinite variance, or attractors, in much the same way as the Gaussian distribution is, for the addition of random variables of finite variance.

Earlier, it was mentioned that the stochastic process underlying a Lévy distribution is much more discontinuous than Brownian motion. This is shown

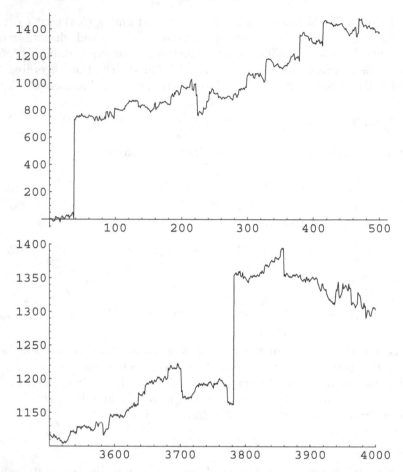

Fig. 5.14. Lévy flight obtained by summing random numbers drawn from a Lévy distribution with $\mu = 3/2$ (*upper panel*). The *lower panel* is a 10-fold zoom on the range (350, 400) and emphasizes the self-similarity of the flight. Notice the frequent discontinuities on all scales

in Fig. 5.14, which has been generated by adding random numbers drawn from a Lévy distribution with $\mu = 3/2$. When compared to a random walk such as Fig. 1.3 or 3.7, the frequent and sizable discontinuities are particularly striking. They directly reflect the fat tails and the infinite variance of the Lévy distribution. When compared to stock quotes such as Fig. 1.1 or 4.5, they may appear a bit extreme, but they certainly are closer to financial reality than Brownian motion.

5.4.4 Non-stable Distributions with Power Laws

Figures 5.10 and 5.11 suggested that the extreme tails of the distributions of asset returns in financial markets decay faster than a stable Lévy distribution would suggest. Here, we discuss two classes of distributions which possess this property: the truncated Lévy distribution where a stable Lévy distribution is modified beyond a fixed cutoff scale, and the Student-t distributions which are examples of probability density functions whose tails decay as power laws with exponents which may lie outside the stable Lévy range $\mu < 2$.

Truncated Lévy Distributions

The idea of truncating Lévy distributions at some typical scale $1/\alpha$ was mainly born in the analysis of financial data [71]. While large fluctuations are much more frequent in financial time series than those allowed by the Gaussian distribution, they are apparently overestimated by the stable Lévy distributions. Evidence for this phenomenon is provided by the S&P500 data in Fig. 5.10 where, especially in the bottom panel, a clear departure from Lévy behavior is visible at a specific scale, $7\ldots8\sigma$, and by the very good fit of the S&P500 variations to a truncated Lévy distribution in Fig. 5.11 (the size of $\alpha \sim 1/2$ is difficult to interpret, however, due to the lack of units in that figure [17]).

A truncated Lévy distribution can be defined by its characteristic function [71, 74]

$$\hat{T}_\mu(z) = \exp\left\{ -a \frac{(\alpha^2 + z^2)^{\mu/2} \cos\left(\mu \arctan \frac{|z|}{\alpha}\right) - \alpha^\mu}{\cos\left(\frac{\pi\mu}{2}\right)} \right\} . \tag{5.52}$$

This distribution reduces to a Lévy distribution for $\alpha \to 0$ and to a Gaussian for $\mu = 2$,

$$\hat{T}_\mu(z) \to \begin{cases} \exp\left(-a|z|^\mu\right) \text{ for } \alpha \to 0 , \\ \exp\left(-a|z|^2\right) \text{ for } \mu = 2 . \end{cases} \tag{5.53}$$

Its second cumulant, the variance, is [cf. (5.17)]

$$c_2 = \sigma^2 = \frac{\mu(\mu-1)a}{|\cos(\pi\mu/2)|}\alpha^{\mu-2} \to \begin{cases} \infty \text{ for } \alpha \to 0 , \\ 2a \text{ for } \mu = 2 . \end{cases} \tag{5.54}$$

The kurtosis is [cf. (5.18)]

$$\kappa = \frac{(3-\mu)(2-\mu)|\cos(\pi\mu/2)|}{\mu(\mu-1)a\alpha^\mu} \to \begin{cases} 0 \text{ for } \mu = 2 , \\ \infty \text{ for } \alpha \to 0 . \end{cases} \tag{5.55}$$

For finite α, the variance and all moments are finite, and therefore the central limit theorem guarantees that the truncated Lévy distribution converges towards a Gaussian under addition of many random variables.

The convergence towards a Gaussian can also be studied from the characteristic function (5.52). One can expand its logarithm to second order in z,

$$\ln \hat{T}_\mu(z) \sim -\frac{a}{2} \frac{\alpha^\mu}{\cos(\pi\mu/2)} (\mu - \mu^2) \frac{z^2}{\alpha^2} + \dots . \tag{5.56}$$

Fourier transformation implies that the Gaussian behavior of the characteristic function for small z translates into Gaussian large-$|x|$ tails in the probability distribution. On the other hand,

$$\ln \hat{T}_\mu(z) \sim -a|z|^\mu \quad \text{for} |z| \to \infty , \tag{5.57}$$

which implies Lévy behavior for small $|x|$. One would therefore conclude that the convergence towards a Gaussian, for the distribution of a sum of many variables, should predominantly take place from the tails. Also, depending on the cutoff variable and due to the stability of the Lévy distributions, the convergence can be extremely slow. As shown in Fig. 5.11, such a distribution describes financial data extremely well.

Notice that one could also use a hard-cutoff truncation scheme, such as [71, 74]

$$T_\mu(x) = L_\mu(x)\Theta(\alpha^{-1} - |x|) . \tag{5.58}$$

While it has the advantage of being defined directly in variable space and avoiding complicated Fourier transforms, the hard cutoff produces smooth distributions only after the addition of many random variables.

Student-t Distribution

A (symmetric) Student-t distribution is defined in variable space by

$$St_\mu(x) = \frac{\Gamma\left[(1+\mu)/2\right]}{\sqrt{\pi}\Gamma(\mu/2)} \frac{A^\mu}{(A^2 + x^2)^{(1+\mu)/2}} . \tag{5.59}$$

A is a scale parameter, $\Gamma(x)$ is the Gamma function, and the definition of the index μ is consistent with Sect. 5.4.3. A priori, there is no restriction on the value of $\mu > 0$. For large arguments, the distribution decays with a power law

$$St_\mu(x) \sim \frac{A^\mu}{|x|^{1+\mu}} \quad \text{for} \quad |x| \gg A , \tag{5.60}$$

that is, formally in the same way as would do a Lévy distribution. Its characteristic function is

$$\hat{St}_\mu(z) = \frac{2^{1-\mu/2}}{\Gamma(\mu/2)} (Az)^{\mu/2} K_{\mu/2}(Az) , \tag{5.61}$$

$$\approx \frac{\sqrt{\pi}}{\Gamma(\mu/2)} \left(\frac{Az}{2}\right)^{(\mu-1)/2} e^{-Az} \quad \text{for} \quad z \to \infty , \tag{5.62}$$

$$\approx 1 + \frac{1}{1-\frac{\mu}{2}} \left(\frac{Az}{2}\right)^2 - \frac{\Gamma(1+\frac{\mu}{2})}{\Gamma(1-\frac{\mu}{2})} \left(\frac{Az}{2}\right)^\mu \quad \text{for} \quad z \to 0 . \tag{5.63}$$

Interestingly, for $\mu > 2$, the dominant term in the expansion of the characteristic function for small z is identical in form to that of a similar expansion of a Gaussian distribution while, for $\mu < 2$, it is identical in form to that of a small-z expansion of a stable Lévy distribution. When $\mu < 2$, the distribution of a sum of many Student-t distributed random variables with index μ will converge to a stable Lévy distribution with the same index, according to the generalized central limit theorem. For example, for $\mu = 1$, the Student-t distribution reduces to the Lorentz–Cauchy distribution

$$St_1(x) = L_1(x) = \frac{1}{\pi} \frac{A}{A^2 + x^2} , \tag{5.64}$$

which also is a stable Lévy distribution. For $\mu > 2$ the central limit theorem requires the distribution of a sum of many Student-t distributed random variables to converge to a Gaussian distribution.

The Student-t distribution is named after the pseudonym "Student" of the English statistician W. S. Gosset and arises naturally when dividing a normally distributed random variable by a χ^2-distributed random variable [44].

5.5 Scaling, Lévy Distributions, and Lévy Flights in Nature

Although the naïve interpretation of the central limit theorem seems to suggest that the Gaussian distribution is the universal attractor for distributions of random processes in nature, distributions with power-law tails arise in many circumstances. It is much harder, however, to find situations where the actual diffusion process is non-Brownian, and close to a Lévy flight [75]–[77].

5.5.1 Criticality and Self-Organized Criticality, Diffusion and Superdiffusion

The asymptotic behavior of a Lévy distribution is, (5.44),

$$p(x) \sim |x|^{-(1+\mu)} . \tag{5.65}$$

The classical example for the occurrence of such distributions is provided by the critical point of second-order phase transitions [78], such as the transition from a paramagnet to a ferromagnet, as the temperature of, say, iron is lowered through the Curie temperature T_c. At a critical point, there are power-law singularities in almost all physical quantities, e.g., the specific heat, the susceptibility, etc. The reason for these power-law singularities are critical fluctuations of the ordered phase (ferromagnetic in the above example) in the disordered (paramagnetic) phase above the transition temperature, resp.

vice versa below the critical temperature, as a consequence of the interplay between entropy and interactions. In general, for $T \neq T_c$, there is a typical size ξ (the correlation length) of the ordered domains. At the critical point $T = T_c$, however, $\xi \to \infty$, and there is no longer a typical length scale. This means that ordered domains occur on all length scales, and are distributed according to a power-law distribution (5.65). The same holds for the distribution of cluster sizes in percolation. The divergence of the correlation length is the origin of the critical singularities in the physical quantities.

Critical points need fine tuning. One must be *extremely close* to the critical point in order to observe the power-law behavior discussed, which usually requires an enormous experimental effort in the laboratory for an accurate control of temperature, pressure, etc. Such fine tuning by a gifted experimentalist is certainly not done in nature. Still, there are many situations where power-law distributions are observed. Examples are given by earthquakes where the frequency of earthquakes of a certain magnitude on the Richter scale, i.e., a certain release of energy, varies as $N(E) \sim E^{-1.5}$, avalanches, traffic jams, and many more [79].

To explain this phenomenon, a theory of *self-organized criticality* has been developed [79]. The idea is that *open, driven, dissipative systems* (notice that physical systems at the critical point are usually in equilibrium!) may spontaneously approach a critical state. An example was thought to be provided by sandpiles. Imagine pouring sand on some surface. A sandpile will build up with its slope becoming steeper and steeper as sand is added. At some stage, one will reach a critical angle where the friction provided by the grains surrounding a given grain is just sufficient to compensate gravity. As sand is added to the top of a pile, the critical slope will be exceeded at some places, and some grains will start sliding down the side of the pile. As a consequence, at some lower position, the critical slope will be exceeded, and more grains will slide. An avalanche forms. It was conjectured, and supported by numerical simulations [79], that the avalanches formed in this way will possess power-law distributions. (Unfortunately, it appears that real sandpiles have different properties.)

Both with critical phenomena, and with self-organized criticality, one looks at statistical properties of a system. Can one observe true "anomalous" distributions, corresponding to Lévy flights, in nature? Diffusion processes can be classified according to the long-time limit of the variance

$$\lim_{t \to \infty} \frac{\sigma^2(t)}{t} = \begin{cases} 0 : \text{subdiffusive} \\ D : \text{diffusive} \\ \infty : \text{superdiffusive} . \end{cases} \tag{5.66}$$

A numerical simulation of superdiffusion, modeled as a Lévy flight in two dimensions with $\mu = 3/2$, is shown in Fig. 5.15. For comparison, Brownian motion was shown in Fig. 3.8. One again notices the long straight lines, corresponding more to flights of the particle than to the short-distance hops associated with diffusion. This is a pictorial representation of superdiffusive motion.

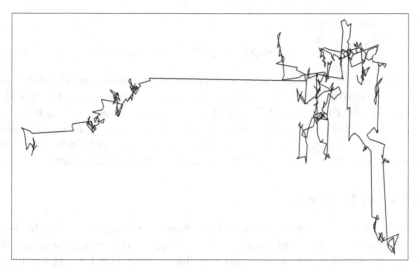

Fig. 5.15. Computer simulation of a two-dimensional Lévy flight with $\mu = 3/2$. Reprinted from J. Klafter, G. Zumofen, and M. F. Shlesinger: in *Lévy Flights and Related Topics,* ed. by M. F. Shlesinger et al. (Springer-Verlag, Berlin 1995) ©1995 Springer-Verlag

Of course, this is in direct correspondence with the continuity/discontinuity observed in the 1D versions, cf. Figs. 1.3 and 5.14.

5.5.2 Micelles

Micelles are long, wormlike molecules in a liquid environment. Unlike polymers, they break up at random positions along the chains at random times, and recombine later with the same or a different strand. The distribution of chain lengths ℓ is

$$p_{\mathrm{mic}}(\ell) \approx \exp(-\ell/\ell_0) . \qquad (5.67)$$

Any fixed-length chain performs ordinary Brownian diffusion with a diffusion coefficient which depends on its length as

$$D(\ell) = D_0\ell^{-2\beta} . \qquad (5.68)$$

In order to observe a Lévy flight, one can attach fluorescent tracer molecules to such micelles [80]. Due to break-up and recombination, any given tracer molecule will sometimes be attached to a short chain, and sometimes to a long chain. When the chain is short, which according to (5.67) happens frequently, it will diffuse rapidly, cf. (5.68). On long chains, it will diffuse slowly.

Using photobleaching techniques, one can observe the apparent diffusion of the tracer molecules and evaluate its statistical properties [80]. One indeed finds the superdiffusive behavior associated with Lévy flights, namely a characteristic function

$$p_{\mathrm{trac}}(q,t) = \exp\left(-D|q|^{\mu}t\right) , \quad \mu = 2/\beta \leq 2 , \tag{5.69}$$

where the precise value of μ depends somewhat on experimental conditions.

Notice, however, that the true physical diffusion process underlying this example is still Brownian (diffusion of the micelles). It is the length dependence of the diffusion constant [again typical of Brownian diffusion, remember Einstein's formula (3.25)] which conveys an apparent superdiffusive character to the motion of the tracer particles when their support is ignored.

5.5.3 Fluid Dynamics

The transport of tracer particles in fluid flows is usually governed both by advection and "normal" diffusion processes. Normal diffusion arises from the disordered motion of the tracer particles, hit by particles from the fluid, cf. Sect. 3.3. Advection of the tracer particles by the flow, i.e., tracer particles being swept along by the flow, leads to enhanced diffusion, but not to superdiffusion. It can be described as an ordinary random walk, but with diffusion rates enhanced over those typical for "normal" diffusion. For long times, therefore, the transport in real fluid flows is normally diffusive.

The short-time limit may be different in some situations. If there are vortices in the system, the tracer particles may stick to the vortices and their transport may become subdiffusive. On the other hand, in flows with coherent jets, the tracer particles may move ballistically for long distances. This process may eventually lead to superdiffusion, and to Lévy flights.

One can now set up an experiment where the flow pattern is composed both of coherent jets and vortices, i.e., sticking and ballistic flights of the tracer particles. The experimental setup is shown in Fig. 5.16 [81, 82]. A 38 weight% mixture of water and glycerol (viscosity $0.03\,\mathrm{cm}^2/\mathrm{s}$) is contained in an annular tank rotating with a frequency of $1.5\,s^{-1}$. In addition, fluid is pumped into the tank through a ring of holes (labeled I in Fig. 5.16) and out of the tank through another ring of holes (O). The radial flow couples to the Coriolis force to produce a strong azimuthal jet, in the direction opposite to the rotation of the tank. Above the forcing rings, there are strong velocity gradients, and the shear layer becomes unstable. As a result, a chain of vortices forms above the outer ring of holes (a similar vortex chain above the inner ring is inhibited artificially). In a reference frame rotating with the vortices, they appear sandwiched between two azimuthal jets going in opposite directions. Such pattern of jets and vortices is shown in the lower panel of Fig. 5.16. Depending on perturbations generated by deliberate axial inhomogeneities of the pattern of the radial flow, different regimes of azimuthal flow can be realized [81].

When a 60 degrees sector has a radial flow less than half of that of the flow between the remaining source and sink holes, a "time-periodic" regime

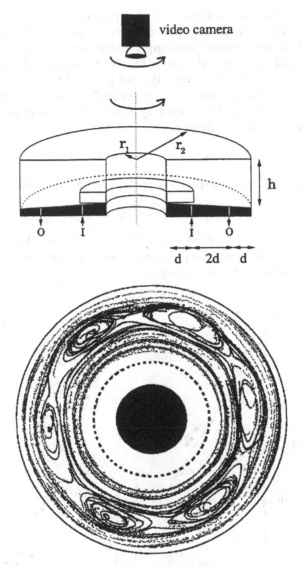

Fig. 5.16. Setup of an experiment with coexisting jets and vortices. *Upper panel*: the rotating annulus. I and O label two rings of holes for pumping and extracting fluid in the sense of the arrows. As explained in the text, under such conditions both jets and vortices form in the tank. *Lower panel*: streaks formed by 90-second-long trajectories of about 30 tracer particles reveal the presence of six vortices sandwiched between two azimuthal jets. The picture has been taken in a reference frame corotating with the vortex chain. By courtesy of H. Swinney. Reprinted with permission from Elsevier Science from T. H. Solomon et al.: Physica D **76**, 70 (1994). ©1994 Elsevier Science

is established. One can then map out the trajectories of passive tracer particles. The motion of these particles, and of the supporting liquid, has periods of flight, where the particles are simply swept along ballistically by the azimuthal jets, of capture and release by the vortices, and diffusion. These processes can be analyzed separately, and probability density functions for various processes can be derived. They generally scale as power laws of their variables. For example, the probability density function of times where the tracer particles stick to vortices behaves as

$$P_s(t) \sim t^{-(1+\mu_s)} , \quad \mu_s \sim 0.6 \pm 0.3 . \tag{5.70}$$

The probability density distribution of the flight times behaves as

$$P_f(t) \sim t^{-(1+\mu_f)} , \quad \mu_f \sim 1.3 \pm 0.2 , \tag{5.71}$$

that is, it carries a different exponent. Figure 5.17 shows the results of such an experiment leading to these power laws. Yet another exponent is measured by the distribution of the flight lengths ℓ

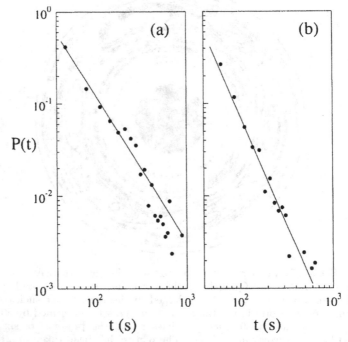

Fig. 5.17. Probability distributions of the sticking times (**a**) and flight times (**b**) for tracer particles moving in a flow pattern composed of two azimuthal jets and vortices. The straight lines have slopes -1.6 ± 0.3 and -2.3 ± 0.2, respectively (cf. text). The tracer particles execute Lévy flights. By courtesy of H. Swinney. Reprinted with permission from Elsevier Science from T. H. Solomon et al.: Physica D **76**, 70 (1994). ©1994 Elsevier Science

$$P(\ell) \sim \ell^{-(1+\mu_\ell)} \ , \quad \mu_\ell \sim 1.05 \pm 0.0.3 \ . \tag{5.72}$$

In these experiments, the fluid and the tracer particles in this system therefore perform a Lévy flight. Pictures of the traces of individual particles are available in the literature [81, 82], and generally look rather similar to the computer simulation shown in Fig. 5.15.

5.5.4 The Dynamics of the Human Heart

The human heart beats in a complex rhythm. Let $B(i) \equiv \Delta t_i = t_{i+1} - t_i$ denote the interval between two successive beats of the heart. Figure 5.18 shows two sequences of interbeat intervals, one (top) of a healthy individual, the

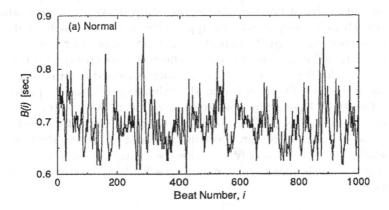

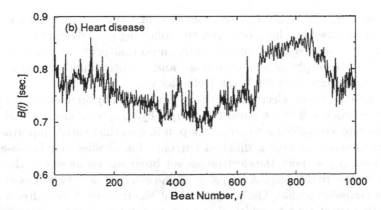

Fig. 5.18. The time series of intervals between two successive heart beats, (**a**) for a healthy subject, (**b**) for a patient with a heart disease (dilated cardiomyopathy). By courtesy of C.-K. Peng. Reprinted from C.-K. Peng et al.: in *Lévy Flights and Related Topics*, ed. by M. F. Shlesinger et al. (Springer-Verlag, Berlin 1995) ©1995 Springer-Verlag

other (bottom) of a patient suffering from dilated cardiomyopathy [83]. For reasons of stationarity, one prefers to analyze the probability for a *variation* of the interbeat interval (in the same way as financial data use returns rather than prices directly). The surprising finding then is that *both* time series lead to Lévy distributions for the increments $I_i = B(i+1) - B(i)$ with the same index $\mu \approx 1.7$ (not shown) [83]. The main difference between the two data sets, at this level, is the standard deviation, which is visibly reduced by the disease.

To uncover more differences in the two time series, a more refined analysis, whose results are shown in Fig. 5.19, is necessary. The power spectrum of the time series of increments, $S_I(f) = |I(f)|^2$ with $I(f)$ the Fourier transform of I_i, for a normal patient has an almost linear dependence on frequency, $S(f) \sim f^{0.93}$. For the suffering patient, on the other hand, the power spectrum is almost flat at low frequencies, and only shows an increase above a finite threshold frequency [83]. To appreciate these facts, note that, for a purely random signal, $S(f) = $ const., i.e., white noise. Correlations in the signal lead to red noise, i.e., a decay of the power spectrum with frequency $S(f) \sim f^{-\beta}$ with $0 < \beta \leq 1$. $1/f$ noise, typically caused by avalanches, is an example of this case. On the other hand, with anticorrelations (a positive signal preferentially followed by a negative one), the power spectrum increases with frequency, $S(f) \sim f^{\beta}$. This is the case here for a healthy patient. With the disease, the small-frequency spectrum is almost white, and the typical anticorrelations are observed only at higher beat frequencies. Also, detrended fluctuation analysis shows different patterns for both healthy and diseased subjects [83].

5.5.5 Amorphous Semiconductors and Glasses

The preceding discussion may be rephrased in terms of a waiting time distribution between one heartbeat and the following one. Waiting time distributions are observed in a technologically important problem, the photoconductivity of amorphous semiconductors and discotic liquid crystals. These materials are important for Xerox technology.

In the experiment, electron–hole pairs are excited by an intense laser pulse at one electrode and swept across the sample by an applied electric field. This will generate a displacement current. Depending on the relative importance of various transport processes, different current–time profiles may be observed. For Gaussian transport, the electron packet broadens, on its way to the other electrode, due to diffusion. A snapshot of the electron density will essentially show a Gaussian profile. The packet will hit the right electrode after a characteristic transit time t_T which shows up as a cutoff in the current profile. Up to the transit time, the displacement current measured is a constant.

In a strongly disordered material, the transport is dispersive, however. Now, electrons become trapped by impurity states in the gap of the semiconductor. They will be released due to activation. The release rates depend

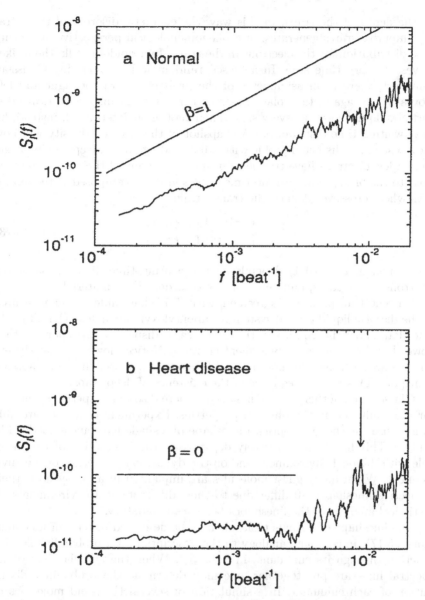

Fig. 5.19. The power spectrum $S_I(f)$ for the time series of increments of the time between two successive heart beats, (**a**) for a healthy individual, (**b**) for a patient with heart disease. The power spectrum of the healthy subject is characteristic of time series with anticorrelations over the entire frequency range, while that of the patient with heart failure is white at low frequencies, and exhibits the anticorrelations only in its high-frequency part. By courtesy of C.-K. Peng. Reprinted from C.-K. Peng et al.: in *Lévy Flights and Related Topics*, ed. by M. F. Shlesinger et al. (Springer-Verlag, Berlin 1995) ©1995 Springer-Verlag

on the depth of the traps. In this way, the energetic disorder of the intra-gap impurity states generates, in a phenomenological perspective, a waiting time distribution for the electrons in the traps. In a random walk, the walker moves at every time step. In a biased random walk, underlying Gaussian transport, there is an asymmetry of the probabilities of the right and left moves which, again, take place at every time step. In dispersive transport, particles do not move at every step. Their motion is determined, instead, by their waiting time distribution. A shapshot of the electron density will now show a strongly distorted profile with a flat leading and a steep trailing edge. Only a few electrons have traveled very far, and many of them are still stuck close to the origin. The current–time curves are now composed of two power laws whose crossover defined the transit time

$$I(t) \sim \begin{cases} t^{-(1-\alpha)} & \text{for } t < t_{\mathrm{T}}, \\ t^{-(1+\alpha)} & \text{for } t > t_{\mathrm{T}}. \end{cases} \tag{5.73}$$

The exponent $\alpha \in (0,1)$ depends on the waiting time distribution of the electrons in the traps, and hence on the disorder in the material.

Current–time profiles in agreement with (5.73) have indeed been measured in the discotic liquid crystal system hexapentyloxytriphenylene (HPAT) [84]. The data show the characteristic structure of dispersive transport, with a power-law decay of the displacement currents. Notice, however, that the exponents are such that the motion is *subdiffusive,* i.e., *slower* than for Gaussian transport. This is a consequence of the existence of deep traps.

Glasses are another class of materials where disorder is, perhaps, the factor most influencing the physical properties. Experimentalists now are able to measure the spectral (optical) lineshape of a single molecule embedded in a glass. This lineshape sensitively depends on the interaction of the molecule with its local environment and on the dynamical properties of the environment. When many guest molecules are implanted in a glassy host, their respective lineshapes all differ due to their different local environments. A statistical analysis of the lineshapes becomes mandatory.

The lineshape of a single molecule may be described in term of its cumulants, (5.17), in complete analogy to the description of a probability density function through its cumulants in Sect. 5.4. When the cumulants of many spectral lines are put together, one may determine the probability distribution of each cumulant. In a simulation of several thousand molecules of terylene embedded in polystyrene, one finds that the first cumulant of the lineshapes is distributed according to a symmetric Lorentz–Cauchy distribution, the second cumulant according to a stable Lévy distribution with $\mu = 1/2$ and a skewness $\beta = 1$ (a maximally skew distribution defined for positive values of the argument only), the third cumulant is drawn from a symmetric Lévy distribution with $\mu = 1/3$, and the fourth cumulant is drawn from an asymmetric distribution with skewness $\beta = 0.61$ and index $\mu = 1/4$ [85].

Another theoretical formulation of the spectral lineshapes of molecules embedded in glasses may even be applied rather straightforwardly to the statistical properties of financial time series [86, 87]. Physically here, each host molecule in the neighborhood of the guest is assumed to shift the guest's delta-function absorption line depending on the individual host–guest interaction by a frequency $\tilde{\nu}(\boldsymbol{R}_n)$. The total lineshape of the molecule then is the superposition of these contributions. The two important factors determining the lineshape are the distance dependence of the host–guest interaction and the density of host molecules around one guest molecule. When the latter is large, a Gaussian lineshape invariably follows. There are so many host molecules around a guest molecule that simply the central limit theorem applies and the details of the interaction process are not sampled in the lineshape. On the other hand, when the density is small, the lineshape sensitively depends on the dependence of the interaction on the spatial separation. With $\tilde{\nu}(\boldsymbol{R}_n) \sim R^{-3/\mu}$, a stable Lévy-like lineshape obtains. We come back to this theory in Chap. 6.

5.5.6 Superposition of Chaotic Processes

Lévy distributions can be generated by superposing specific chaotic processes [88]. Chaotic processes are defined by non-linear mappings of a variable X

$$X_{n+1} = f(X_n) , \tag{5.74}$$

and are often used to model non-linear dynamical systems. If the mapping function

$$f(X_n) = \frac{1}{2} \left(X_n - \frac{1}{X_n} \right) \tag{5.75}$$

is used in (5.74), and many processes corresponding to different initial conditions of the variable X are superposed, the probability distribution of the variable X iterated and superposed in this way will converge to

$$p(X) \to \frac{1}{\pi(1 + X^2)} , \tag{5.76}$$

that is, the Lorentz–Cauchy distribution. This is a special Lévy distribution with $\mu = 1$. More general Lévy distributions can be obtained from the mapping function

$$f(X_n) = \left| \frac{1}{2} \left(|X_n|^\alpha - \frac{1}{|X_n|^\alpha} \right) \right|^{1/\alpha} \operatorname{sign} \left(X_n - \frac{1}{X_n} \right) . \tag{5.77}$$

If this mapping is iterated, and many processes corresponding to different initial conditions are superposed, the probability density of X converges to

$$p(X) \to \frac{\alpha}{\pi} \frac{|X|^{\alpha-1}}{\left(1 + |X|^{2\alpha} \right)} , \tag{5.78}$$

which has Lévy behavior with $\mu = \alpha$. This is shown in Fig. 5.20.

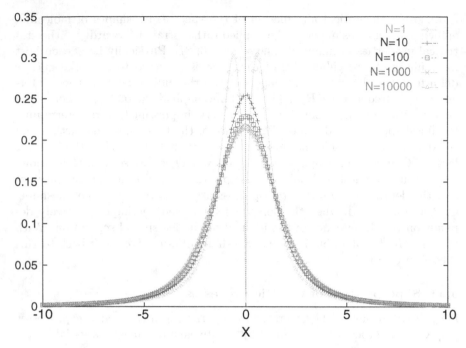

Fig. 5.20. Superposition of many chaotic processes, as described in the text. $\alpha = 3/2$ has been used, and N denotes the number of processes with different initial conditions. By courtesy of K. Umeno. Reprinted from K. Umeno: Phys. Rev. E **58**, 2644 (1998), ©1998 by the American Physical Society

5.5.7 Tsallis Statistics

There are many properties which the Gaussian and the stable Lévy distributions share: both are fixed points, or attractors, for the distributions of sums of independent random variables, in both cases guaranteed by a central limit theorem, and both of them describe the probability distributions associated with certain stochastic processes. The main difference is that the Gaussian distribution is of finite variance, while the power-law decay of the Lévy distributions leads to infinite variance.

From a physics perspective, we could derive the Gaussian distribution from a maximization of the entropy, subject to constraints, cf. Sect. 5.4.2. Is it possible to generate stable Lévy distributions in a similar way?

There are two possibilities for achieving this. One is to keep the definition of the Boltzmann–Gibbs entropy, (5.36), unchanged but to introduce a constraint different from (5.37) for the variance of the distribution function. This requires rather complicated constraints. The alternative is to change the definition of the entropy. It also requires a change of the variance constraint, but a rather simple one only. This is the way taken by Tsallis et al. [77, 89].

They generalize the entropy to

$$S_q[p(x)] = k_B \frac{1 - \int_{-\infty}^{\infty} d\left(\frac{x}{\sigma}\right) [\sigma p(x)]^q}{q - 1} . \tag{5.79}$$

This reduces to the familiar Boltzmann–Gibbs entropy (5.36) in the limit $q \to 1$. The probability distribution of the variable x characterizing the equilibrium state of maximal entropy can then be determined by maximizing $S[p]$ subject to the constraints

$$\int_{-\infty}^{\infty} dx\, p(x) = 1 , \quad \langle x^2 \rangle_q = \int_{-\infty}^{\infty} d\left(\frac{x}{\sigma}\right) x^2 [\sigma p(x)]^q = \sigma^2 , \tag{5.80}$$

generalizing (5.37). With these generalizations, one can essentially derive the thermodynamic relations, such as $T^{-1} = \partial S/\partial U$, but the thermodynamics and statistical mechanics are no longer extensive. This has given rise to the name of "non-extensive statistical mechanics", an area of rather intense research currently.

The probability distributions maximizing S now depend on q. For $q \le 5/3$, the Gaussian obtains. For $5/3 < q < 3$, one finds the stable Lévy distributions with index $\mu = (3 - q)/(q - 1)$ varying from 2 to zero as q increases from $5/3$ to 3. For $q > 3$, there is no solution [77, 89].

The stationary probability distributions obtained in this way are

$$p_q(x) = \frac{1}{Z_q} \left[1 - \tilde{\beta}(1 - q)U(x) \right]^{1/(1-q)} , \tag{5.81}$$

where $\tilde{\beta} = 1/k_B T$ is the inverse temperature, Z_q is the partition function, and $U(x)$ is a "potential" [90]. The label "potential" is to be taken in a generalized sense which is explained below.

One may now ask: what dynamics equations could lead to such stationary distributions? The question can be posed at a macroscopic level, i.e., how must the evolution equation of $p_q(x)$ be structured in order to produce stationary solutions of the form (5.81)? In ordinary Boltzmann–Gibbs statistical mechanics, this is the question about the appropriate Fokker–Plank equation. One could also search for time-dependent solutions of these Fokker–Planck equations, but we will not pursue this further here. On the other hand, one can ask for the evolution equation for the stochastic variable, i.e., take a microscopic view. This is the question for the appropriate Langevin-type equation.

We first turn to the macroscopic level. In ordinary statistical physics for Markov processes, the evolution of the probability density function is governed by a Fokker–Planck equation

$$\frac{\partial p(x,t)}{\partial t} = -\frac{\partial}{\partial x}\left[D^{(1)}(x,t)p(x,t) \right] + \frac{1}{2}\frac{\partial^2}{\partial x^2}\left[D^{(2)}(x,t)p(x,t) \right] , \tag{5.82}$$

where $D^{(1)}$ and $D^{(2)}$ are the drift and diffusion "coefficients", respectively. For constant $D^{(2)}$ and time-independent $D^{(1)}(x)$, the stationary solution is

$$p(x) = N \exp\left[-\tilde{\beta} U(x)\right] \quad \text{with} \quad D^{(1)}(x) = -\frac{\partial U(x)}{\partial x} . \tag{5.83}$$

N is a normalization factor. The "potential" is thus defined via the drift coefficient $D^{(1)}$. The drift is a result of a force acting on the particles.

In the macroscopic view of non-extensive statistical mechanics, one can imagine arriving at (5.81) either from a Fokker–Planck equation linear in the probability density $p_q(x)$, such as the preceding one [91], or one containing non-linear powers of $p_q(x)$ [92]. In the linear framework, our approach is to derive special relations between $D^{(1)}(x)$ and $D^{(2)}(x)$ by equating the general stationary solution of the Fokker–Planck equation [37]

$$p(x) = N \exp\left\{2 \int dx \frac{1}{D^{(2)}(x)} \left[D^{(1))}(x) - \frac{1}{2}\frac{\partial D^{(2)}(x)}{\partial x}\right]\right\} \tag{5.84}$$

with the stationary distribution (5.81) obtained from entropy maximization [91]. This equation is solved when the condition

$$\frac{2}{D^{(2)}(x)} \left[D^{(1)}(x) - \frac{1}{2}\frac{\partial D^{(2)}(x)}{\partial x}\right] = \frac{-\tilde{\beta}}{1 - \tilde{\beta}(1 - q)U(x)} \frac{\partial U}{\partial x} \tag{5.85}$$

is satisfied, where $U(x)$ is defined in (5.83). On the other hand, a Fokker–Planck equation implies a Langevin equation for the evolution of the stochastic variable [37]

$$\frac{dx}{dt} = D^{(1)}(x) + \sqrt{D^{(2)}(x)}\eta(t) , \tag{5.86}$$

where $\eta(t)$ is white noise. For any given $U(x)$, (5.85) thus determines a family of microscopic Langevin equations which give rise to non-extensive statistical mechanics on the macroscopic level [91]. It is the special interplay of the deterministic drift $D^{(1)}$ and the stochastic diffusion coefficients $D^{(2)}$ which determines the steady-state distribution of the system, and not so much the particular form of the coefficients.

Using (5.81), (5.85) can be rewritten as

$$\frac{dp}{p^q} = -\tilde{\beta}(Z_q)^{q-1}U(x)dx , \tag{5.87}$$

which is identical to the stationary solution of non-linear Fokker–Planck equations. The non-linear Fokker–Planck equation is [92]

$$\frac{\partial f(x,t)}{\partial t} = -\frac{\partial}{\partial x}\left[D^{(1)}(x)f^\rho(x,t)\right] + \frac{1}{2}\frac{\partial^2}{\partial x^2}\left[D^{(2)}(x)f^\nu(x,t)\right] . \tag{5.88}$$

ν and ρ are real numbers characterizing the non-linearity. This equation is equivalent to a Langevin equation

$$\frac{dx}{dt} = D^{(1)}(x) + \sqrt{D^{(2)}(x)}f(x,t)^{(\nu-\rho)/2}\eta(t) . \tag{5.89}$$

Here, $f(x,t)$ is an auxiliary distribution and not the physical probability distribution $p(x,t)$. The physical distribution is $p(x,t) = f^{\rho}(x,t)$.

The important feature of (5.89) is the dependence of the effective diffusion coefficient $D^{(2)}f^{(\nu-\rho)/2}$ acting on the microscopic level on the probability density $f^{(\nu-\rho)/2} = p^{(\nu-\rho)/2\rho}$ realized at the macroscopic level [92]. The non-linear Fokker–Planck equation (5.88) and the Langevin equation (5.89) no longer are equivalent, complementary descriptions of a stochastic system but here turn into a system of coupled equations. The feedback of the macroscopic into the microscopic level apparently is the prerequisite to turn ordinary Boltzmann–Gibbs statistical mechanics into non extensive statistical mechanics. When an interpretation in terms of Brownian motion is sought, one would conclude that the amplitude of the shocks the Brownian particle picks up from its environment depends on the frequency of its visits to specific regions of space. This might lead to a cleaving of phase space.

The scaling properties of the variance of the system variable

$$\langle x^2(bt) \rangle = b^{2/(3-q)} \langle x^2(t) \rangle \quad \text{with} \quad \frac{1-\nu}{1+\nu} = \frac{q-1}{3-q} \quad \text{at} \quad \rho = 1 \qquad (5.90)$$

demonstrate that non-extensive statistical mechanics describes anomalous diffusion [92]. This equation suggests that the Hurst exponent H of this system is $H = 1/(3-q)$. It might even suggest that the processes are related to fractional Brownian motion, (4.42). When the Hurst exponent is calculated in the way it was originally defined [33], one finds, however, that, for the anomalous diffusion described in non-extensive statistical mechanics, the Hurst exponent is $H = 0.5$ as for ordinary Brownian motion and independent of q while, for fractional Brownian motion, it is different. The underlying reason is that the stochastic process of non-extensive statistical mechanics described by (5.89) is uncorrelated in time. Fractional Brownian motion, (4.42), on the other hand, possesses long-range temporal correlations which are at the origin of the non-trivial Hurst exponent.

A less formal and more intuitive approach starts from the ordinary Langevin equation

$$\frac{dx}{dt} = -\gamma x + \sigma \eta(t) . \qquad (5.91)$$

The identification with (5.86) is made through $D^{(1)}(x) = -\gamma x$ and $D^{(2)}(x) = \sigma^2$. Our emphasis here is not on the dependences of the drift and diffusion coefficients on the stochastic variable but rather on the possibility that they slowly fluctuate in time. A specific assumption is that $\beta = \gamma/\sigma^2$ is χ^2-distributed with degree n [93], i.e., that

$$p(\beta) = \frac{1}{\Gamma(n/2)} \left(\frac{n}{2\beta_0} \right)^{n/2} \beta^{n/2-1} \exp\left(-\frac{n\beta}{2\beta_0} \right) . \qquad (5.92)$$

A variable which is the sum of the squares of n Gaussian distributed random variables is distributed according to a χ^2-distribution of degree n. $\beta_0 = \langle \beta \rangle$ is the average of β.

If the time scale on which β fluctuates is much longer than $1/\gamma$, the time scale of the stochastic variable, the conditional probability of x on β, is

$$p(x|\beta) = \sqrt{\frac{\beta}{2\pi}} \exp\left(-\frac{\beta x^2}{2}\right) . \tag{5.93}$$

The marginal probability of x then is

$$p(x) = \int p(x|\beta)p(\beta)\mathrm{d}\beta = \frac{\Gamma(\frac{n+1}{2})}{\Gamma(n/2)} \sqrt{\frac{\beta_0}{\pi n}} \left(1 + \frac{\beta_0}{n}x^2\right)^{-(n+1)/2} . \tag{5.94}$$

Comparison with (5.81) shows that the distribution found for this system with slowly fluctuating drift and diffusion coefficients is a stationary distribution of non-extensive statistical mechanics provided one identifies $q = 1 + 2/(n + 1)$ and $\tilde{\beta} = 2\beta_0/(3 - q)$. The potential is $U(x) = x^2/2$, as appropriate for ordinary diffusion. Also, $p(x)$ is identical to a Student-t distribution, (5.59), with index $\mu = n$ and scale parameter $A = \sqrt{n/\beta}$.

Non-linear Langevin equations may be studied in the same way. As expected from (5.82) with the potential $U(x)$ defined in (5.83), a power-law dependence in the drift coefficient will translate into a non-trivial power-law dependence on x in the probability distribution. We postpone the application of this theory to a physical example, hydrodynamic turbulence, and to financial markets, to the next chapter.

An aspect which has not been clarified satisfactorily yet is the scope of application of non-extensive statistical mechanics. Thermodynamics and statistical mechanics usually treat systems in or close to equilibrium. The experimental results discussed above, where Lévy-type scaling was found, to a varying degree depart from equilibrium situations. The example of micelles certainly is close to equilibrium, but the rotating fluid containing jets and vortices is a stationary state rather far away from equilibrium. What about turbulence, the subject of the next chapter? And social systems or financial markets? Does the non-extensive statistical mechanics describe situations both close to and far away from equilibrium? Where in the theory could we nail down this opening to non-equilibrium physics? Attempts to answer these questions are just beginning to appear [94]. The state of the art in this field of research is summarized in the proceedings of a conference on Tsallis statistics [95].

5.6 New Developments: Non-stable Scaling, Temporal and Interasset Correlations in Financial Markets

The assumption of statistical independence of subsequent price changes, made by the geometric Brownian motion hypothesis, is apparently rather well satisfied by stock markets, both concerning the decay of return correlation functions, and the use of correlations in practical trading rules. On the contrary, the distribution of returns of real markets is far from Gaussian, and Sect. 5.3.3 suggested that returns were drawn from distributions which either were stable Lévy distributions, or variants thereof with a truncation in their most extreme tails.

5.6.1 Non-stable Scaling in Financial Asset Returns

There were, however, observations in the economics literature which could raise doubts about the simple hypothesis of stable Lévy behavior. As an example, it appeared that the Lévy exponent μ somewhat depended on the time scale of the observations, i.e., if intraday, daily, or weekly returns were analyzed [96]. This is not expected under a Lévy hypothesis because the distribution is stable under addition of many IID random variables. Returns on a long time scale obtain as the sum of many returns on short time scales, and therefore must carry the same Lévy exponent.

Lux examined the tail exponents of the return distributions of the German DAX stock index, and of the individual time series of the 30 companies contained in this index by applying methods from statistics and econometrics [97]. Interestingly, he found his results consistent with stable Lévy behavior for the majority of stocks and for the DAX share index, with exponents in the range $\mu \approx 1.42, \ldots, 1.75$.

A counter-check, using an estimator of the tail index introduced in extreme-value theory, led to different conclusions, however. It turned out that all stocks, and the DAX index, were characterized by tail exponents $2 < \mu \leq 4$, i.e., outside the stable Lévy regime. In most cases, even the 95% confidence interval did not overlap with the regime required for stability, $\mu \leq 2$. Moreover, statistical tests could not reject the hypothesis of convergence to a power law.

The estimator used is more sensitive to extremal events in the tails of a distribution than a standard power-law fit. It deliberately analyzes the tail of large events where, e.g., in the bottom panel of Fig. 5.10, deviations of the data from Lévy power laws become visible. It would indicate that a power-law tail with an exponent $\mu > 2$ is more appropriate than an exponential truncation scheme.

These conclusions are corroborated by an investigation using both two years of 15-second returns and 15 years of daily returns of the DAX index. The corresponding price charts are given in Figs. 5.5 and 1.2. Figure 5.21 displays the normalized returns of the DAX high-frequency data presented earlier, in double-logarithmic scale [59, 60]. The figure is essentially independent of whether positive, negative, or absolute returns are considered, and the last possibility has been chosen. Again, we find approximately straight behavior for large returns, suggesting power-law behavior and fat tails.

Using the Hill estimator of extreme-value theory [98, 99] to estimate the asymptotic distribution for $|\delta s_{15''}| \to \infty$, a tail index $\mu \approx 2.33$ for a power-law distribution

$$p(\delta s_\tau) \sim |\delta s_\tau|^{-1-\mu} \qquad (5.95)$$

is determined. This power law is shown as the dotted line in Fig. 5.21. The solid line in Fig. 5.21 is a one-parameter fit to a Student-t distribution (5.59), where the exponent derived from the Hill estimator was taken fixed and only the scale parameter A of the distribution was fitted. The index $\mu \approx 2.33$ is significantly bigger than Mandelbrot's 1963 value and outside the range of stable Lévy distributions, but roughly in line with Lux's result using data on a longer time scale [97].

Both the curvature of the data *away* from the straight line in Fig. 5.21 and the convergence, with a finite slope, of the Hill estimator to its infinite fluctuation limit suggest that the probability distribution of extreme returns is *not* a pure power law but rather contains multiplicative corrections varying more slowly than a power law. The existence of such (e.g., logarithmic) corrections to power-law properties is well known in statistical physics in the vicinity of critical points. The idea of slowly varying corrections to power laws

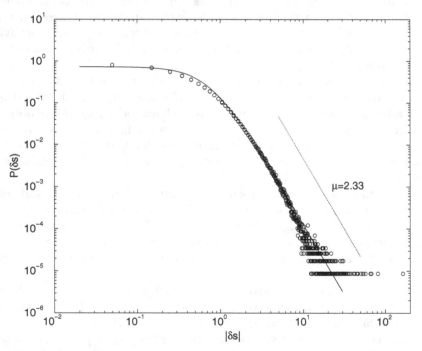

Fig. 5.21. Probability density function of 15-second DAX returns. The *straight dotted line* indicates a power law with its index $\mu = 2.33$ derived from extreme-value theory. The *solid line* is a fit to a Student-t distribution using the exponent determined independently. From S. Dresel: *Modellierung von Aktienmärkten durch stochastische Prozesse*, Diplomarbeit, Universität Bayreuth, 2001, by courtesy of S. Dresel

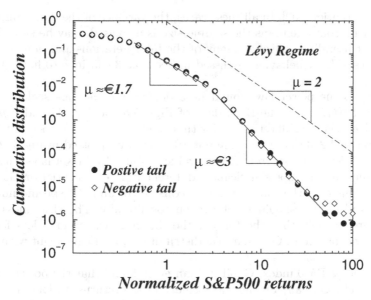

Fig. 5.22. Cumulative distribution function of normalized returns of the S&P500 share index. Three different regimes can be distinguished: small returns much less than a standard deviation, where no analysis was performed; an intermediate regime of returns $\approx 0.4 \dots 3\sigma$, where a stable Lévy power law is appropriate; and a large-fluctuation regime of power-law return with an exponent $\mu = 3$ outside the stable Lévy range. The *dashed line* $\mu = 2$ is the limit of Lévy stability. By courtesy of P. Gopikrishnan. Reprinted from P. Gopikrishnan et al.: Phys. Rev. E **60**, 5305 (1999) ©1999 by the American Physical Society

is already contained in work by Cont [100] and a recent paper by LeBaron [101].

A tail exponent of $\mu = 3$ is also found for other stock markets, such as the S&P500, the Japanese Nikkei 225, and the Hong Kong Hang Seng indices [62]. Figure 5.22 shows the cumulative probability

$$P_>(\delta S) = \int_{\delta S}^{\infty} \mathrm{d}(\delta S')\, p(\delta S') \qquad (5.96)$$

for the normalized returns of the S&P500 index. Clearly, the tails containing the extreme events follow a power law

$$P_>(\delta S) \sim (\delta S)^{-\mu} \qquad (5.97)$$

with an exponent $\mu = 3$, beyond the limit of stability of Lévy laws. However, we also recognize that a stable Lévy law with $\mu = 1.7$ is a good description in an intermediate range of returns $0.4\sigma \leq \delta S \leq 3\sigma$. This power law had been emphasized in earlier studies and was discussed in Sect. 5.3.3.

Tail exponents $2 < \mu \leq 6$ have also been found in a variety of other markets, most notably foreign exchange markets, interbank cash interest rates, and commodities [102]. In all these cases, the variance of the data sample exists, but its convergence as the sample size is increased, may be slow. While no longer literally applicable, many of the interpretations and practical consequences of Lévy behavior discussed in Sect. 5.3.3 continue to hold qualitatively.

How do the power laws found here depend on the time scale τ of the returns $\delta S_\tau(t)$? Using the DAX data of Figs. 5.5 and 1.2, the index $\mu(\tau)$ is determined via the Hill estimator for time lags varying in powers of four from a quarter to $1\,243\,136$ minutes, about 10 years, and plotted in Fig. 5.23 [59, 60]. The index increases from 2.33 to values around 10. Power laws with such high exponents are not significantly different from exponential or Gaussian distributions over the range of values considered, and the specific numbers for the tail indices should not be taken too literally. The clear message of the figure, namely that the tails of the distributions become less fat and gradually converge to Gaussian-like distributions, is in agreement with other studies [62].

For the S&P500 index, Gopikrishnan et al. found that the power laws do not depend essentially on the time scale τ of the returns, so long as $\tau \leq 4d$ [62]. Only for returns evaluated on scales above four days does the shape of

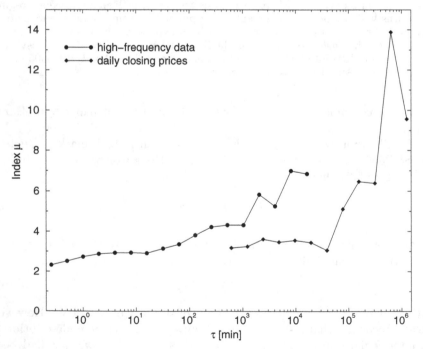

Fig. 5.23. Dependence of the index μ of the power laws, (5.95), on the time scale of the returns. From S. Dresel: *Modellierung von Aktienmärkten durch stochastische Prozesse,* Diplomarbeit, Universität Bayreuth, 2001, by courtesy of S. Dresel

the cumulative probability depend significantly on τ, not quite in agreement with the rather gradual increase in the DAX data shown in Fig. 5.23. For the S&P500, both inspection of the cumulative probabilities on longer time scales, as well as an analysis of the scaling of the moments of the distribution, indicate that it becomes more Gaussian as the time scales τ are increased.

5.6.2 The Breadth of the Market

A market index is a weighted sum of many individual share prices. How can non-stable power-law probability distributions arise as weighted sums of some other probability distributions? What are these probability distributions of individual shares, underlying a market index? How can we effectively characterize the individual variations of securities traded in a financial market on a given day, when we summarize by saying that the market went up/down, e.g., 2%?

Comprehensive studies of the statistical properties of individual share price variations were undertaken by the Boston group based on data taken from a variety of databases with different historical extension, data frequency, and market breadth [103, 104]. In one study, the variation of market capitalization (equal to the share price multiplied by the number of outstanding shares) has been investigated rather than the share returns themselves. Variation of market capitalization is a good proxy for share-price variation when the number of outstanding shares varies on a much slower time scale than the share prices. This has not been studied but, for simplicity, we will neglect this subtlety here.

By and large, the price statistics of individual companies is rather similar to that of a market index [103, 104]. The cumulative distribution functions in a log–log plot against asset returns are straight lines, implying power-law return distributions. The exponents realized slightly depend on the companies considered: on a five-minute time scale, most of them fall into the range $2 \leq \mu \leq 6$; only very few companies possess a tail index $\mu < 2$ in the stable Lévy range. A histogram of the returns peaks at $\mu = 3$. When the returns of a stock are normalized by its standard deviation on the corresponding time scale, the cumulative distribution functions collapse onto a single master curve. This master curve has a slope of approximately -3 in log–log representation [103]. More specifically, regression fits produce significantly different tail indices $\mu_+ = 3.10 \pm 0.03$ and $\mu_- = 2.84 \pm 0.12$ for the positive and negative tails, respectively. On the other hand, using the Hill estimator [98] produces lower values $\mu_+ = 2.84 \pm 0.12$ and $\mu_- = 2.73 \pm 0.13$, which are essentially the same, given the error bars [104]. When the time scale increases, these indices increase gradually up to values of the order $\mu_+ \approx 5 \ldots 6$ for scales of the order of four years. Apparently, the asymmetry between positive and negative returns increases: the indices μ_- remain of order $3 \ldots 3.5$ even at the largest time scales. $\mu_+ > \mu_-$ implies that rallies are less severe than crashes but, given the positive global averages of the markets, also implies that positive returns have more weight than negative returns in the range of moderate variations. When changing from databases with high-frequency data to those containing daily data, a break similar to that in Fig. 5.23 is observed.

These findings, however, leave us with a puzzle: why is the probability distribution apparently (almost?) form-invariant under addition of random variables for the short time scales although the underlying distributions are not stable? Why does convergence towards a Gaussian occur only beyond four days? Or why is it so slow, if we refer to the more gradual convergence of the DAX returns? Why do stock indices have the same power-law behavior in their return probability density functions as have individual stocks, although the basic probability distributions are not stable and many individual stock returns are added to produce that of the index? The answer is not truly established at present although it is likely that it has to do with correlations – both temporal and interasset correlations. Some elements of an answer, and much more information on the structure of financial time series, is provided by higher-order correlations in the returns, or in the volatility. Other elements are provided by studying the correlation matrix of the shares traded in one or several markets. Before addressing these problems, we briefly turn to the homogeneity of the markets.

The power-law tails do not inform us directly of the width of the distribution of the stock returns in a given market on a give time horizon, say one day. A stock index may rise by 1% in a day. However, there may be days where this 1% return is generated by moderate rises of almost all stocks, and other days where half of the stocks rise, perhaps even by 10%, or so, and the other half fall by almost the same amount. This effect is not captured by the market index, neither in its return nor in its volatility, which is a property of the index time series. It will not show up either in the power-law exponents directly. On the other hand, such information on the inhomogeneity of the price movements in a market may be valuable both from a fundamental point of view and for investors.

Let $\gamma = 1, \ldots, N$ label a specific stock in a market, and consider one-day returns $\delta S_{1d}^{(\gamma)}(t)$ only. (For the remainder of this discussion, we drop the time-scale subscript.) On each trading day, the returns of the ensemble of stocks will be random variables, and a probability distribution $p[\delta S^{(\gamma)}(t)]$ can be attributed to them. This probability distribution has been determined for the 2188 stocks traded at the New York Stock Exchange from January 1987 to December 1998 [105].

In general, the time series of the individual stocks have different widths and somewhat different shapes. They are transformed to random variables with zero mean and unit variance by subtracting their temporal mean, and dividing by the standard deviation of the time series. In log-scale, their central part is approximately triangular, i.e., the variables are drawn from a Laplace distribution $p(\delta s^{(\gamma)}) \propto \exp(-a|\delta s^{(\gamma)}|)$ [105]. Furthermore, the distribution in crash periods is very different from that of normal days where it displays an approximately constant form. In crash periods, the distribution is significantly broader and asymmetric. The crash of October 1987 (Black Monday), another event in early 1991, the October crash in 1997 (Asian crisis), and the crash of August 1998 (Russian debt crisis, for the latter two cf. Fig. 1.1 for the German DAX index) clearly stand out from the remainder of the distribution. However, negatively skewed distributions at the crash (or the day thereafter) are often followed by positively skewed rebounds shortly after the crash, which is at the origin of the apparently symmetric shape of the 1987 crash.

A more quantitative and condensed description of the market-return distribution is obtained from its first moments. The average and standard deviation are

$$\mu(t) = \frac{1}{N} \sum_{\gamma=1}^{N} \delta S^{(\gamma)}(t) , \quad \Sigma(t) = \sqrt{\frac{1}{N} \sum \left[\delta S^{(\gamma)}(t) - \mu(t) \right]^2} . \quad (5.98)$$

$\mu(t)$ is the average return of the market on a given trading day and, apart from weight factors, should be equal to the return of the market index $\delta S_{\mathrm{market}}(t)$. $\Sigma(t)$ gives the width of the return distribution of the market on each trading day. Lillo and Mantegna have proposed to call this quantity *variety* of the ensemble. It measures the inhomogeneity of the market on a given day and should be clearly distinguished from the volatility of the market, which measures the day-to-day variations (for this reason, we have chosen the capital letter Σ).

The probability distribution of the mean $\mu(t)$ of the 3032 trading days in the period studied is non-Gaussian and approximately Laplacian. The probability distribution of the daily mean of each of the 2188 stocks has a similar shape but is much narrower [105]. The probability distribution of the variety $\Sigma(t)$ is positively skewed in log–log scale, while that of the volatility σ_i is negatively skewed. A quadratic approximation in the central parts would give log-normal distributions, although the accuracy of such an approximation is questionable in their tails, due to the skewness. Similar to the returns and the volatility of the market, the mean $\mu(t)$ is essentially uncorrelated in time, while the variety $\Sigma(t)$ possesses long-time power-law correlations with an exponent of the order 0.23, comparable in order of magnitude to the exponents which describe the volatility correlations [105].

To what extent can one understand these results in a picture where a market provides a collective dynamics, and the individual companies execute additional ("idiosyncratic") fluctuations around the market dynamics? Such a one-factor model (one collective driver of dynamics) can be written as

$$\delta S^{(\gamma)}(t) = \alpha^{(\gamma)} + \beta^{(\gamma)} \delta S_{\mathrm{market}}(t) + \epsilon^{(\gamma)}(t) . \quad (5.99)$$

$\alpha^{(\gamma)}$ is the stock-specific deviation of mean returns with respect to the market return, and $\epsilon^{(\gamma)}(t)$ describes the zero-mean idiosyncratic fluctuations of the stock γ with respect to the market dynamics. $\beta^{(\gamma)}$ is a measure of the correlation of the stock γ with the market. The market return $\delta S_{\text{market}}(t)$ can be taken as an actual market time series, e.g., the S&P500. In this way, the one-factor model can generate surrogate time series for the market. It turns out that the probability density of the mean $\mu(t)$ of the return distribution of such surrogate data is in good agreement with the probability density of the real market. On the other hand, the probability density of the variety $\Sigma(t)$ of the surrogate data is different from that of the market data: it is almost symmetric and much narrower than the market distribution [105]. Also, the one-factor model cannot describe correctly changes in the symmetry of the ensemble return distributions during crash and rally periods [106].

5.6.3 Non-linear Temporal Correlations

In Sect. 5.3.2, we saw that *linear* correlations of the returns of three assets, sampled on a 5-minute time scale, (5.3), decayed to zero within 30 minutes. For the S&P500, the linear correlation function of 1-minute returns decays to zero even faster: within 4 minutes, it reaches the noise level [62]. The correlations of the DAX high-frequency data decay to zero within 10 minutes [59, 60]. This, however, is not true for non-linear correlations which persist to much longer times.

One can consider various higher-order correlation functions, e.g.,

$$C_{\text{abs},\tau}(t - t') \quad \sim \langle |\delta S_\tau(t)||\delta S_\tau(t')|\rangle \,, \tag{5.100}$$
$$C_{\text{square},\tau}(t - t') \sim \langle [\delta S_\tau(t)]^2 [\delta S_\tau(t')]^2 \rangle \,, \tag{5.101}$$

$$\ldots$$

where $\delta S_\tau(t)$ is defined in (5.1). $C_{\text{abs},\tau}$ measures the correlations of the absolute returns, and $C_{\text{square},\tau}$ those of the returns squared. Both are related to volatility correlations, $C_{\text{square},\tau}$ perhaps in a more direct way. Various higher-order correlation functions can also be defined and evaluated.

Geometric Brownian motion assumes the volatility to be a constant. This is true over rather short time scales, at best. Empirical volatilities vary strongly with time and suggest considering volatility as a stochastic variable. This fact has led to the development of the ARCH and GARCH models [48, 49], briefly mentioned in Sect. 4.4.1. The probability distribution of volatilities of the S&P500 is close to log-normal [107], a fact which also holds for various other markets, in particular foreign exchange [108]. However, the probability distribution does not exhaust stochastic volatility: in fact, volatility is strongly correlated over time!

Figure 5.24 displays the correlation function of the absolute returns of the S&P500 [62]. The absolute correlations decay very slowly with time,

$$C_{\text{abs},\tau} \sim |t - t'|^{-0.3} \,. \tag{5.102}$$

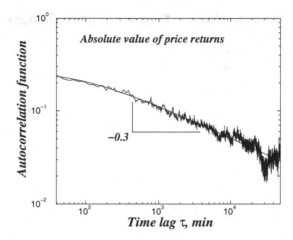

Fig. 5.24. Correlation function of the absolute returns of the S&P500 index. The correlations decay as a power law with an exponent −0.3. By courtesy of P. Gopikrishnan. Reprinted from P. Gopikrishnan et al.: Phys. Rev. E **60**, 5305 (1999) ©1999 by the American Physical Society

Decay of correlations with a power law is so slow that no characteristic correlation time can be defined. Correlations of the absolute returns therefore extend infinitely far in time!

The same is true for correlations of the returns squared. Figure 5.25 shows the volatility correlations of S&P500 index futures: they again decay as power laws

$$C_{\text{square},\tau} \sim |t - t'|^{-0.37} \tag{5.103}$$

with an exponent −0.37, rather similar to that of the absolute returns. Again, no characteristic time scale can be defined, and the correlations extend infinitely far in time.

The analysis of the Hurst exponent H of a probability density function, (4.41), also supports long-time correlations in absolute and square returns. Lux reports such an analysis for the German stock market (DAX share index and its constituent stocks individually) [110], and finds exponents in the range $H = 0.7, \ldots, 0.88$ for the absolute returns, and $H = 0.62, \ldots, 0.77$ for the square returns. For comparison, purely random behavior leads to $H = 1/2$, and is rather well obeyed by the returns. Further evidence for long-time correlation is also available for the US and UK stock markets [111], and for foreign exchange markets [102]. Quite generally, the correlations are the stronger, the lower the power of the returns taken [102].

Zipf analysis, too, points towards serial correlations in financial time series. In this method, taken from statistical studies of languages, one studies the rank dependence of the frequency of "words". Rank denotes the position of a "word" after ordering according to frequency. "Word" is taken literally in linguistics, but any sequence of up- or down-moves of a stock price may be

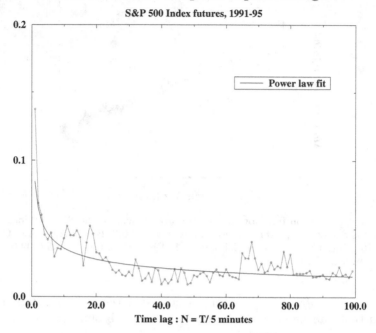

Fig. 5.25. Correlations of the squared returns of S&P500 index futures. The solid line is a fit to a power law with an exponent -0.37. From [74] courtesy of R. Cont

decomposed into characteristic words whose frequency in the entire pattern is then evaluated. From this kind of analysis, significant correlations have been discovered, e.g., in the chart of Apple stock [112]. Similar results have also been obtained for foreign exchange rates [113].

Writing the returns of an asset as

$$\delta S_\tau(t) = \text{sign}\left[\delta S_\tau(t)\right] |\delta S_\tau(t)| \,, \tag{5.104}$$

the absence of linear correlations and presence of long-time non-linear correlations in financial time series implies that the time series of the sign changes of the returns is uncorrelated or short-range-correlated, while the long-time correlations are embodied in the amplitudes of the returns.

This decomposition can be pursued further and suggests an interesting analogy with diffusion [114]. On a rather long time scale τ, the asset return $\delta S_\tau(t)$ is aggregated from a number N_τ of individual returns δS_i in the time interval $[t, t + \tau]$:

$$\delta S_\tau(t) = \sum_{i=1}^{N_\tau} \delta S_i \,. \tag{5.105}$$

This equation also applies to a 1D diffusion problem, where $\delta S_\tau(t)$ and δS_i would correspond to the distance traveled by a test particle in the time interval τ and the distance traveled as a consequence of each of the N_τ individual shocks. (We emphasize that the subscript i here is used to number individual shocks on a particle/individual transactions in a market, in a small time interval τ.) When a measurement on an actual financial time series or a diffusing particle is made, all quantities in (5.105), $\delta S_\tau(t)$, δS_i, and N_τ, and the times between two shocks, turn out to be random. We may thus inquire about their statistical properties.

In ordinary diffusion, the probability distribution $p(\delta S_\tau)$ is Gaussian with variance $\langle \delta S_\tau^2 \rangle = N_\tau \langle \delta S_i^2 \rangle_\tau = D\tau$, and D is the diffusion constant. The distribution of the number of shocks in a given time interval (attempt frequency) $p(N_\tau)$ is narrow Gaussian, and the attempt frequencies only have short-time exponential correlations. Looking at the distribution of variances of the individual shocks sampled over the interval τ, $p(\langle \delta S_i^2 \rangle)$ again is a narrow Gaussian, and these variances are short-time- correlated only. One can then introduce an effective variable,

$$\epsilon(t) = \frac{\delta S_\tau(t)}{\sqrt{N_\tau \langle \delta S_i^2 \rangle_\tau}} . \tag{5.106}$$

In diffusion, $\epsilon(t)$ is uncorrelated and Gaussian distributed. Of course, this discussion refers to equilibrium, and diffusion in a stirred environment would have different statistical properties.

Financial markets are very different from this classical diffusion problem: for an ensemble of 1000 stocks, $p(N_\tau)$ is not Gaussian but possesses a power-law tail with an exponent -4.4 [114]. The correlations $\langle N_\tau(t)N_\tau(t') \rangle \sim |t - t'|^{-0.3}$, i.e., show a power-law decay with a rather small exponent similar to those observed above for the volatility. The distribution of $\langle \delta S_i^2 \rangle_\tau$ is a power law with an exponent -3.9, but this variable is essentially uncorrelated in time. Finally, $\epsilon(t)$ turns out to be uncorrelated and Gaussian distributed, as in ordinary diffusion. Putting everything together again, we find that an asset return can be written as

$$\delta S_\tau(t) = \epsilon(t)\sqrt{\langle \delta S_i \rangle_\tau N_\tau} . \tag{5.107}$$

As announced above, $\epsilon(t)$ being Gaussian distributed and uncorrelated plays a role similar to the sign of the return, while the square root essentially is the amplitude of the return, and contains the long-time correlations. Alternatively, in the perspective of stochastic volatility, ARCH and GARCH models, we can say that the price changes are drawn from an uncorrelated Gaussian variable with an instantaneous variance $N_\tau \langle \delta S_i \rangle_\tau$ which contains long-range correlations. The tails of the distribution of the price changes come from the tails in $\langle \delta S_i \rangle_\tau$, and the long-time correlations originate in those of N_τ. The similarity in the exponents of the volatility correlations of financial time series and the correlations of $\langle N_\tau(t)N_\tau(t') \rangle$ therefore is not accidental but, on the contrary, causal [114].

We finally turn to a more complicated kind of correlation known in financial markets, the leverage effect [116]: the volatility of the returns of an asset tends to increase when its price drops. In option markets, these negative correlations induce a negative skew in the return distributions on longer time scales [17]. The leverage correlation function is defined as [117]

$$\mathcal{L}(t - t') = Z^{-1} \langle [\delta S_{1d}(t)]^2 \, \delta S_{1d}(t') \rangle \,, \tag{5.108}$$

that is, a third-order correlation function between volatility and returns (for simplicity, we assumed that $\langle \delta S_{1d} \rangle = 0$). Our discussion will be limited to daily returns. Consequently, we temporarily drop the subscript 1d. The normalization constant is chosen as $Z = \langle [\delta S(t)]^2 \rangle^2$. 10-year daily closing prices of 437 US stocks have been analyzed. The leverage effect is significant and negative for $t > t'$ while it essentially vanishes for $t > t'$. This implies that falling prices cause increased volatilities, and not vice versa. An exponential fit to

$$\mathcal{L}(t - t') = -A \exp(-|t - t'|/T) \tag{5.109}$$

gives a satisfactory description of the data. The best fit is generated with $A = 1.9$ and $T = 69$ days.

A similar analysis can be performed for stock indices [117]. An exponential function again gives a reasonable fit, however, with very different parameters: $A = 18$ and $T = 9.3$ days, i.e., a significantly increased amplitude and a much shorter correlation time. Moreover, there are some significant positive correlations for $t - t' < -4$ days, i.e., the volatility increases a couple of days *before* the indices rally. Possibly, these correlations are related to rebounds shortly after a strong market increase causes increased volatility.

A retarded return model, eventually extended by a stochastic volatility, can account for some of the effects observed [117]. Write the change in asset price (we use ΔS for the absolute price change to distinguish from the return) over the fixed time scale of one day as

$$\Delta S(t) = S^R(t) \sigma(t) \epsilon(t) \,, \tag{5.110}$$

where $\sigma(t)$ is the (possibly time-dependent) volatility, $\epsilon(t)$ is a random variable with unit variance, and the retarded price $S^R(t)$ is defined as

$$S^R(t) = \sum_{t-t'=0}^{\infty} \mathcal{K}(t - t') S(t') \,. \tag{5.111}$$

$\mathcal{K}(t - t')$ is a kernel normalized to unity with a typical decay time T. This retarded model interpolates between an additive stochastic process when the decay time T tends to infinity, and a purely multiplicative process when $T \to 0$. The argument for considering such a model is that the proportionality of return to share price should hold on the longer time scales of investors. On shorter time scales where traders rather than investors operate, the prices are more determined by limit orders which are given in absolute units of money.

Evaluating this model for constant volatility and in the limit of small price fluctuations over the decay time T ($\sigma\sqrt{T}$ $ll1$), one obtains for the small-time limit of the leverage function $\mathcal{L}(t - t' \to 0) = -2$. Stochastic volatility fluctuations could increase the magnitude of this term. This limit is satisfied by the individual stocks analyzed, as well as similar data from European and Japanese markets. In the perspective of the retarded model, the leverage effect would just be a consequence of a different market structure, or of different market participants, determining the price variations on different time scales.

It is surprising then that the leverage of stock market indices is much bigger, and decays on a much shorter time scale, than that of individual stocks [117]. The index being an average of a number of stock prices, one would expect rather similar properties than for the single stocks. Apparently, an additional panic effect is present in indices, which leads to significantly more severe volatility increases following a downward price move which, however, would persist only over time scales of one to two weeks.

The leverage effect has also been observed in a 100-year time series of the daily closing values of the Dow Jones Industrial Average [118]. The effect there is about one order of magnitude smaller than the individual-stock effect discussed above, and more than two orders of magnitude smaller than that of the stock indices just discussed. Also, the decay time of the effect here is about $20 \ldots 30$ days, somewhat intermediate between the stock and index decay times of Bouchaud et al. [117]. Perelló and Masoliver [118] show that stochastic volatility models, even without retardation, are able to explain the effect observed.

5.6.4 Stochastic Volatility Models

The preceding sections have demonstrated that the assumption of constant volatility underlying the hypothesis of geometric Brownian motion in financial markets is at odds with empirical observations. Volatility is a random variable drawn from a distribution which is approximately log-normal and which possesses long-time correlations in the form of a power law. The question then is to what extent stochastic volatility should be explicitly included in the model of asset prices.

Two standard models with stochastic volatility were briefly described in Sect. 4.4.1. In the ARCH(p) and GARCH(p,q) processes, (4.46) and (4.48), the volatility depends on the past returns and (for the GARCH process) the past volatility, i.e., these models are examples of conditional heteroskedasticity. These models have been analyzed extensively in the financial literature. Another popular class of stochastic volatility models considers the volatility as an independent variable driving the return process. The starting point formally is geometric Brownian motion, (4.53), with a time-dependent volatility

$$\mathrm{d}S(t) = \mu S(t)\mathrm{d}t + \sigma(t)S(t)\mathrm{d}z_1 \ . \tag{5.112}$$

$dz_1(t)$ describes a Wiener process. With $v(t) = \sigma^2(t)$, the time-dependent variance again follows a stochastic process

$$dv(t) = m(v)dt + s(v)dz_2 \ . \tag{5.113}$$

Several popular models use different specifications for $m(v)$ and $s(v)$ [10]:

$$
\begin{aligned}
& m(v) = \gamma v \ , && s(v) = \kappa v && \text{(Rendleman} - \text{Bartter model),} \\
& m(v) = \gamma(\theta - v) \ , && s(v) = \kappa && \text{(Vasicek model),} \\
& m(v) = \gamma(\theta - v) \ , && s(v) = \kappa\sqrt{v} && \text{(Cox} - \text{Ingersoll} - \text{Ross model).}
\end{aligned}
\tag{5.114}
$$

In the Vasicek and Cox–Ingersoll–Ross models, the volatility is mean-reverting with a time constant γ^{-1} and an equilibrium volatility of θ.

The leverage effect suggests that the volatility and return processes may be correlated in addition:

$$dz_2(t) = \rho_{r-v}dz_1(t) + \sqrt{1 - \rho_{r-v}^2}dZ(t) \ , \tag{5.115}$$

where $dZ(t)$ describes a Wiener process independent of $dz_1(t)$. Recently, the Cox–Ingersoll–Ross model with a finite return-volatility correlation ρ_{r-v} has been solved for its probability distributions [119], extensively using Fokker–Planck equations. The logarithmic probability distributions for log-returns on short time scales (1 day) are almost triangular in shape, while they become more parabolic for longer time scales, e.g., 1 year.

For long time scales $\gamma\tau \gg 1$, the probability distribution of $x_\tau(t) = \delta S_\tau(t) - \langle \delta S_\tau(t) \rangle$ takes the scaling form

$$P(x_\tau) = N_\tau e^{-p_0 x_\tau} P_*(z) \ , \quad P_*(z) = K_1(z)/z \ . \tag{5.116}$$

N_τ is a time-scale-dependent normalization constant, p_0 is a constant depending on the return-volatility correlations and the parameters of the volatility process, and $K_1(z)$ is the modified Bessel function. The argument z is of the schematic form $z^2 = (ax_\tau + b)^2 + c^2$ [119]. In the limit of large returns, $\ln P(x_\tau) \sim -p_0 x_\tau - (\ldots)|x_\tau|$, i.e., the tails of the probability distribution of the returns are exponential with a different slope for the positive and negative returns. These slopes, however, do not depend on the time scale τ in this long-time-scale limit. The exponential tails are reminiscent of some variants of the truncated Lévy distributions discussed in Sect. 5.3.3. In the limit of small returns at long time scales, a skewed Gaussian distribution of returns is obtained. When the solutions are compared to 20 years of Dow Jones data, an excellent collapse onto a single master curve is obtained for time scales from 10 days to 1 year *with four fitting parameters only*, γ, θ, κ, μ. Independently, the correlation coefficient ρ_{r-v} has been found to vanish [119]. These four parameters are summarized in Table 5.3, where they are given both in daily and annual units.

Table 5.3. Parameters of the stochastic volatility model obtained from the fit of the Dow Jones data. In addition to the parameters listed, $\rho = 0$ for the correlation coefficient and $1/\gamma = 22.2$ trading days for the relaxation time of the variance are found

Units	γ	θ	κ	μ
1/day	4.50×10^{-2}	8.62×10^{-5}	2.45×10^{-3}	5.67×10^{-4}
1/year	11.35	0.022	0.618	0.143

5.6.5 Cross-Correlations in Stock Markets

With the exception of the Black–Scholes analysis where we used the correlations in price movements between an option and its underlying security, we have not yet considered possible correlations between financial assets. However, it would be implausible to assume that the price movements of a set of stocks in a market are completely uncorrelated. There are periods where a large majority of stocks moves in one direction, and thus the entire market goes up or down. On the other hand, in other periods, the market as a whole moves quite little, but sectors might move against each other, or within an industry share values of different firms could move against each other, either as a result of changing market share, or due to more psychological factors.

Can correlations between different stocks be quantified, or those between stocks and the market index be quantified? As will become apparent in Chap. 10, knowing such correlations accurately is a prerequisite for good risk management in a portfolio of assets. Unfortunately, it turns out that many of these correlations are hard to measure.

Correlations between the prices or returns of two assets γ and δ are measured by the correlation matrix

$$C(\gamma, \delta) = \frac{\langle [\delta S^{(\gamma)}(t) - \langle \delta S^{(\gamma)}(t) \rangle] [\delta S^{(\delta)}(t) - \langle \delta S^{(\delta)}(t) \rangle] \rangle}{\sigma^{(\gamma)} \sigma^{(\delta)}} \qquad (5.117)$$

$$\equiv \frac{1}{T} \sum_{t=1}^{T} \delta s^{(\gamma)}(t) \delta s^{(\delta)}(t) . \qquad (5.118)$$

A time scale $\tau = 1$ day has been assumed for the returns, and the corresponding subscript has been dropped, $\delta S_{1d}^{(\gamma)}(t) \equiv \delta S^{(\gamma)}(t)$. We also assume stationary markets, i.e. $C(\gamma, \delta)$ is time-independent. The returns $\delta S^{(\gamma)}(t)$ have been defined in (5.1), $\sigma^{(\gamma)}$ are their standard deviations, the normalized returns $\delta s^{(\gamma)}$ were defined in (5.2), and the averages $\langle \ldots \rangle$ are taken over time. Uncorrelated assets have $C(\gamma, \delta) = \delta_{\gamma, \delta}$. In finance, the label β is reserved for the correlation of a stock γ (or a portfolio of stocks) with the market [10]:

$$\beta = C(\gamma, \text{market}) . \qquad (5.119)$$

In order to appreciate the subsequent discussion, let us look at two un-correlated time series $\delta s^{(1)}(t)$ and $\delta s^{(2)}(t)$, each of length T (and zero mean, unit variance, of course). From (5.117), we have

$$C(1,2) = \frac{1}{T} \sum_{t=1}^{T} \delta s^{(1)}(t) \delta s^{(2)}(t) \ . \tag{5.120}$$

$C(1,2)$ is the sum of T random variables with zero mean. Despite the absence of correlations (by construction) between the two time series, for finite T, $C(1,2)$ is a random variable itself and different from zero. $C(1,2)$ is drawn from a distribution with zero mean and a standard deviation decreasing as $1/\sqrt{T}$. Only in the limit $T \to \infty$ will $C(1,2) \to 0$, as is appropriate for uncorrelated random variables. The finite time scale T, over which the correlations between the two time series are determined, produces a *noise dressing* of the correlation coefficient. More specifically, for two independent time series of length T of normally distributed random numbers $\varepsilon_i(t)$ with zero mean and unit variance, the correlation coefficient again is a random number [120]

$$\langle \varepsilon_i(t)\varepsilon_j(t) \rangle = \delta_{ij} + \sqrt{\frac{1+\delta_{ij}}{T}}\varepsilon_i(t) \ . \tag{5.121}$$

The finite-length autocorrelation is a random normally distributed variable with mean unity and variance $2/T$, and the cross-correlation is a random normally distributed variable with zero mean and variance $1/T$.

For correlation *matrices* where many time series enter, noise dressing may be a severe effect. N time series with T entries each may be grouped into an $N \times T$ random matrix $\mathbf{M}$, and the correlation matrix is written as $\mathbf{C} = T^{-1} \mathbf{M} \cdot \overline{\mathbf{M}}$ where $\overline{\mathbf{M}}$ is the transpose of $\mathbf{M}$. In the same way as noise dressing for finite T produced an artificial finite random value for $C(1,2)$, for finite T, noise dressing will produce artificial finite random entries $C(\gamma, \delta)$ in the correlation matrix. Figure 5.26 demonstrates this effect: the correlation matrix $\mathbf{C}$ of 40 uncorrelated time series is random when the time series is only 10 steps long (left panel). The absence of correlations $C(\gamma, \delta) = \delta_{\gamma,\delta}$ is well visible for 1000 time steps (right panel). The two panels of Fig. 5.26 are consistent with (5.121). For $T = 10$, the autocorrelation is a Gaussian variable with mean unity and standard deviation 0.48, and the cross-correlation coefficients are Gaussians with mean zero and standard deviations of 0.32. For $T = 1000$, the mean values are the same but standard deviations have decreased by one order of magnitude. Roughly, for N time series, $T \gg N$ time steps are required in the series in order to produce statistically significant correlation matrices.

Random matrix theory predicts the spectrum of eigenvalues λ of a random matrix (of the type appropriate for financial markets [121, 122]) to be bounded and distributed according to a density

$$\rho(\lambda) = \frac{Q}{2\pi\sigma^2} \frac{\sqrt{(\lambda_{\max} - \lambda)(\lambda - \lambda_{\min})}}{\lambda} \ ,$$

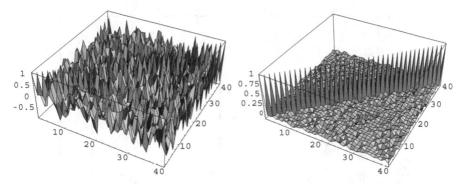

Fig. 5.26. Noise dressing of a correlation matrix. The correlation matrix of 40 uncorrelated time series is shown for a length of 10 steps (*left panel*) and 1000 steps (*right panel*)

$$\lambda_{\min}^{\max} = \sigma^2 \left(1 + \frac{1}{Q} \pm 2\sqrt{\frac{1}{Q}} \right) , \qquad (5.122)$$

where $Q = T/N \geq 1$ is the ratio of time series entries to assets. This density is shown as the dotted line in Fig. 5.27.

Recently, two groups calculated the correlation matrices of large samples of stocks from the US stock markets [121, 122], and compared their results to predictions from random matrix theory. This is partly done with reference to the complexity of a real market (a detailed analysis of all correlation coefficients would not be useful) and partly in order to compare empirical correlations with a null hypothesis (purely random correlations, the alternative null hypothesis of zero correlations being rather implausible). Random matrix theory was developed in nuclear physics in order to deal with the energy spectra of highly excited nuclei in a statistical way when the complexity of the spectra made the task of a detailed microscopic description hopeless [123] – a situation reminiscent of financial markets.

Figure 5.27 displays the eigenvalue density of the correlation matrix of 406 firms out of the S&P500 index based on daily closes from 1991 to 1996 [121]. Similar results are available also for other samples of the US stock market [122]. A very large part of the eigenvalue spectrum is indeed contained in the density predicted by random matrix theory, and therefore noise-dressed.

There are some eigenvalues falling outside the limits of (5.122), however, which contain more structured information [121, 122]. The most striking is the highest eigenvalue $\lambda_1 \approx 60$. Its eigenvector components are distributed approximately uniformly over the companies, demonstrating that this eigenvalue represents the market itself. Another 6% of the eigenvalues fall outside the random matrix theory prediction for the spectral density but lie close to its upper end. An evaluation of the inverse participation ratio of the eigenvectors [122] suggests that there may be a group of about 50 firms with definitely non-random correlations which are responsible for these eigenvalues.

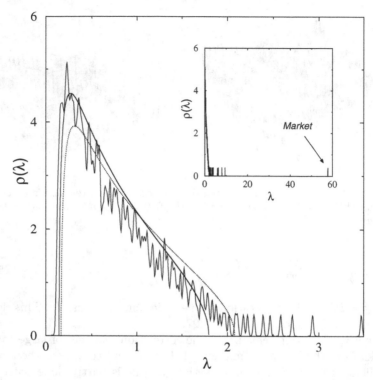

Fig. 5.27. Density of eigenvalues of the correlation matrix of 406 firms out of the S&P500 index. Daily closing prices from 1991 to 1996 were used. The *dotted line* is the prediction of random matrix theory. The *solid line* is a best fit with a variance smaller than the total sample variance. The inset shows the complete spectrum including the largest eigenvalue which lies about 25 times higher than the body of the spectrum. By courtesy of J.-P. Bouchaud. Reprinted from L. Laloux et al.: Phys. Rev. Lett. **83**, 1467 (1999), ©1999 by the American Physical Society

Interestingly, high inverse participation ratios are also found for some very small eigenvalues. While they apparently fall inside the spectral range of random matrix theory, the high values found here seem to give evidence for possibly small groups of firms with strong correlations [122]. However, these groups would not have significant cross-group correlations.

Kwapién et al. have shown that drawing 451 time series of length 1948 each out of a Gaussian distribution produces a remarkably good approximation to (5.122) [124]. For fixed N, Q increases with T, and λ_{max} and λ_{min} approach each other and both approach σ^2 ($\sigma = 1$ in our case). We therefore recover an N-fold-degenerate eigenvalue 1, as expected for uncorrelated variables.

The empirical properties of the S&P500 correlation matrices can be clarified further using a model of group correlations [125]. Here, one assumes that industries cluster in groups (labeled by g while the individual firms are labeled by γ), and that the return of a stock contains both a "group component" and an "individual component"

$$\delta s^{(\gamma)}(t) = \sqrt{\frac{w_{g_\gamma}}{1 + w_{g_\gamma}}} f_{g_\gamma}(t) + \sqrt{\frac{1}{1 + w_{g_\gamma}}} \varepsilon_\gamma(t) . \tag{5.123}$$

$f_{g_\gamma}(t)$ and $\varepsilon_\gamma(t)$ are both random numbers and represent the synchronous variation of the returns within a group, and the individual component with respect to the group, respectively. The relative weight of the group dynamics with respect to the individual dynamics is measured by the weight factor w_{g_γ}. In the model, there may also be a number of companies which do not belong to a group. They formally obtain a weight factor $w = 0$. This is a straightforward generalization of the one-factor model (5.99) introduced when discussing variety. There is no built-in correlation between industries. With infinitely long time series, the correlation matrix of the model without in-group randomness $[\varepsilon_\gamma(t) \equiv 0]$ is a block diagonal matrix. It is a direct product of $N_g \times N_g$ matrices whose entries are all unity (N_g is the size of group g). These blocks have one eigenvalue equal to N_g, and $N_g - 1$ eigenvalues equal to zero. When the time series are finite, and the firms have an individual random component in their returns, the eigenvalues will be changed. The influence on the eigenvalue N_g will be minor so long as the individual randomness is not too strong. However, the most important effect will be a splitting of the $(N_g - 1)$-fold-degenerate zero eigenvalues into a finite spectral range. Under special circumstances, one may also observe high inverse participation ratios for small eigenvalues [125]. This happens when the noise strength of a group is small, i.e., when the variance of the "individual firm contribution" to the returns is small compared to the variance of the "group contribution". This effect is also seen in numerical simulations [125].

A nice feature of this model is that its correlation coefficients can be de-termined analytically for finite times series lengths T [120] when the price dynamics is governed by geometric Brownian motion (returns normally dis-tributed). From (5.117) and using (5.121), we find

$$
\begin{aligned}
C(\gamma, \delta; T) = & \sqrt{\frac{w_{g_\gamma}}{1 + w_{g_\gamma}}} \sqrt{\frac{h_{g_\delta}}{1 + w_{h_\delta}}} \left(\delta_{g_\gamma h_\delta} + \sqrt{\frac{1 + \delta_{g_\gamma h_\delta}}{T}} \varepsilon_{g_\gamma h_\delta} \right) \\
& + \sqrt{\frac{1}{1 + w_{g_\gamma}}} \sqrt{\frac{1}{1 + w_{h_\delta}}} \left(\delta_{\gamma\delta} + \sqrt{\frac{1 + \delta_{\gamma\delta}}{T}} \varepsilon_{\gamma\delta} \right) \\
& + \sqrt{\frac{w_{g_\gamma}}{1 + w_{g_\gamma}}} \sqrt{\frac{1}{1 + w_{h_\delta}}} \frac{1}{\sqrt{T}} \varepsilon_{g_\gamma\delta} \\
& + \sqrt{\frac{h_{g_\delta}}{1 + w_{h_\delta}}} \sqrt{\frac{1}{1 + w_{g_\gamma}}} \frac{1}{\sqrt{T}} \varepsilon_{\gamma h_\delta} \tag{5.124}
\end{aligned}
$$

to leading order in $\sqrt{T}$. The indexation of the four random numbers ε is meant to indicate that they are different and independent, but is irrelevant else.

Moreover, the model can be simulated numerically quite easily. When comparable parameters are used, an eigenvalue spectrum similar to Fig. 5.27 is obtained. This is demonstrated in Fig. 5.28. In that simulation it was assumed that assume that, among $N = 508$ stocks considered, there are six correlated groups $g = 1, \ldots, 6$ with sizes growing as 2^{g+1} and weights $w_g = 1 - 2^{-g-1}$. The sizes increase from 4 to 128 companies, and weight factors increase from 0.75 to 0.99 [120]. The remaining 256 stocks were supposed to be uncorrelated.

For a time series length of $T = 1,650$, the spectrum in the top left panel of Fig. 5.28 is rather similar to 5.27. When the length of the time series is increased to $T = 5,000$ and on to $T = 50,000$, the structure of the eigenvalue spectrum of the correlation matrix is changed. The bulk of the spectrum first develops a bimodal structure and subsequently splits into two distinct and clearly separated spectra, one centered around $\lambda = 0.5$ and the other spectrum centered around $\lambda = 1$. In addition, we still have the large eigenvalues discussed in the analysis of the S&P500 data.

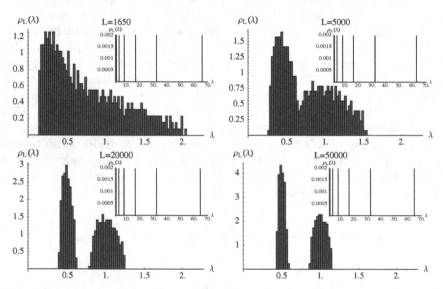

Fig. 5.28. Spectral densities $\rho_T(\lambda)$ of simulated correlation matrices. The length of the time series increases from *top left* to *bottom right* as $T = 1,650,\ 5,000,\ 20,000,\ 50,000$. The densities are split into two regions, $0 \leq \lambda \leq 2.2$ (main body of each panel) and $2.2 \leq \lambda \leq 70$ (inset of each panel). The densities are given in units of N. By courtesy of B. Kälber. Reprinted from T. Guhr and B. Kälber: J. Phys. A: Math. Gen. **36**, 3009 (2003), ©2003 by the Institute of Physics

Extending Noh's argument [125], we can attribute the three groups of spectra to different mechanisms. The large eigenvalues outside the spectrum described by random matrix theory, consist of the market component and the large eigenvalues of each individual industry. The eigenvalues centered around $\lambda = 0.5$ represent intra-industry correlations. For every industry, there is an almost $N_g - 1$-fold degenerate eigenvalue at $\lambda = 1/(1 + w_g)$ which, with w_g-factors in the range $0.75 \ldots 0.99$, lies close to $\lambda = 0.5$. (The N_g^{th} eigenvalue of the industry group is among the "large" eigenvalues.) These eigenvalues descend from the $N_g - 1$-fold degenerate zero eigenvalue obtained in the simplified problem where all entries of the intra-industry correlation matrix equal unity. Finally, the group of eigenvalues around $\lambda = 1$ represents the trivial autocorrelation of those companies which do not belong to any industry group.

The detailed understanding of the T-scaling of the entries of the correlation matrix, (5.124), in the Noh model [125] allows to formulate a heuristic method called power mapping, to identify instrinsic correlations in a broad eigenvalue spectrum such as that shown in Fig. 5.27. Power mapping is equivalent to artificially extending the length T of the time series underlying the correlation matrix [120]. Power mapping is achieved by raising every element of the correlation matrix to its qth power

$$C^{(q)}(\gamma, \delta; T) = \text{sign}\left[C(\gamma, \delta; T)\right] |C(\gamma, \delta; T)|^q \ . \tag{5.125}$$

Notice that the power-mapped matrix $C^{(q)}(\gamma, \delta; T)$ is different from the qth power of the correlation matrix $[\mathbf{C}(\gamma, \delta; T)]^q$. Now consider the influence of this mapping on the three different types of contributions to $C(\gamma, \delta; T)$. The diagonal terms

$$C(\gamma, \gamma; T) \sim 1 + \frac{b_1}{T^{1/2}} \to C^{(q)}(\gamma, \gamma; T) \sim 1 + q\frac{b_1}{T^{1/2}} \ , \tag{5.126}$$

where b_1 is a constant. The intra-industry off-diagonal terms $g = h$ but $\gamma \neq \delta$ are mapped as

$$C(\gamma, \delta; T) \sim a + \frac{b_2}{T^{1/2}} \to C^{(q)}(\gamma, \gamma; T) \sim a^q + q\frac{b_2}{T^{1/2}} \ , \tag{5.127}$$

with constants $0 < a < 1$ and b_2. The terms off-diagonal both in industry and in company index, on the other hand, behave as

$$C(\gamma, \delta; T) \sim \frac{b_3}{T^{1/2}} \to C^{(q)}(\gamma, \gamma; T) \sim \left(\frac{b_3}{T^{1/2}}\right)^q \sim T^{-q/2} \ . \tag{5.128}$$

When $q > 1$, the decay of these terms is accelerated by power-mapping with respect to the diagonal or intra-industry off-diagonal terms. It is for this suppression of off-diagonal noise-induced correlation coefficients that power-mapping is equivalent to a prolongation of the time series.

Numerical simulations of the Noh model confirm that power mapping with $q > 1$ acts to reduce the noise dressing of the correlation matrix. With $q = 1.5$, a clear two-peak structure in the eigenvalue spectrum is visible when the original ($q = 1$) spectrum looked similar to Fig. 5.27. All three components of the eigenvalue spectrum, intra-industry correlations, isolated companies, and industry and market collective contributions are readily apparent. However, it turns out that the range of powers q where the mapping separates the spectral components, is actually quite limited. When q increases, the a^q-constant in the intra-industry off-diagonal terms are strongly suppressed with respect to the equivalent term of size unity in the diagonal terms. Consequently, the intra-industry correlation structure is distorted significantly, and the two-peak structure in the eigenvalue spectrum of $C^{(q)}(\gamma, \delta; T)$ is lost. Apparently, $q = 1.5$ is the optimal value for the power-mapping approach [120].

A variant of this model allows to perform a mean-field analysis of the correlations in a stock market [126]. The dynamical equation is written as

$$S^\alpha(t+1) = (1 - \epsilon_M - \epsilon_g)\left[S^\alpha(t) + \varepsilon^\alpha(t)\right]$$

$$+ \frac{\epsilon_M}{N}\sum_{\beta=1}^{N}\left[S^\beta(t) + \varepsilon^\beta(t)\right] + \frac{\epsilon_g}{N_g}\sum_{\gamma \in g}\left[S^\gamma(t) + \varepsilon^\gamma(t)\right] . \quad (5.129)$$

N is the number of stocks in the market, and N_g is the size of the industry group which a particular stock belongs to. ϵ_M and ϵ_g are coupling constants (weight factors) parameterizing the correlation of the price movement of the stock S^α with the market and the industry group. One important difference to (5.123) is the explicit presence of the market mode. This is typical of mean-field approaches in statistical physics. Its appearence in (5.129) does not have an immediate financial interpretation. (However, one might think about the benchmark-driven fund managers of today's mutual fund industry.) The other important difference to (5.123) becomes apparent when regrouping the terms in (5.129) in a different manner (we set $\epsilon_M = 0$ for simplicity)

$$\delta S^\alpha(t) + \epsilon_g\left[S^\alpha(t) - \frac{1}{N_g}\sum_{\gamma \in g}S^\gamma(t)\right] = \varepsilon^\alpha(t) - \epsilon_g\left[\varepsilon^\alpha(t) - \frac{1}{N_g}\sum_{\gamma \in g}\varepsilon^\gamma(t)\right]$$

$$(5.130)$$

The coupling to the stocks of the same industry is implemented through the difference terms which measure the deviation of the current stock price S^α from the "industry mean-field" $\sum_\gamma S^\gamma$, and similarly for the price changes ε^α. The coupling to the market mode is realized with the same structure [126].

(5.129) can be rewritten as a continuity equation

$$\delta S_1(t) = S(t+1) - S(t) = \varepsilon(t) + \Delta \cdot [S(t) + \varepsilon(t)] . \quad (5.131)$$

$\Delta = \Delta_M + \Delta_g$ is a Laplace-type operator which describes flows due to the presence of gradients from the market and industry modes over an underlying

network. The gradients due to the intra-industry correlations are exhibited
by the difference terms in (5.130), and the gradients from market correlations
have similar structure. The elements of Δ_M and Δ_g are functions of ϵ_M and
ϵ_g, respectively [126]. The picture embedded is that of a network whose nodes
are formed by the labels of the stocks in the market, where a part of the price
changes is generated by flows induced by the correlations.

Setting $\epsilon_g = 0$ and $\epsilon_M = 1$ produces a mean-field limit where the correlation matrix can be calculated analytically. Its entries are [126]

$$C(\gamma, \delta; T \to \infty) = \begin{cases} \dfrac{a(\epsilon_M)}{[1 - a(\epsilon_M)]\, N + a(\epsilon_M)} & \text{if } \gamma \neq \delta \\ 1 & \text{if } \gamma = \delta \end{cases} \tag{5.132}$$

with $a(\epsilon_M) = \epsilon_M(3 - 2\epsilon_M)/(2 - \epsilon_M)$. The largest eigenvector of this correlation
matrix

$$\lambda_M = \frac{N}{[1 - a(\epsilon_M)]\, N + a(\epsilon_M)} \approx \frac{2 - \epsilon_M}{(1 - \epsilon_M)^2} \text{ as } N \to \infty. \tag{5.133}$$

The eigenvalue of the market component diverges quadratically as the coupling strength $\epsilon_M \to 1$ in the large-N limit. This divergence, which is reminiscent of critical phenomena as the fully correlated state is approached, is
confirmed by numerical simulations. The actual position of the market eigenvalue can be used to calibrate the coupling constant ϵ_M of the model. When,
at the next stage, the industry groups and the coupling constants ϵ_g are determined, one obtains good fits to the eigenvalue spectra shown in Fig. 5.27.
In particular, the fits produce the very large market eigenvector λ_M, several
large eigenvalues due to industry correlations above the spectral range of random matrix theory, and significant spectral weight at or below the lower edge
of the random matrix theory spectrum [126]. As has been shown above based
on the model (5.123), this weight is the necessary counterpart to the large
intra-industry eigenvalues. Based on the eigenvalues, a rather detailed picture
of correlations and industry groups for financial markets can be derived.

Approaches developed for cross-correlations in markets can also be adapted to search for temporal correlation structures in one time series [124, 127].
Take a high-frequency time series of the DAX such as that shown in Fig.
5.5, and transform to normalized returns, (5.2). Now divide the history into
N days, and let T denote the length of the intraday time series recorded in
15-second intervals. One now can form a correlation matrix $C(n_i, n_j)$ where
n_i denotes the n_ith day of the history. Averaging is done over the intraday
recordings. $C(n_i, n_j) = 1$ would imply that the time series of days n_i and n_j
were identical. Of course, $\mathbf{C}$ again is a random matrix, and one can proceed
as above.

From about three years of DAX high-frequency data a spectrum quite
similar to Fig. 5.27 is found where two eigenvalues of the order 4 fall outside
the spectrum of random matrix theory, and are thus statistically significant

[124, 127]. They can be interpreted by generating a weighted return time series

$$\delta s_{15''}^{(\lambda_k)}(t) = \sum_{\alpha=1}^{N} \text{sign}(v_\alpha^k)|v_\alpha^k|^2 \delta s_{15''}^{(\alpha)}(t) \, . \tag{5.134}$$

For the eigenvalue λ_k, the weights are determined by the corresponding eigenvectors v^k. These two time series show one prominent spike each. The spike of one time series is positive and located at 2:30 p.m. This is the local time in Germany when the financial news release in the United States starts. Interestingly, it is one hour before the opening of Wall Street which is not clearly detectable, and there is a significant weighted positive return at that time. The other spike is negative and located at 5 p.m., which corresponds to the closing of the German market.

There are other ways of representing correlations in a financial market. The preceding discussion may be thought of, roughly speaking, as an ensemble view containing (*all* correlation coefficients of) a correlation landscape built on a regular lattice (the indices of the correlation matrix entries), containing all fine details in a kind of grayscale (all values between -1 and 1 represented). An alternative representation could be a view where only the highest elevations in a landscape are connected (maximal correlations involving a stock emphasized, irrespective of its position in an index), and contrast is enhanced to black and white (all subdominant correlation coefficients dropped). In this way, the mountain ranges of the landscape become correlation clusters of stocks in a market or market indices in the global financial systems. A taxonomy of stock markets is built [128]–[130], which emphasizes the topology of correlations. This taxonomy is similar in structure though different in detail from the one derived from the model of coupled random walks [126].

We slightly simplify the discussion of the actual analysis, which proceeds by using elements of spin-glass theory such as ultrametric spaces. Let $C(\gamma, \delta)$ defined in (5.117) be the correlation coefficient between the assets γ and δ and define a "distance"

$$d(\gamma, \delta) = \sqrt{2\left[1 - C(\gamma, \delta)\right]} \, . \tag{5.135}$$

Highly correlated assets have a small distance in this representation. In this way, a hierarchical structure of asset clusters can be formed, and their evolution with time can be monitored. When, e.g., country indices of stock markets are analyzed, three distinct clusters, North America, Europe, and the Asia–Pacific region, emerge [129]. The participation of countries in these clusters evolves with time, however. The North American cluster including the Dow Jones Industrial Average, the S&P500, the Nasdaq 100, and the Nasdaq Composite is stable over time. The European cluster contains, in the late 1980s, the Amsterdam AEX, the Paris CAC40, the DAX, and the London FTSE. In the mid-1990s, the Madrid General and Oslo General indices have joined the European cluster. Other countries, most notably Italy, stayed outside this cluster. A similar expansion is observed for the Asia–Pacific cluster, where Japan remains a poorly linked important economy in the cluster region.

A similar analysis can be performed for the stocks within one market [128, 130]. When the best linked stocks of the New York Stock Exchange are graphed, a rather fractal structure emerges. Branches of this cluster often can be identified as industries. Interestingly, the stock of General Electric forms a natural center of this network.

Also, a connection to graph and network theory can be derived, in rather close analogy to such a taxonomy [131]. After defining a reduced variable (again for a one-day time horizon) $\delta S^{(\gamma)} - N^{-1} \sum_{\delta=1}^{N} \delta S^{(\delta)}(t)$ by subtracting the one-day return of the entire market, one readily calculates the correlation matrix of this reduced variable for the N assets of a market whose structure roughly is comparable to that of $C(\gamma, \delta)$. The correlation coefficients are assigned to the edges of fully connected graphs. (A fully connected graph is a graph generated from an ensemble of points/vertices by connecting each point to all other points.) The sum of all edges connecting to one particular vertex is the influence strength of this vertex, i.e., a measure of how well connected this vertex is to the rest of the system. Using data from the 500 companies of the S&P500, it turns out that the distribution of these influence strengths follows a power law with an exponent -1.8, i.e., the network formed from the cross-correlations of the S&P500 is scale-free with a fat-tailed influence-strength distribution [131]. A comparable analysis of other scale-free networks, such as the world wide web or the metabolic network, produces exponents systematically larger than two. These systems apparently possess less fat tails in their influence-strength distribution than financial markets.

With a somewhat related procedure, one can also map the correlations in a stock market on a liquid [132]. Here, however, the idea is to search for a quantity satisfying the axiomatic properties of a distance in Euclidean space. To do this, introduce an instantaneous stock price conversion factor $P_{\gamma\delta}$ by

$$S^{(\gamma)}(t) = P_{\gamma\delta}(t) S^{(\delta)}(t) \,. \tag{5.136}$$

The three equations for three stocks can only be satisfied when $P_{\beta\delta}(t) = P_{\beta\gamma}(t) P_{\gamma\delta}(t)$. These relations can be defined on an ensemble of H time horizons $T_1 < T_2 < \ldots < T_H$. Finally, logarithmic variations of these conversion factors with time are defined as

$$d_{\gamma\delta}^{\alpha}(t) = \frac{1}{T_\alpha} \ln \left[\frac{P_{\gamma\delta}(t)}{P_{\gamma\delta}(t - T_\alpha)} \right] \,. \tag{5.137}$$

Interestingly, the H-component vector $\boldsymbol{d}_{\gamma\delta} \equiv (d_{\gamma\delta}^1, \ldots, d_{\gamma\delta}^N)$ has all the properties required for an oriented distance vector between the assets γ and δ and, for any norm in Euclidean space, $||\boldsymbol{d}_{\gamma\delta}||$ is a well-defined distance between γ and δ. Assets with a small distance behave as strongly correlated.

Summing over one index, i.e., all shares in the market, generates position vectors

$$x_\gamma(t) = \frac{1}{N} \sum_{\delta=1}^{N} d_{\gamma\delta}(t) , \quad x_\gamma - x_\delta = d_{\gamma\delta} . \tag{5.138}$$

The temporal fluctuations of the assets translate into fluctuations of the conversion factors $P_{\gamma\delta}(t)$, their distances $d_{\gamma\delta}(t)$, and their positions $x_\gamma(t)$. We would thus obtain a mapping of the financial assets on the positions of particles in a gas or a liquid [132].

The standard deviation of the positions is

$$\sigma = \frac{1}{N} \sqrt{\sum_{1\leq\gamma<\delta\leq N} ||d_{\gamma\delta}||^2} \tag{5.139}$$

and, within this formalism, it plays a role reminiscent of the variety discussed above, and is a measure of the linear extension of the system. Transposing to financial markets, it is a measure of the heterogeneity of the market at a given time.

Using the time dimension, one can construct a temperature. To this end, the linear system size is scaled to unity through $r_\gamma = x_\gamma/\sigma$, and a velocity is defined as $v_\gamma(t) = [r_\gamma(t) - r_\gamma(t-T_1)]/T_1$. A temperature then is defined via $T = \langle\langle v_\gamma^2(t)\rangle_\gamma\rangle_t/H$. Finally, from the two-point pair correlation function, one can derive a pair potential for the particles. This potential possesses a long-range attractive tail and a short-range repulsive core. The long-range attractive tail confines the particles in a finite volume σ^H. Therefore, so long as σ is finite, they behave as a droplet of liquid. Such a mapping of financial market on droplets of liquids is possible both for small ensembles of assets such as the 30 stocks composing the DAX [132] as well as for larger ensembles, e.g., 2800 stocks traded at the New York Stock Exchange [133].

6. Turbulence and Foreign Exchange Markets

The preceding chapter has shown that, when looking at financial time series in fine detail, they are more complex than what would be expected from simple stochastic processes such as geometric Brownian motion, Lévy flights or truncated Lévy flights. One of the main differences to these stochastic processes is the *heteroscedasticity of financial time series,* i.e., the fact that their volatility is not a constant. While this has given rise to the formulation of the ARCH and GARCH processes [48, 49] briefly mentioned in Chap. 4.4.1, we here pursue the analogy with physics and consider phenomena of increased complexity.

6.1 Important Questions

The flow properties of fluids are such an area. In this chapter, we will discuss the following questions:

- How do fluid flows change as, e.g., their velocity is increased?
- Is there a phase transition between a slow-flow (laminar) and a fast-flow (turbulent) regime?
- What are the hallmarks of turbulence? What are its statistical properties?
- Are there models of turbulence?
- Are there similarities in the time series and in the statistical properties between turbulence and financial assets?
- Are the models of turbulence useful to formulate models for financial markets?
- Are there benchmark financial assets which are particularly well suited to study statistical and time series properties?
- Is there a relation to geometrical constructions such as fractals and multi-fractals, and is it useful?

6.2 Turbulent Flows

A good introduction to the field of turbulence has been written by Frisch [134]. We first introduce turbulence in a phenomenological way. In a second step, we discuss time series analysis of turbulent signals.

6.2.1 Phenomenology

The basic question is: how do fluids flow? The answer is not clear-cut, and depends on a control parameter, the Reynolds number

$$R = \frac{Lv}{\nu} \; . \tag{6.1}$$

Here, L is a typical length scale, v a typical velocity, e.g., $\sqrt{\langle v^2(\boldsymbol{x}) \rangle}$, and ν the kinematic viscosity. For incompressible flows in fixed geometry, the Reynolds number R is the only control parameter.

Then, in the limit $R \to 0$, laminar flow obtains. In the opposite limit, $R \to \infty$, one has turbulent flow. What happens in between is much less clear. Apparently, it is not clear to what extent the transition to turbulence is sharp or smooth, and even less the critical value R_c at which it might take place.

To illustrate this point, we consider a uniform flow with velocity $\boldsymbol{v} = v\hat{x}$, past a cylinder of diameter L, oriented along $\hat{z}$ [134]. In this simple case, the quantities L and v directly enter the numerator of the Reynolds number. A pictures of the resulting flow at small Reynolds number is shown in the upper panel of Fig. 6.1. $R = 1.54$ is typical for the laminar flow in the small-R limit. The fluid flows along the cylinder surface on both sides and closes behind the cylinder. As the Reynolds number is increased, say in the range $R \sim 10, \ldots, 20$, the flow detaches from the cylinder walls at the rear, and forms two countercirculating eddies. The bottom panel of Fig. 6.1 shows a rather extreme case ($R = 2300$) of the opposite limit. At very large Reynolds numbers, eddies of all sizes form in irregular structures behind the cylinder. The situation is rather similar to the bottom panel of Figs. 6.1. This picture shows a turbulent water jet emerging from a nozzle at $R \approx 2300$, and has been preferred for its photographic quality.

The basic equation for fluid flow is the Navier–Stokes equation

$$\partial_t \boldsymbol{v} + \boldsymbol{v} \cdot \boldsymbol{\nabla} \boldsymbol{v} = -\boldsymbol{\nabla} P + \nu \boldsymbol{\nabla}^2 \boldsymbol{v} \; . \tag{6.2}$$

The various terms have an immediate interpretation. The left-hand side is the total derivative $d\boldsymbol{v}/dt$ including two contributions: the explicit acceleration of fluid molecules within a small volume, and the change in velocity due to the flow, i.e., molecules entering and leaving the small reference volume with different velocities. The first term on the right-hand side is the external force (pressure gradient), and the second term represents friction. An incompressible fluid, in addition, has

$$\boldsymbol{\nabla} \cdot \boldsymbol{v} = 0 \; . \tag{6.3}$$

It is believed that these two equations are sufficient to describe turbulence. The problem is that there are no explicit solutions, and almost no exact information on their properties. Much of our information therefore comes from computer simulations.

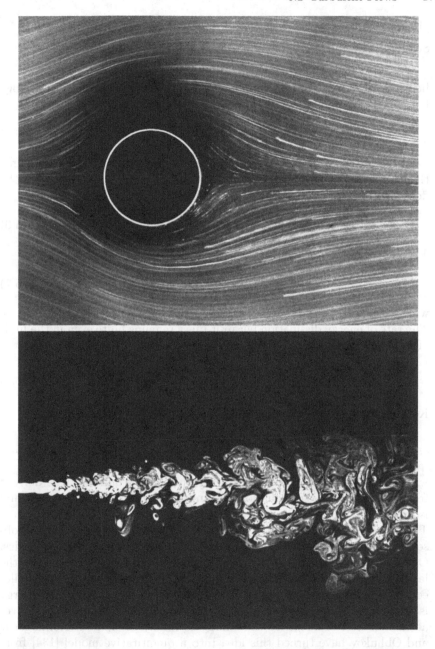

Fig. 6.1. Flow past a circular cylinder at $R = 1.54$ (*top panel*, photograph by S. Taneda). Turbulent water jet at $R \approx 2300$ (*bottom panel*, photograph by Dimotakis, Lye, and Papatoniou). Reprinted from M. Van Dyke (ed.): *An Album of Fluid Motion*, ©1982 Parabolic Press, Stanford

Here are a few important facts:

- Scaling: With

$$\delta v_{\parallel}(\ell) = [\boldsymbol{v}(\boldsymbol{r} + \boldsymbol{\ell}) - \boldsymbol{v}(\boldsymbol{r})] \cdot \hat{\boldsymbol{\ell}} \tag{6.4}$$

being the difference of the component of the velocity parallel to the flow, between two points along the flow direction, the structure function

$$S_2(\ell) = \langle \delta v_{\parallel}^2(\ell) \rangle \sim \ell^{2/3} \tag{6.5}$$

has power-law scaling with the distance of the points. At the same time, the same function involving the velocity components perpendicular to the flow direction scales with the same exponent

$$\langle \delta v_{\perp}^2(\ell) \rangle \sim \ell^{2/3} . \tag{6.6}$$

The energy spectrum then scales as

$$E(k) \sim k^{-5/3} , \tag{6.7}$$

where $k \sim \ell^{-1} \sim \omega$ is the wavenumber.

- The rate of energy dissipation per unit mass remains finite even in the limit of vanishing viscosity:

$$\frac{\mathrm{d}\varepsilon}{\mathrm{d}t} > 0 \ \text{ even for } \nu \to 0 . \tag{6.8}$$

- Kolmogorov theoretically derived

$$\langle \delta v_{\parallel}^3(\ell) \rangle = -\frac{4}{5} \varepsilon \ell \ \text{ for } R \to \infty , \tag{6.9}$$

which represents one of the few exact results on turbulence. It is derived from the Navier–Stokes equation, assuming in addition homogeneity and isotropy.

- The *cascade idea* is illustrated in Fig. 6.2. Here, one starts from the observation that turbulence generates eddies at many different length scales. One now assumes that external energy is injected into the eddies at the largest scale of the problem (injection scale). Eddies break up into smaller eddies which themselves break up into smaller eddies, etc., and energy is transferred from the big eddies into the small eddies, until one arrives at the smallest scale where the energy is finally dissipated. Kolmogorov and Obhukov have turned this idea into a quantitative model [134] from which, e.g., the scaling exponents of the various moments of the velocity differences, (6.5), (6.6), or (6.9), can be derived.

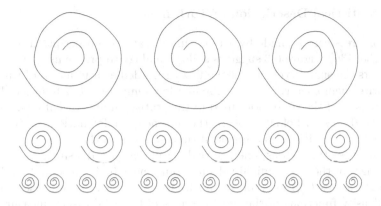

Fig. 6.2. The cascade idea. Energy is injected at the biggest length scale. Eddies at that scale break up into smaller eddies, transferring energy to smaller and smaller scale, until it is dissipated at the smallest scale, the dissipation scale

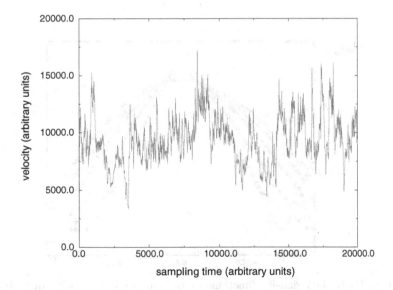

Fig. 6.3. Time series of a turbulent flow. The local velocity of a helium jet at low temperature has been recorded with a hot-wire anemometer. Data provided by J. Peinke, Universität Oldenburg

6.2.2 Statistical Description of Turbulence

Some progress can be made by attempting a statistical description of turbulence [135, 136]. Figure 6.3 suggests a close analogy to problems of finance: it represents the signal (velocity of the flow) recorded as a function of time, by a hot-wire anemometer, a local probe in a low-temperature helium jet. These data are part of the time series used in the statistical analysis of Chabaud et al., to be discussed below [135]. In the absence of information, it would be difficult to decide if this is a financial time series or not!

From these time series, one can deduce probability density functions for the changes of the longitudinal velocity component (6.4) measured on different length scales ℓ_i [135, 136]. In much the same way, we discussed probability density functions of the price changes of financial assets, measured on different time scales, e.g., Fig. 5.10. Figure 6.4 displays such a set of distribution functions. For large scales, the probability densities are approximately Gaussian, while they approach a more exponential distribution as the length scales are reduced.

Do these probability densities show scaling? We may rescale the distributions empirically as (the index $\parallel$ on δv will be dropped from now on)

$$P_\ell(\delta v) \rightarrow \frac{1}{\sigma_\ell} P_\ell \left(\frac{\delta v}{\sigma_\ell} \right) \, , \tag{6.10}$$

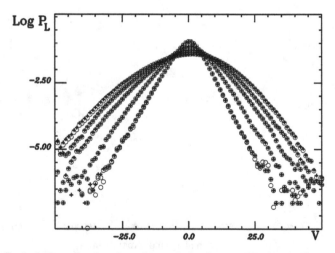

Fig. 6.4. Probability density functions of the longitudinal velocity in turbulent flows at different length scales, decreasing from top to bottom in the wings. $\ell_i = 424, 224, 124, 52, 24$ from top to bottom with $\ell_0 = 1024$. Circles are data points, and crosses have been obtained by iteration with the experimentally determined conditional probability density functions, starting at ℓ_0. By courtesy of J. Peinke. Reprinted from R. Friedrich and J. Peinke: Phys. Rev. Lett. **78**, 863 (1997), ©1997 by the American Physical Society

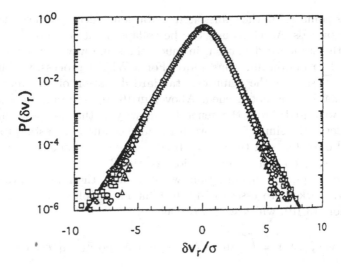

Fig. 6.5. Rescaled probability density function for the longitudinal velocity changes of turbulent flows. The *solid line* is a fit explained in the text. By courtesy of J. Peinke. Reprinted from B. Chabaud, et al.: Phys. Rev. Lett. **73**, 3227 (1994), ©1994 by the American Physical Society

where σ_ℓ is the empirical standard deviation at length scale ℓ. As shown in Fig. 6.5, all data now more or less collapse onto a single master curve, demonstrating that they obey the same basic laws, and are only distinguished by the different length scales of the measurement. The master curve has some similarity to a Gaussian in its center and is more like an exponential distribution in the wings. While there is no simple expression, it can be described as an integral over a continuous family of Gaussians [135]

$$\frac{1}{\sigma_\ell} P_\ell \left(\frac{\delta v}{\sigma_\ell} \right) = \int_{-\infty}^{\infty} d\ln\sigma \, G_\ell(\ln\sigma) \frac{1}{\sigma} P_{\ell_0} \left(\frac{\delta v}{\sigma} \right) \qquad (6.11)$$

with

$$G_\ell(\ln\sigma) = \frac{1}{\sqrt{2\pi\lambda(\ell)}} \exp\left(-\frac{\ln^2 \sigma/\sigma_{\ell_0}}{2\lambda(\ell)} \right). \qquad (6.12)$$

Notice that the empirical distribution function at the largest length scale, P_{ℓ_0}, is nearly Gaussian. The probability distribution at a smaller length scale $\ell < \ell_0$ is therefore represented by a weighted integral over Gaussians whose standard deviations are log-normally distributed. This integral over Gaussians describes the curves extremely well. The standard deviations σ_ℓ are scale dependent through the width $\lambda(\ell)$ of their distribution, and generate themselves from those on bigger length scales. This directly implements the cascade idea. However, Kolmogorov's theory predicts that $\lambda \propto -\ln(\ell/\ell_0)$ which is not observed in the experiment.

At this point, recall the Gaussian distribution which describes the Wiener stochastic process. All Gaussians can be collapsed onto each other by rescaling with the standard deviation, in much the same way as we rescaled the empirical distribution functions above. For a Wiener process, we know that $\sigma \propto \sqrt{T-t}$, so that the empirical standard deviation of an eventually incomplete data set is not needed. Above, in (6.10), the empirical standard deviation was used. From the general similarity of the two procedures, one may wonder if the similarity between turbulence and stochastic processes is superficial only. Or: can turbulence be described as a stochastic process in length scales, instead of (or in addition to) time?

In order to pursue this question, we first check the Markov property, cf. Sect. 4.4.1. Markov processes satisfy the Chapman–Kolmogorov–Smoluchowski equation, (3.10), which we rewrite as

$$p(\delta\tilde{v}_2, \lambda_2 | \delta\tilde{v}_1, \lambda_1) = \int_{-\infty}^{\infty} d\tilde{v}_3 \, p(\delta\tilde{v}_2, \lambda_2 | \delta\tilde{v}_3, \lambda_3) p(\delta\tilde{v}_3, \lambda_3 | \delta\tilde{v}_1, \lambda_1) \ . \qquad (6.13)$$

λ has been defined above, and the velocities have been rescaled as

$$\delta\tilde{v} = \delta v \sqrt[3]{\ell_0/\ell} \ . \qquad (6.14)$$

This form of rescaling is suggested by theoretical arguments [134, 136]. Under the assumption that the eddies are space-filling, and that the downward energy flow is homogeneous, on can derive a scaling relation $\xi_n = n/3$ between the exponents ξ_n of the structure functions, and their order n which is rather well satisfied by the data at least for small n, cf. Fig. 6.7 below.

Indeed, the empirical data satisfy the Chapman–Kolmogorov–Smoluchowski equation: one can superpose the conditional probability density function $p(\delta\tilde{v}_2, \lambda_2 | \delta\tilde{v}_1, \lambda_1)$ derived from the experimental data, with that calculated according to (6.13), using experimental data on the right-hand side of the equation. The result very well matches the $p(\delta\tilde{v}_2, \lambda_2 | \delta\tilde{v}_1, \lambda_1)$ measured directly. As a consequence, one is allowed to iterate the experimental probability density function for ℓ_0 (not shown in Fig. 6.4) with (6.13) and the experimentally determined conditional probability distributions. The results of this procedure (shown as crosses in Fig. 6.4) exactly superpose the experimental data on scales $\ell < \ell_0$, shown as circles.

When the Chapman–Kolmogorov–Smoluchowski equation is fulfilled, one may search for a description of the scale-evolution of the probability density functions in terms of a Fokker–Planck equation. Quite generally, one can convert the convolution equation (6.13) into a differential form by a Kramers–Moyal expansion [37],

$$\frac{\partial p(\delta\tilde{v}_2, \lambda_2 | \delta\tilde{v}_1, \lambda_1)}{\partial \lambda_2} = \sum_{n=1}^{\infty} \left[-\frac{\partial^n}{\partial(\delta\tilde{v}_2)^n} D^{(n)} \left(\delta\tilde{v}_2, \lambda_2 \right) p(\delta\tilde{v}_2, \lambda_2 | \delta\tilde{v}_1, \lambda_1) \right] \ .$$

$$(6.15)$$

The Kramers–Moyal coefficients are defined as

$$D^{(n)}(\delta\tilde{v}_2, \lambda_2) = \frac{1}{n!} \lim_{\lambda_3 \to \lambda_2} \frac{1}{\lambda_3 - \lambda_2} \int_{-\infty}^{\infty} d(\delta\tilde{v}_3)(\delta\tilde{v}_3 - \delta\tilde{v}_2)^n p(\delta\tilde{v}_3, \lambda_3|\delta\tilde{v}_2, \lambda_2) .$$

$$(6.16)$$

If all $D^{(n)}$ with $n > 2$ vanish, one obtains a Fokker–Planck partial differential equation

$$\frac{\partial p(\delta\tilde{v}_2, \lambda_2|\delta\tilde{v}_1, \lambda_1)}{\partial \lambda_2} = \left[-\frac{\partial^n}{\partial(\delta\tilde{v}_2)} D^{(1)}(\delta\tilde{v}_2, \lambda_2) \right. \tag{6.17}$$

$$\left. + \frac{\partial^2}{\partial(\delta\tilde{v}_2)^2} D^{(2)}(\delta\tilde{v}_2, \lambda_2) \right] p(\delta\tilde{v}_2, \lambda_2|\delta\tilde{v}_1, \lambda_1) ,$$

and the stochastic process is completely characterized by the drift and diffusion "constants" $D^{(1)}$ and $D^{(2)}$. It turns out that, within experimental accuracy, this is the case, and

$$D^{(1)}(\delta\tilde{v}, \lambda) \sim -\delta\tilde{v} , \tag{6.18}$$

$$D^{(2)}(\delta\tilde{v}, \lambda) \sim a(\lambda) + b(\lambda)(\delta\tilde{v})^2 . \tag{6.19}$$

Great care must be taken when estimating these quantities from actual data. In particular, for a discrete, finite sampling interval contributions from the drift terms $D^{(1)}$ may contaminate the esimators of $D^{(2)}$ and lead to incorrect estimates of the parameters in (6.19) [137]. This may have affected actual estimates [136, 137], but our general conclusions are robust. A related observation has been made from the perspective of integration of stochastic differential equations by Timmer [138].

In this perspective, turbulence is described as a stochastic process over a hierarchy of length scales. The drift term contains the systematic downward flow of energy postulated by the cascade model. The diffusion term describes the fluctuations around the otherwise deterministic cascade [136], and shows that there is a strong random component in this energy cascade. This is connected with the indeterminacy of the number and size of the smaller eddies produced from one big eddy, as one drifts down the cascade.

6.2.3 Relation to Non-extensive Statistical Mechanics

When the evolution of the probability density of a stochastic process is described by a Fokker–Planck equation, an equivalent stochastic differential equation for the stochastic variable can be found, and it takes the form of a Langevin equation, (5.86) [37]. A general form of a non-linear Langevin equation is [93]

$$\frac{dx}{dt} = -\gamma F(x) + \sigma\eta(t) . \tag{6.20}$$

It is not necessary to consider explicitly a non-linearity in the diffusion term, as it can be reduced to a constant by a transformation of variables [37]. $F(x) = -\partial U(x)/\partial x$ is the non-linear force. If $U(x) = C|x|^{2\alpha}$, the non-linear

Langevin equation generates one of the power-law probability distributions of non-extensive statistical mechanics, cf. Sect. 5.5.7. The parameters are identified as

$$q = 1 + \frac{2\alpha}{\alpha n + 1} , \quad \tilde{\beta} = \frac{2\alpha}{1 + 2\alpha - q}\beta_0 . \tag{6.21}$$

As in Sect. 5.5.7, $\tilde{\beta} = 1/k_B T$ is the inverse temperature and n the number of degrees of freedom of the χ^2-distribution used to describe the slow temporal fluctuations of the parameters of the Langevin equation [93].

Assume now that a test particle in a turbulent flow moves for a while in a region with a (fluctuating) energy-dissipation rate ϵ_r on scale r. τ is a typical time scale during which energy is dissipated, typically the time of sojourn in the region with ϵ_r. Then $\beta = \epsilon_r \tau \Lambda$ is a fluctuating quantity, and as a model we may assume that it is χ^2-distributed. Λ is a constant necessary to adjust the dimensions of β. At the smallest scale, the dissipation scale η, $\beta = u_\eta^2 \Lambda$, where u_η is a fluctuating velocity. In the simplest model, the three components of u_η would fluctuate independently and would be drawn from Gaussian distributions with mean zero. This would suggest that $n = 3$, and with $\alpha \approx 1$ (weak non-linearity of the forcing potential) would give $q \approx 3/2$. These values are in very good agreement with experimental data on velocity differences of a test particle over very small time scales in turbulent Taylor–Couette flow with a Reynolds number $R = 200$ [93, 139]. The probability distributions observed there are rather similar to those depicted in Figs. 6.4 and 6.5. At the dissipation scale, the fluctuations of $\epsilon_{r=\eta}$ can be viewed in terms of the ordinary diffusion of a particle of mass M which is subject to noise of a temperature $T = 1/k_B\tilde{\beta}$. The changes of the distributions as the spatial scale of the experiments is varied are embodied in different values of n, q, and $\tilde{\beta}$. Tsallis statistics allows us to relate one to another. This discussion suggests that Tsallis statistics is applicable to systems with fluctuating energy-dissipation rates.

6.3 Foreign Exchange Markets

6.3.1 Why Foreign Exchange Markets?

Foreign exchange markets are extremely interesting for statistical studies because of the number and quality of data they produce [102, 140]. The markets have no business-hour limitations. They are open worldwide, 24 hours a day including weekends, except perhaps a few worldwide holidays. Trading is essentially continuous, the markets (at least for the most frequently traded currencies) are extremely liquid, and the trading volumes are huge. Daily volumes are of the order of US\$ 10^{12}, approximately the gross national product of Italy. Typical sizes of deals are of the order of US\$ 10^6–10^7, and most of the deals are speculative in origin. As a consequence of the liquidity, good databases contain about 1.5 million data points per year, and data have been collected for many years.

6.3.2 Empirical Results

Ghashghaie et al. [141] analyzed high-frequency data consisting of about 1.5×10^6 quotes of the US$/DEM exchange rate taken from October 1, 1992 until September 30, 1993. The probability density function for price changes δS_τ over a time scale τ is shown in Fig. 6.6. The time scale τ of the returns increases from top to bottom, and the curves have been displaced vertically, for clarity. Both the fat tails characteristic of financial data, and the similarity to the distributions in turbulence, e.g., Fig. 6.4, are apparent. Specifically, one can notice a crossover from a more tent-shaped probability density function at short time scales to a more parabolic (Gaussian) one at longer scales. This would imply that the probability density function is *not* form-invariant under rescaling, as was found, at least for not too long time scales, in the analysis of stock market data [62] discussed in Sect. 5.6.1.

There are more analogies between foreign exchange markets and turbulence. For example, one can investigate the scaling of the moments of the

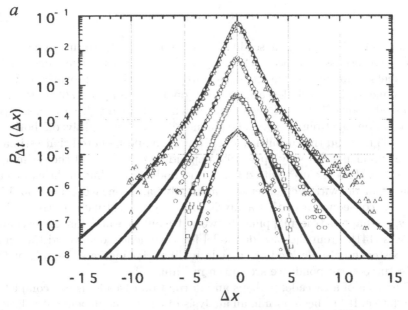

Fig. 6.6. Probability density function for variations of the US$/DEM exchange rate for time delays $\Delta t \equiv \tau = 640\,\mathrm{s}$, $5120\,\mathrm{s}$, $40\,960\,\mathrm{s}$, and $163\,840\,\mathrm{s}$ (*top to bottom*). The *full lines* are fits using integrals over Gaussian distributions. The identification of the legends with our text is: $\Delta x \equiv \delta S_\tau$ and $\Delta t \equiv \tau$. By courtesy of W. Breymann. Reprinted by permission from Nature **381**, 767 (1996) ©1996 Macmillan Magazines Ltd.

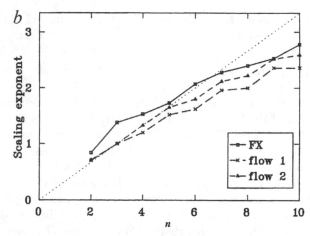

Fig. 6.7. Dependence of the scaling exponents ξ_n of the nth moment of the probability densities on its order for foreign exchange markets and turbulent flows. The *dotted line* is $\xi_n = n/3$. By courtesy of W. Breymann. Reprinted by permission from Nature **381**, 767 (1996) ©1996 Macmillan Magazines Ltd.

distribution function with time scale (referring to the financial data)

$$\langle |\delta S_\tau|^n \rangle \sim \tau^{\xi_n} \ . \tag{6.22}$$

Examples of the equivalent scaling behavior in turbulence, involving $\delta v(\ell)$, have been discussed in Sect. 6.2.1. Figure 6.7 shows the dependence of the exponents found in foreign exchange rates and turbulence on the order of the moment. The turbulence data start on the line $\xi_n = n/3$ at small n, discussed above [141], and then bend downward, in rough agreement with a prediction by Kolmogorov. The financial data are slightly off both the $\xi_n = n/3$ line and the turbulence data, but it should be noted that estimates of the exponents can vary up to 30% depending on details of the estimation procedure. However, even with different methods, the scaling of the exponents of the moments with their order systematically has a concave shape [140, 142].

As a consequence of this analysis, one would postulate a strong similarity, perhaps a true mapping, between turbulence and foreign exchange markets [141]. From the cascade model for turbulence, one would then infer the existence of some kind of cascade in financial markets. Details of this conjectured correspondence are shown in Table 6.1.

The idea of a cascade, perhaps an information cascade, is not completely speculative. It has been born in an analysis of time-scale-dependent volatility in FX and commodity markets in the economics literature [143], and has also been hypothezized for the S&P500 stock market index [144]. As we have seen in the previous chapter, volatility is a long-time-correlated variable. It therefore can be predicted, in principle. Obviously, the better the stochastic

Table 6.1. Postulated correspondence between fully developed three-dimensional turbulence and foreign exchange markets. Adapted from [86]

Hydrodynamic Turbulence	Foreign Exchange Markets
Energy	Information
Spatial distance	Time delay
Intermittency (laminar periods interrupted by turbulent bursts)	Volatility clustering
Energy cascade in space hierarchy	Information cascade in time hierarchy
$\langle|\delta v|^n\rangle \sim \ell^{\xi_n}$	$\langle|\delta S|^n\rangle \sim \tau^{\xi_n}$

volatility process and its driving mechanisms are understood, the better a prediction one can hope to generate.

In a heterogeneous market, the different types of traders present, e.g., long-term investors, day traders, etc., in general act with different time horizons. A day trader will observe market volatility on a very short scale. On the other hand, a long-term investor will not watch the market often enough to even perceive short-term volatility. The question of how statistics reflects the various types of operators in the marketplace then is reduced to the correlations between the volatilities characterizing the various actors. In FX and commodity markets, Müller et al. [143] have studied the correlation of finely defined volatility with coarsely defined volatility.

We define the finely and coarsely defined volatilities by absolute values of return and, to be specific, use a one-week time scale

$$\sigma^{\text{fine}}(N) = \frac{1}{5}\sum_{i=1}^{5}|\delta S_{1\text{d}}(N,i)| \quad \text{and} \quad \sigma^{\text{coarse}}(N) = |\delta S_{1\text{w}}(N)| . \quad (6.23)$$

The fine volatility is the sum of daily volatilities while the coarse volatility is the weekly return directly. N labels weeks and, where necessary, i labels the business days of the week. A lagged correlation

$$\rho_\tau = \frac{\langle[\sigma^{\text{coarse}}(N+\tau) - \langle\sigma^{\text{coarse}}(N+\tau)\rangle][\sigma^{\text{fine}}(N) - \langle\sigma^{\text{fine}}(N)\rangle]\rangle_N}{\sqrt{\text{var}[\sigma^{\text{coarse}}(N)]\,\text{var}[\sigma^{\text{fine}}(N)]}} \quad (6.24)$$

measures the correlation of the coarse volatility with the fine volatility τ weeks earlier. Empirically, it turns out that $\rho_\tau - \rho_{-\tau} < 0$ for $\tau > 0$ quite generally [143]. This implies that the coarse volatility predicts the fine volatility better than vice versa. This result is observed both for daily data (assumed in the equations above) and for high-frequency intraday data. It can be explained by a hypothesis of heterogeneous markets: coarse volatility matters both for a long-term investor and for a day trader. It will set the overall scale for the latter, and a day trader will take different positions depending on the level of volatility. On the other hand, short-term volatility is only important for the short-term trader.

More formally, in complete analogy to turbulence, one can search for a stochastic process across time *scales* in foreign exchange markets. It may not surprise the reader that indeed the probability density functions of the US\$/DEM exchange rate satisfy the Chapman–Kolmogorov–Smoluchowski equation and allow one to reduce it to a Fokker–Planck equation [145]. Differences are only found in details, such as the precise functional form of the rescaling of δS_τ. The drift and diffusion constants are found to be

$$D^{(1)}(\delta S, \tau) = -0.93\,\delta S\,, \tag{6.25}$$

$$D^{(2)}(\delta S, \tau) = 0.016\tau + 0.11(\delta S)^2\,. \tag{6.26}$$

The numerical prefactor of τ in (6.26) is given in units of days. The comparison of the scale dependence of the "experimental" probability density function with the one obtained by solving the Fokker–Planck equation using the appropriate empirical probability density function for long time scales is shown in Fig. 6.8.

In fact, the numbers given in (6.25) and (6.26) above are not the original results of Friedrich et al., but have been corrected by improved data analysis and a more robust fitting procedure, based on conditional instead of unconditional probability distributions and accounting explicitly for possible observational noise [146]. Firstly, one may calculate the power spectrum of the time series

$$\Pi(\omega) = \int_{-\infty}^{\infty} dt\, e^{i\omega t} \langle S(t)S(0)\rangle\,. \tag{6.27}$$

For approximately two decades in frequency, it decreases as ω^{-2} before leveling off to a constant, i.e., white noise, at high frequency. The presence of white noise suggests that the signal may be composed of two components, the intrinsic signal with $\Pi(\omega) \sim \omega^{-2}$, and white observational noise. The presence of observational noise in similar financial data has been shown independently in an investigation using more traditional approaches of time-series analysis [147]. However, there observational noise is found neither in the prices nor in the returns but in the time series of squared returns.

As a consequence of observational noise, one should work with a smoothed signal where this observational noise has been averaged out. The width of the averaging window defines a minimal time scale of 4 minutes. With this analysis, the expressions for $D^{(1)}(\delta S, \tau)$ and $D^{(2)}(\delta S, \tau)$ are obtained. The tail index μ of the unconditional probability distribution can also be calculated from the drift and diffusion coefficients. A value of $\mu = 4.2 \pm 0.8$ is obtained, quite in the range of the data analyzed in Sect. 5.6 [146, 148]. It is not clear to what extent this improved analysis still is affected by possible errors in $D^{(2)}$ related to the finite time scales used [137, 138].

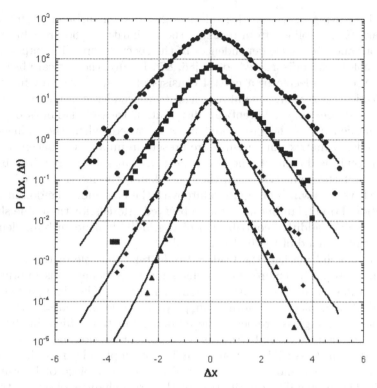

Fig. 6.8. Probability densities for variations of the US$/DEM exchange rates (*dots*) compared to the solutions of a Fokker–Planck equation (*solid lines*) with the initial distribution taken as the one for $\tau = 40\,960$ s. Time delays were $\tau = 5120$ s, $10\,240$ s, $20\,480$ s, and $40\,960$ s. Both the data set and the notation are the same as in Fig. 6.6. By courtesy of J. Peinke. Reprinted from Friedrich et al.: Phys. Rev. Lett. **84**, 5224 (2000), ©2000 by the American Physical Society

Quite some time before this work, possible analogies between hydrodynamic turbulence and financial time series [141] have been questioned because of different (and, in some instances, less well defined) power-law scaling in the S&P500 index and air-flow data at $R = 1500$ [149]. One interesting aspect of this work is that the power spectrum of the S&P500 data is ω^{-2} in the entire frequency range considered, and no crossover to observational noise is observed. This may be related to the time scale $\tau = 1$ hour of the S&P500 returns analyzed.

With a Fokker–Planck equation for the probability distributions of financial data at hand, it would be interesting to search for improvements, e.g., in the theory of option pricing, etc., to include the effects of non-Gaussian statistics. This will be pursued further in Sect. 7.6 below.

An interesting phenomenological analogy between turbulence and financial markets also follows from realizing the similarity of the probability distributions and their scale dependences to the spectroscopic lineshapes of impurity molecules in disordered solids [86, 87]. Ideally, the optical absorption spectrum of a molecule in a crystal consists of a series of delta functions. Imperfections is real systems always lead to a broadening. There, a change of the lineshape from a Lorentzian distribution to a Gaussian is observed when the density of the disordered units is varied: when the influence of the disordered matrix units (which are present in important concentrations) on a molecule is dominant, the lineshape is a Gaussian, as required by the central limit theorem. When, on the other hand, the interaction of certain two-level systems which are quite dilute with the host molecule dominate its absorption, Lorentzian lineshapes are observed. Models for these line shapes usually assume additive contribution of the individual perturbing elements in the neighborhood of the molecules probed.

In a financial market, the traders would take the role of the dye molecules in glasses. The environment influencing their behavior is information which becomes available at various moments of time. The time passed since the arrival of a piece of information plays the role of spatial distance in the molecule-in-a-glass problem. The influence function which, in the spectroscopy problem, is taken by the dipole–dipole interaction, becomes a memory function $af(t - t')$ in a market. a is the amplitude, t is the time of a trading decision, and t' is the time of arrival of a piece of information [86, 87]. The probability distribution of the price changes observed then is determined by the functional form of $f(t - t')$. If the frequency of information arrival is large with respect to the inverse time scale of the returns under consideration, the precise form of $f(t - t')$ does not matter: the central limit theorem requires that the resulting probability distribution will be Gaussian, independently of the details of the memory functions. On the other hand, when the frequency of information arrival is low or the time scale of the returns short enough, the functional form of $f(t - t')$ matters. For example, for an exponential memory kernel, the short-time probability distributions have very flat wings with a pronounced spike at zero return. Such spikes are not observed in real markets, but they were generated in numerical simulations of artificial financial markets to be discussed in Sect. 8.3.2. On the other hand, for a stretched-exponential decay in the memory kernel, a set of time-scale-dependent probability distribution functions similar to Fig. 6.6 with a truncation in the wings were obtained. Finally, for an algebraic memory function, the probability distributions at short times were of the form of the truncated Lévy distributions discussed in Sect. 5.4.4. The interesting conclusion from this work is that, in terms of fundamental analysis, traders would account for, resp. the market would reflect, information with a memory which is scale-free (stretched-exponential or power-law memory function) [86, 87]. In turbulence, the role of the dye molecule/trader would be played by the measurement device (anemometer), and that of the perturber would be taken by the eddies depicted in Fig. 6.1.

6.3.3 Stochastic Cascade Models

The idea that turbulent flows or foreign exchange markets are described by a stochastic cascade across spatial or time scales can be formalized. Here, we restrict ourselves to capital markets [144, 150]. Our discussion in the preceding section is equivalent to postulating for returns on a scale τ

$$\delta S_\tau(t) = \sigma_\tau(t)\varepsilon(t) . \tag{6.28}$$

Here, $\varepsilon(t)$ is a scale-independent random variable, and $\sigma_\tau(t)$ is a positive random variable depending on the scale, and identified with the standard deviation on that scale τ. There is a hierarchy of scales $\tau_0 = T > \cdots > \tau_k > \cdots > \tau_N$. If the cascade is purely multiplicative, the σs are related by

$$\sigma^{(k)}(t) = a^{(k)}(t)\sigma^{(k-1)}(t) , \quad \sigma^{(k)}(t) \equiv \sigma_{\tau_k}(t), \tag{6.29}$$

with time-dependent random factors $a^{(k)}(t)$. If our discussion of turbulent signals by a cascade in Sect. 6.2.2, and a similar analysis of foreign exchange quotes underlying the solid lines in Fig. 6.6, are rephrased in terms of (6.29), the probability distribution of the $a^{(k)}(t)$ is log-normal with a k-dependent width, cf. (6.12). This gives for the volatility at scale τ_m

$$\sigma^{(m)} = \sigma^{(0)} \prod_{k=1}^{m} a^{(k)}(t) . \tag{6.30}$$

One particularly simple realization would be a geometric progression in the inverse scales, e.g., $\tau_k = \tau_{k-1}/2$, and to associate two random numbers $a_1^{(k)}$ and $a_2^{(k)}$ with the passage from one level to the next lower [144].

In this model, there is a definite direction for the net flow of information from large to small scales. Namely, one can calculate the cross-correlation coefficient [144]

$$C_{\tau_m,\tau_n}(\Delta t) = \frac{\langle \ln \sigma^{(m)}(t) \ln \sigma^{(n)}(t + \Delta t)\rangle}{\mathrm{var}(\ln \sigma^{(m)})\mathrm{var}(\ln \sigma^{(n)})} . \tag{6.31}$$

One finds that $C_{\tau_m,\tau_n}(\Delta t) > C_{\tau_m,\tau_n}(-\Delta t)$ if $\tau_m > \tau_n$ and $\Delta t > 0$. This can be interpreted as a flow of information contained in $\ln \sigma^{(n+\Delta)}(t)$ to $\ln \sigma^{(n)}(t + \Delta t)$.

More sophisticated updating schemes for the random numbers $a^{(k)}$, relating the volatilities on neighboring time scales, have been devised [150]. At t_0, one draws the $a^{(k)}(t_0)$ from a log-normal distribution with k-dependent width. In later time steps, the factors at the top of the hierarchy, $a^{(1)}(t_{n+1})$, are updated with a certain probability, again from a log-normal distribution. If this factor is updated, all lower-level factors $a^{(k>1)}(t_{n+1})$ are also updated. If the top-level factor was not updated at t_{n+1} the next-level factor will be updated only with a certain, level-dependent probability, and so on. These level-dependent probabilities are small near the top level, leading to very few updates, and increase as one descends the cascade to give rather frequent updates there. As shown in Fig. 6.9, when the parameters of the model are suitably fixed, a numerical simulation can reproduce very well the observed probability distributions of the US\$–Swiss Franc exchange rates over time scales from 1 hour to 4 weeks [150]. With optimized parameters, the model also reproduces other important features of the data set such as, e.g., the slow decay of the autocorrelation function of absolute returns [150].

An alternative to cascade models is provided by a variant of the ARCH processes, the HARCH process [143]. Applying it to FX data, it turns out that seven market components, each with a characteristic time scale τ_n, are both necessary and sufficient to provide an adequate description of the lagged coarse–fine volatility correlations.

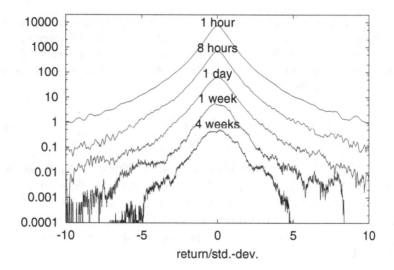

Fig. 6.9. *Dots*: distribution of returns of the US\$–Swiss Franc exchange rate for time horizons ranging from 1 hour to 4 weeks. *Full lines*: simulation of the stochastic cascade model described in the text, with optimized parameters. Data are offset for clarity. By courtesy of W. Breymann. Reprinted from Breymann et al.: Int. J. Theor. Appl. Financ. **3**, 357 (2000), ©2000 by World Scientific

6.3.4 The Multifractal Interpretation

Fractals and Multifractals

Geometry is the most popular area for thinking about fractals [33]. While ordinary macroscopic bodies such as spheres, cubes, cones, etc., are characterized by a small surface-to-volume ratio (surface scales as L^2 and volume as L^3, with L the linear dimension of the system), there are other objects with large surface-to-volume ratio. They look porous, ragged, hairy, and often play a fundamental role in natural phenomena. Examples are sponges, the human lung, the landmass on earth, dendrites, the fault structure of the earth's crust, or river basins. These systems are fractals. Fractals are scale-invariant over several orders of magnitude of size, i.e., their observed volume depends on the resolution with a power law. On the contrary, regular bodies are not scale-invariant, and their observed volume does not essentially depend on resolution.

We introduce a grid with cell size ℓ. Then, the observed volume of a fractal is the number of cells filled (partially or totally) by the object. With resolution defined as $\varepsilon = \ell/L$, the observed volume scales as $N(\varepsilon) \sim \varepsilon^{-D_0}$, where D_0 is the fractal dimension of the object [151]. The simplest mathematical fractals are built by the repetitive application of a generator to an initiator: the Cantor set, e.g., takes as initiator the interval $[0, 1]$, and the generator wipes out the central third of this line, yielding $\{[0, 1/3], [2/3, 1]\}$ in the first stage, $\{[0, 1/9], [2/9, 1/3], [2/3, 7/9], [8/9, 1]\}$ in the second stage, etc. The Cantor set has $D_0 = \ln 2/\ln 3$, and is an example of a deterministic monoscale fractal.

Three successive generalizations lead us from geometry to stochastic time series [151]. The first one is the introduction of several scales, producing multiscale fractals: the initiator is now divided into unequal parts. For the example of the Cantor set, we can construct a first stage as $\{[0, r_1], [r_2, 1]\}$. The second one is fractal functions. These are simple functions of an argument (perhaps time) which are nowhere differentiable. Their graph is a fractal curve. An example is the Weierstrass–Mandelbrot function [33]

$$C(t) = \sum_{n=-\infty}^{\infty} \frac{1 - \cos(\gamma^n t)}{\gamma^{(2-D_0)n}}. \tag{6.32}$$

The third generalization is randomness. For a multiscale fractal, one might choose randomly which rule to apply from a menu of choice. Randomness can also be introduced into fractal functions. An example is the fractional Brownian motion introduced in Sect. 4.4.1. The fractal dimension of the graph is related to the Hurst exponent H by $D_0 = 2 - H$.

A physical process on a fractal support may generate a stationary distribution. A fractal measure is a fractal with a time-independent distribution attached to it [151], e.g., the voltage distribution of a random resistor network. Such distributions can be used to analyze the fractal by opening fractal subsets, e.g., by selecting the subset which gives the dominant contribution to the nth moment of the distribution. If this is the case, the system will be called multifractal. Returning to the example of the Cantor set, instead of eliminating the central third of the initiator, we may attach two different probabilities p_1 and $p_2 = 1 - p_1$ to the extremal and central thirds of the initiator, respectively. By iterating this rule an infinite number of times, a probability distribution which is discontinuous everywhere, a multifractal measure, is generated.

Multifractal Time Series

The generation of a multifractal time series is best illustrated with a specific example devised for financial markets. One multifractal model for the time series of asset returns has been defined by [22], [152]–[155]

$$\delta S_\tau(t) = B_H\left[\Theta(t)\right] . \tag{6.33}$$

Here, B_H describes fractional Brownian motion with a Hurst exponent H, cf. (4.42), and $\Theta(t)$ is a multifractal time deformation.

A non-fractal time series $x(t_n) = x_0 + \sum_{m=0}^{n} \Delta x(t_n)$, say Brownian motion, is constructed by adding increments $\Delta x(t_n) = \varepsilon(t_n)\sqrt{\Delta t}$ with $\Delta t = t_n - t_{n-1} = \text{const.}$, or the corresponding continuum limit, cf. (4.22) and (4.23). This can be generalized to increments scaling as $\Delta x(t) = \varepsilon(t)(\Delta t)^H$, at least in the sense of expectation values (cf. Sect. 4.4.1 for the case of ordinary Brownian motion). Fractional Brownian motion corresponds to a non-trivial Hurst exponent $H \neq 1/2$ [155]. A generalization allowing for non-constant t-dependent exponents $H(t)$ then defines the increments of a *multifractal time series*

$$\Delta x_{\text{mf}}(t) = \varepsilon(t)(\Delta t)^{H(t)} . \tag{6.34}$$

$H(t)$ may be a deterministic or a random function.

Take as the initiator again the line interval $[0,1]$. In a binomial cascade, divide the interval into two subintervals of equal length and assign fractions p_1 and $p_2 = 1 - p_1$ of the total probability mass to the subintervals. Then repeat this process *ad infinitum*. In a log-normal cascade, at each iteration step, p_1 is random and is drawn from a log-normal distribution. The results of this cascade, with each subinterval interpreted as a time step, define a multifractal time series. In the model formulated by Mandelbrot, Calvet, and Fisher, this time series is used as a transformation device from chronological time t_n to a multifractal time $\Theta(t)$. The values of the final iteration of the cascade are interpreted as the increments of a (positive-valued) stochastic time process $\Delta\Theta(t)$. $\Theta(t)$, which is an irregularly increasing function of chronological time t, can be interpreted as a trading time [155]. Tick-by-tick data of real financial markets show that the trading activity is very non-stationary, and that there are periods of hectic trading alternating with quiescent periods. $\Theta(t)$ would increase very quickly in periods of heavy trading, and more slowly when the tick-to-tick interval is rather large. The existence of such a Θ-time has been demonstrated empirically in foreign exchange markets [156] and used for asset-pricing theories [157].

As a last step, the multifractal model of asset prices (6.33) uses this deformed time as the driver of a fractional Brownian motion process [152]–[155]. Empirically, however, the statistical evidence for a non-trivial Hurst exponent $H \neq 1/2$ seems to be rather weak, and one may as well inject the multifractal Θ-time series into ordinary Brownian motion, $H = 1/2$ [158].

At the level of multifractal stochastic processes, it is important to notice one important difference to ordinary (non-fractal and monofractal) processes. In an ordinary stochastic process, the subsequent value of the random variable can be determined from the past time series and an "innovation", a new random increment, and the time series can be continued as long as one wishes. For multifractal time series as they have been formulated to date, the entire time series is constructed in one shot. In the case above, this applies in particular to the cascade generating the multifractal Θ-time, while the stochastic process driven by Θ-time obeys the usual rules.

Our discussion very much emphasized the construction of a multifractal stochastic process. Alternatively, one can simply define a multifractal stochastic process by its statistical properties, as do Mandelbrot, Calvet, and Fisher [152]–[154]: *a stochastic process $\delta S_\tau(t)$ is called multifractal if it satisfies the scaling property*

$$\langle |\delta S_\tau(t)|^n \rangle = c(n)\, \tau^{\xi_n} \, . \tag{6.35}$$

This brings us to the statistical properties of multifractals.

Multifractal Statistics

Let us return to the multifractal generated by attaching probabilities p_1 and $p_2 = 1 - p_1$ to the generator of the Cantor set. We first ask which regions of $[0, 1]$ give the main contribution to the total probability. The locus of these regions (boxes) defines a fractal subset with a dimension f_1 [151]. The number of such boxes scales with ε as $N_1(\varepsilon) \sim \varepsilon^{-f_1}$. For the specific case of the binomial cascade, we have $f_1 = -(2p_1 \ln p_1 + p_2 \ln p_2)/\ln 3$. The dominant contributions to the nth moment of the distribution define different fractal subsets, each with its specific fractal dimension f_n, which can be calculated [151]. f_n describes the nth fractal subset of the multifractal, i.e., the support of a distribution, but not the distribution itself.

This probability distribution is described by another set of exponents, α_n, called crowding indices or Hölder exponents. We write the probability in a specific box in the lth iteration as $P_m = p_1^m p_2^{l-m}$. Then, the dominant contribution to the total probability defining the first fractal subset comes from regions with a single specific index m_1. P_{m_1} scales with box size as

$P_{m_1} \sim \varepsilon^{\alpha_1}$, defining α_1. The idea in this procedure is straightforward (assume $p_1 > p_2$). On the one hand, the maximal probability is in the cell with p_1^l, but the weight of this cell is $1/l$ and thus negligible. On the other hand, the most rarified boxes are numerous, but their probability mass is too small. The dominant contribution will thus come from cells with some intermediate values of m, and it turns out that, for $l \to \infty$, only a single index m_1 contributes [151]. The higher α_n then are defined through a similar procedure using the higher fractal subsets. Eliminating n from f_n and α_n generates a relation $f(\alpha)$. The spectra (f_n, α_n) or the $f(\alpha)$ spectrum characterize a given multifractal. The relation $\alpha_n = 1/H_n$ relates the Hölder exponents α_n to the generalizations H_n of the Hurst exponent, often also called Hölder exponents.

The $f(\alpha)$ spectrum also determines the structure function $\chi_n(\varepsilon)$, which is defined as

$$\chi_n(\varepsilon) = \sum_j^{N(\varepsilon)} P_j^n , \qquad (6.36)$$

that is, the sum over the nth-power box probabilities. Using the scaling laws found above, this can be rewritten as

$$\chi_n(\varepsilon) = \sum_j N_n(\varepsilon, f_k)(\varepsilon^{\alpha_k})^n \sim \sum_k \varepsilon^{\alpha_k n - f(\alpha_k)+1} . \qquad (6.37)$$

Since $\varepsilon \ll 1$, the last sum will be dominated by those values of k for which the exponent of ε is minimal, leading to

$$\chi_n(\varepsilon) \sim \varepsilon^{\xi_n} \quad \text{with} \quad \xi_n = \min_\alpha \left[n\alpha - f(\alpha) \right] + 1 . \qquad (6.38)$$

In complete analogy, the empirical study of multifractal return time series $\delta S_\tau(t)$ of a financial asset proceeds via the scaling of its moments, i.e., the estimation of its structure function

$$\chi_n(\tau) = \langle |\delta S_\tau(t)|^n \rangle . \qquad (6.39)$$

As in (6.22), we expect a scaling

$$\chi_n(\tau) \sim \tau^{\xi_n} . \qquad (6.40)$$

ξ_n in general is a concave function of n and satisfies $\xi_0 = 0$. Both for the binomial and for the log-normal cascades, the spectra $f(\alpha)$, and thus the scaling exponents ξ_n, are known [158],

$$
\begin{aligned}
f_{\text{bin}}(\alpha) \quad &= -\frac{\alpha_{\text{max}} - \alpha}{\alpha_{\text{max}} - \alpha_{\text{min}}} \log_2 \left(\frac{\alpha_{\text{max}} - \alpha}{\alpha_{\text{max}} - \alpha_{\text{min}}} \right) \\
&\quad - \frac{\alpha - \alpha_{\text{min}}}{\alpha_{\text{max}} - \alpha_{\text{min}}} \log_2 \left(\frac{\alpha - \alpha_{\text{min}}}{\alpha_{\text{max}} - \alpha_{\text{min}}} \right) ,
\end{aligned}
\qquad (6.41)
$$

$$f_{\text{log-nor}}(\alpha) = 1 - \frac{(\alpha - \lambda)^2}{4(\lambda - 1)} . \qquad (6.42)$$

For the binomial cascade with $p_1 > 1/2$, $\alpha_{\min} = -\log_2 p_1$ and $\alpha_{\max} = -\log_2(1 - p_1)$, while for the log-normal cascade, the logarithms of the multipliers are drawn from a normal distribution with mean $-\lambda$ and variance $2(\lambda - 1)/\ln 2$. These expressions characterize the multifractal properties of the cascade generating $\Theta(t)$. Assuming that the return process in chronological time is ordinary Brownian motion, the $f(\alpha)$ spectrum of the compound return process is $f_{\delta S}(\alpha) = f_{\Theta}(2\alpha)$ [158].

Figure 6.7 shows examples for the dependence of the scaling exponents of the moments on their order, taken from FX markets and from two turbulent flows [141]. All three data sets display a concave bend downward away from a straight line $\xi_n = n/3$, corresponding to the Kolmogorov hypothesis for turbulence. One can, in principle, derive the $f(\alpha)$ spectrum from such a scaling behavior, inverting (6.38). Numerous analyses of turbulent flows in terms of multifractal properties have been performed following the pioneering work of Mandelbrot [159]. We will not discuss them here. Some of the most recent work, e.g., finds evidence for multifractal atmospheric cascades from global scales down to about 1 km from the analysis of satellite cloud pictures at visible and infrared wavelengths [160].

Qualitatively similar though quantitatively different behavior has been found in 14 years of daily data of the French Franc (FRF) against the Swiss Franc (CHF), the US Dollar (USD), the Great Britain Pound (GBP), and the Japanese Yen (JPY) [142, 161]. Firstly, the slope of the small-n approximations is rather close to $1/2$, instead of $1/3$ as above, for the high-frequency USD/DEM rates. $1/2$ is the slope expected for Brownian motion, so one may wonder if the appearance of this slope may be related to the longer time scale analyzed. Secondly, while again one observes a systematic concavity of the ξ_n versus n curves, it is particularly weak for the JPY and particularly pronounced for the DEM exchange rate. The case of FRF against GBP is revealing because, during the last two years of the sampling interval, the GBP entered the European Exchange Mechanism which allowed a maximal deviation of 12% from a preset reference value: imposing this restriction leads to a significant increase of the concave downward bend in the ξ_n versus n curves, rather similar to the FRF/DEM curves, while before the behavior was more akin to FRF/USD or FRF/CHF [161]. If confirmed, this finding would imply that unregulated and regulated markets can be discriminated by the concavity of their $\xi_n(n)$ curves.

The behavior of the exponents of the lowest moments can be interpreted in simple pictures [162]. $\xi_1 = H$, the Hurst exponent, describes the roughness of the path described by the time series: $\xi_1 > 1/2$, a persistent time series gives a more ragged path than Brownian motion while an antipersistent time series ($\xi_1 < 1/2$) gives a smoother path. A sparseness coefficient C_1 can be defined by taking ξ_n as a continuous function of n, and taking the derivative $C_1 = -\mathrm{d}(\xi_n/n)/\mathrm{d}n|_{n=1}$. The sparseness describes the intermittency, or temporal concentration, of the signals. For $C_1 = 0$, i.e., $\xi_n \propto n$, $\xi_1 = 0$ describes white noise and $\xi_1 = 1$ describes differentiable functions, with Brownian motion midway in between. On the other hand, for $\xi_1 = 0$, there is an evolution from white noise at $C_1 = 0$ to Dirac delta functions at $C_1 = 1$. Quite generally, one then can locate various signals in a C_1 versus ξ_1 diagram. Analyzing a multitude of foreign exchange rates, Vandewalle and Ausloos have found that they scatter over a rather large part of the diagram, perhaps with the exception of the corners [162].

Many of these studies are based on graphical superposition of data analyzed and theoretical predictions of multifractal models. These methods can fail, however, as demonstrated by the multifractal analysis of a simulated monofractal stochastic process [163]. Here, the apparent multiscaling is a consequence of a crossover phenomenon at an intermediate time scale in the process. As an alternative, statistical hypothesis tests can also be used to assess the significance or "explanatory power" of multifractal cascade models [158]. No parametric tests for multifractal models are available, but one can turn around this problem by setting up a Monte Carlo simulation of the stochastic multifractal process with the estimated parameters, and then apply a Kolmogorov–Smirnov test [44]. This test evaluates the probability of the null hypothesis that both sets of data are drawn from the same underlying probability distribution. This test program was carried out by Lux using daily data for the DAX stock index, the New York Stock Exchange Composite Index, the USD/DEM exchange rate, and the gold price [158]. With only one adjustable parameter, p_1 for the binomial or λ for the log-normal cascades, the null hypothesis cannot be rejected at the 95% significance level. The tests perform equally well for both types of cascades, and the parameter estimates for the four time series are rather similar. The p_1 estimates fall into the range $p_1 = 0.63 \ldots 0.69$, while $\lambda = 1.04 \ldots 1.12$ is estimated for the log-normal cascade. On the contrary, the description of the empirical probability distributions by a GARCH(1,1) process are significantly worse, and drawing the random increments in GARCH(1,1) from a Student-t distribution only partially improves the situation. This would suggest that a multifractal model indeed can capture some important elements of the return dynamics of financial assets.

7. Derivative Pricing Beyond Black–Scholes

In the two preceding chapters, we have observed that the price dynamics of real-world securities differs significantly from geometric Brownian motion, most importantly by fat tails in the return distributions and by volatility correlations. The fundamental assumptions behind the Black–Scholes theory of option pricing and hedging do not hold in real markets. More general methods which include these stylized facts are called for.

7.1 Important Questions

This leads us to the following important questions concerning derivative pricing.

- Can the Black–Scholes theory of option pricing and hedging be worked out for non-Gaussian markets?
- Can we formulate a theory of option pricing which does not make any assumptions on the properties of the stochastic process followed by the underlying security, and for which Black–Scholes obtains as a special limit?
- Are analytic expressions for option prices available when the underlying returns are taken from a stable Lévy distribution?
- Are path-integral methods from physics useful in the elaboration of option pricing schemes for non-Gaussian markets, and can we formulate a quantum theory of financial markets?
- How are American-style options priced?
- Can option prices and hedges be simulated numerically?

7.2 An Integral Framework for Derivative Pricing

In Chap. 4, we determined exact prices for derivative securities. In particular, we derived the Black–Scholes equation for (simple European) options. Our derivation relied on the construction of a risk-free portfolio, i.e., a perfect hedge of the option position was possible.

The derivation was subject, however, to a few unrealistic assumptions: (i) security prices performing geometric Brownian motion, (ii) continuous adjustment of the portfolio, (iii) no transaction fees. That (i) is unrealistic was demonstrated at length in Chap. 5. It is clear that transaction fees forbid a continuous adjustment of the portfolio. Also liquidity problems may prevent this. Both factors imply that a portfolio adjustment at discrete time steps is more realistic. However, both with non-Gaussian statistics, and with discrete-time portfolio adjustment, a complete elimination of risk is no longer possible.

A generalization of the Black–Scholes framework, using an integral representation of global wealth balances, was formulated by Bouchaud and Sornette [17, 164]. To explain the basic idea, we take the perspective of a financial institution writing a derivative security. In order to hedge its risk, it uses the underlying security, say a stock, and a certain amount of cash. In other words, it constitutes a portfolio made up of the short position in the derivative, the long position in the stock, and some cash. The stock and cash positions are adjusted according to a strategy which we wish to optimize. The optimal strategy, of course, should minimize the risk of the bank (it can't eliminate it completely). However, in a non-Gaussian world, this strategy will depend on the quantity used by the bank to measure risk, and in contrast to the Black–Scholes framework, where the risk is eliminated instantaneously, here one can minimize the global risk, incurred over the entire time interval to maturity. While the Black–Scholes theory was differential, this method is integral.

To formalize this idea, we establish the wealth balance of the bank over the time interval $t = 0, \ldots, T$ up to the maturity time T of the derivative. The unit of time is a discrete subinterval of length $\Delta t = t_{n+1} - t_n$. The asset has a price S_n at time t_n, it is held in a (strategy dependent) quantity $\Phi(S_n, t_n) \equiv \Phi_n$ and has a return μ. The amount of cash is B_n, and its return is the risk-free interest rate r. At $t = t_n$, the wealth of the bank then is

$$W_n = \Phi(S_n, t_n)S_n + B_n . \tag{7.1}$$

How does it evolve from $n \to n + 1$? The updated cash position is

$$B_{n+1} = B_n e^{r\Delta t} - S_{n+1}(\Phi_{n+1} - \Phi_n) . \tag{7.2}$$

The first term accounts for the interest, and the second term is due to the portfolio adjustment $\Phi_n \to \Phi_{n+1}$, due to stock price changes $S_n \to S_{n+1}$. The difference in wealth between t_n and t_{n+1} is then

$$W_{n+1} - W_n = \Phi_n(S_{n+1} - S_n) + B_n \left(e^{r\Delta t} - 1\right) . \tag{7.3}$$

B_n can be eliminated from this equation by using (7.1), the resulting equation can be iterated, and the wealth of the bank after n time steps can be expressed in terms of the stock position alone:

$$W_n = W_0 e^{rn\Delta t} + \sum_{k=0}^{n-1} \Phi_k e^{r(n-k-1)\Delta t} \left(S_{k+1} - S_k e^{r\Delta t}\right) . \tag{7.4}$$

The term in parentheses is the stock price change discounted over one time step, and its prefactor in the sum is the cost of the portfolio adjustment.

7.3 Application to Forward Contracts

As a simple application, we consider a forward contract. In a forward, the underlying asset of price S_N is delivered at maturity $T = N\Delta t$ for the forward price F, to be fixed at the moment of writing the contract. As we have seen in Sect. 4.3.1, there are no intrinsic costs associated with entering a forward contract because the contract is binding for both parties. The value of the bank's portfolio at any time before maturity therefore is

$$\Pi_n = W_n \text{ at } t_n < T = N\Delta t \,. \tag{7.5}$$

At maturity, it becomes

$$\Pi_N = W_N + F - S_N \text{ at } T = N\Delta t \,. \tag{7.6}$$

The bank delivers the asset for S_N and receives the forward price F.

Using (7.4), it is possible to rewrite the resulting equation so that the stock price S_k only appears in the form of differences $S_{k+1} - S_k$, and of the initial stock price S_0

$$\Pi_N = F + W_0 e^{rT} - S_0 - S_0(e^{r\Delta t} - 1) \sum_{k=0}^{N-1} \Phi_k e^{r(N-1-k)\Delta t}$$

$$+ \sum_{k=0}^{N-1} (S_{k+1} - S_k) \tag{7.7}$$

$$\times \left(\Phi_k e^{r(N-1-k)\Delta t} - \left[e^{r\Delta t} - 1\right] \sum_{l=k+1}^{N-1} \Phi_l e^{r(N-1-l)\Delta t} - 1 \right) \,.$$

The idea behind this complicated rewriting is that the only term representing risk in this equation is the evolution of the stock price from one time step to the next, $S_{k+1} - S_k$. If its prefactor can be made to vanish, the risk will be eliminated completely. (As we know, this must be possible for a forward contract because the contract is not traded and binding to both parties.) This gives the conditions

$$\left(\Phi_k e^{r(N-1-k)\Delta t} - \left[e^{r\Delta t} - 1\right] \sum_{l=k+1}^{N-1} \Phi_l e^{r(N-1-l)\Delta t} - 1 \right) = 0 \tag{7.8}$$

at every time step. This equation can be iterated backwards, starting at $k = N - 1$,

$$\Phi_{N-1} - 1 = 0 \,. \tag{7.9}$$

In order to completely hedge its risk in the short forward position, the bank must hold one unit of stock at the last time step before the delivery of the stock is due at maturity. In the second-last time step, we have

$$\Phi_{N-2}e^{r\Delta t} - \left[e^{r\Delta t} - 1\right]\Phi_{N-1} - 1 = 0 \Rightarrow \Phi_{N-2} = 1, \qquad (7.10)$$

where $\Phi_{N-1} = 1$ has been used. This process can be continued,

$$\Phi_n = 1 \text{ for all } n. \qquad (7.11)$$

The portfolio need not be adjusted in the case of a forward contract, and a perfect hedge of the short forward position is possible by going long in the underlying security at the time of writing the contract.

The sum in (7.8) is a geometric series which can be summed, and the final value of the portfolio is

$$\Pi_N = F + W_0 e^{rT} - S_0 e^{rT}. \qquad (7.12)$$

No arbitrage is possible if this is equal to the wealth of the bank in the absence of the forward contract

$$\Pi_N = W_0 e^{rT}. \qquad (7.13)$$

Then, the value of the contract is the same for the long and the short positions. This gives the forward price

$$F = S_0 e^{rT} \qquad (7.14)$$

already derived in Sect. 4.3.1. This is not surprising. By construction of the forward contract, a perfect hedge does not require portfolio adjustment, and our derivation of the forward price (4.1) in Sect. 4.3.1 did not make any reference to the statistics of price changes.

7.4 Option Pricing (European Calls)

The situation is very different for option positions, however. The value of the portfolio at the maturity of a European call is

$$\Pi_N = W_0 e^{rT} + C e^{rT} - \max(S_N - X, 0) + \sum_{k=0}^{N-1} \Phi_k e^{r(N-1-k)\Delta t}\left(S_{k+1} - S_k e^{r\Delta t}\right).$$
$$(7.15)$$

The first and the last terms on the right-hand side have been discussed in the preceding section. The second term is the price of the option which the bank receives up front, compounded by interest, and the third term is the amount it has to pay to the long position at maturity. As this term is nonlinear in S_N, the risk can no longer be eliminated completely.

A fair price for the option, C, can now be fixed from the requirement that the expected change in the value of the bank's portfolio, over its initial value compounded by the riskless rate r, vanishes,

$$\langle \Delta W \rangle = \langle \Pi_N - W_0 e^{rT} \rangle = 0 \ , \tag{7.16}$$

which can be solved for the call price

$$C = e^{-rT} \left[\langle \max(S_N - X, 0) \rangle - \sum_{k=0}^{N-1} \left\langle \Phi_k e^{r(N-1-k)\Delta t} \left(S_{k+1} - S_k e^{r\Delta t} \right) \right\rangle \right] . \tag{7.17}$$

This price, a priori, is strategy dependent (Φ_k appears and cannot be eliminated). Moreover, since even the optimal strategy carries a residual risk, a risk premium can be added to the call price C.

The price changes during $k \to k+1$, $S_{k+1} - S_k$, are statistically independent of the fraction of stock held at t_k, Φ_k. Then $S_{k+1} - S_k e^{r\Delta t}$ is also statistically independent of Φ_k, and one can separate

$$\left\langle \Phi_k \left(S_{k+1} - S_k e^{r\Delta t} \right) \right\rangle = \langle \Phi_k \rangle \left\langle \left(S_{k+1} - S_k e^{r\Delta t} \right) \right\rangle \tag{7.18}$$

in (7.17). If $r \ll 1$, the exponential can be set to unity. If the stock price is then drift-free,

$$\langle S_{k+1} - S_k \rangle = 0 \ . \tag{7.19}$$

Alternatively, in a risk-neutral world, the same conclusion would obtain without making the assumptions on the smallness of r and the martingale property of S_k. A priori, however, the notion of a risk-neutral world is tied to geometrical Brownian motion, and should be used with much care here. Then

$$C = e^{-rT} \langle \max(S_N - X, 0) \rangle = e^{-rT} \int_X^\infty dS \, (S - X) p(S, N | S_0, 0) \ . \tag{7.20}$$

One recovers the expectation value pricing formula for option prices (4.95) which reduces to the Black–Scholes expression (4.85) for a log-normal distribution. The result is a direct consequence of the assumed martingale property (7.19) of the stock price which also had to be made to derive (4.95). Of course, in this limit, the option price comes out strategy-independent.

If the stochastic process of the stock price is not a martingale, the full expression (7.17) must be used. The drift in the second term will then partly compensate the drift in the first term. Both terms will drift because the historical price densities are used in the calculation of the expectation values in (7.17).

Then, the optimal hedging strategy $\{\Phi_k^\star\}$ must be designed so as to minimize the risk of the bank. One possible definition of the risk R in this framework is to minimize the variance of the (integral) wealth balance

$$R^2 = \langle (\Delta W)^2 \rangle - \langle \Delta W \rangle^2 = \langle (\Delta W)^2 \rangle \ . \tag{7.21}$$

This is minimized by equating to zero the functional derivative

$$0 = \frac{\delta R^2}{\delta \Phi_k} \tag{7.22}$$

$$= \frac{\delta}{\delta \Phi_k} \left\{ \sum_{k=0}^{N-1} \left\langle \Phi_k^2 \right\rangle e^{2r(N-1-k)\Delta t} \left\langle \left(S_{k+1} - S_k e^{r\Delta t} \right)^2 \right\rangle \right. \tag{7.23}$$

$$\left. - 2 \sum_{k=0}^{N-1} \left\langle \max(S_N - X, 0) \left(S_{k+1} - S_k e^{r\Delta t} \right) \Phi_k \right\rangle e^{r(N-1-k)\Delta t} \right\} .$$

Here, terms independent of Φ_k have already been dropped. Moreover, price changes have been assumed to be independent, $\langle \delta S_k \delta S_l \rangle = \langle (\delta S_k)^2 \rangle \delta_{kl}$, and terms proportional to $\left\langle \Phi_k \left(S_{k+1} - S_k e^{r\Delta t} \right) \right\rangle$ have been neglected with the same assumptions as above.

A rather subtle problem concerns the use of probability density functions in the various expectation values. The strategy Φ_k is determined by the stock price S_k. Therefore, $p(S_k, k | S_0, 0)$ is the appropriate distribution for the first expectation value. The price changes $S_{k+1} - S_k$ are governed by $p(S_{k+1}, k + 1 | S_k, k)$, which must be used in the second expectation value. Finally, in the third expectation value, $p(S_N, N | S_{k+1}, k + 1)$ must be introduced for the payoff of the option. Also, in this expectation value, only those variations of $S_{k+1} - S_k$ must be allowed which end up at S_N after N time steps. For IID random variables, all intermediate steps contribute the same amount, and [17]

$$\langle S_{k+1} - S_k \rangle_{(S_k, k) \to (S_N, N)} = \frac{S_N - S_k}{N - k} . \tag{7.24}$$

Using this result, (7.22) becomes

$$0 = \frac{\delta}{\delta \Phi_k} \left\{ \sum_{k=0}^{N-1} \int_{-\infty}^{\infty} \mathrm{d}S \, \Phi_k^2(S) p(S, k | S_0, 0) e^{2r(N-1-k)\Delta t} \left\langle \left(S_{k+1} - S_k e^{r\Delta t} \right)^2 \right\rangle \right.$$

$$- 2 \sum_{k=1}^{N-1} \int_{-\infty}^{\infty} \mathrm{d}S \, \Phi_k(S) p(S, k | S_0, 0) e^{r(N-1-k)\Delta t} \tag{7.25}$$

$$\left. \times \int_X^{\infty} \mathrm{d}S' \, (S' - X) p(S', N | S k) \langle S_{k+1} - S_k \rangle_{(S_k, k) \to (S_N, N)} \right\}$$

$$= 2\Phi_k e^{2r(N-1-k)\Delta t} p(S_k, k | S_0, 0) \langle (S_{k+1} - S_k)^2 \rangle$$

$$- p(S_k, k | S_0, 0) e^{r(N-1-k)\Delta t} \tag{7.26}$$

$$\times \int_X^{\infty} \mathrm{d}S' \, (S' - X) p(S', N | S_k, k) \langle S_{k+1} - S_k \rangle_{(S_k, k) \to (S_N, N)} .$$

This can be solved to determine the optimal strategy

$$\Phi_k^{\star}(S_k) = \frac{e^{-r(N-1-k)\Delta t}}{\langle (S_{k+1} - S_k)^2 \rangle} \int_X^{\infty} \mathrm{d}S' \, (S' - X) \frac{S' - S_k}{N - k} p(S', N | S_k, k) , \tag{7.27}$$

which should be inserted into (7.17) to provide the correct option price. If $p(S', N|S_k, k)$ is taken from either a Gaussian or a log-normal distribution, and if one takes the continuum limit for time, one can show that the optimal strategy reduces to the Δ-hedge of Black, Merton, and Scholes. In general, however, $\Phi_k^\star$ will give a different strategy, and more importantly, a residual risk

$$R^2\left[\{\Phi_k^\star\}\right] \neq 0 \qquad (7.28)$$

will remain.

A pedagogical example is provided by assuming that returns are IID random variables drawn from a Student-t distribution $St_\mu(\delta S)$ as defined in (5.59) [165]. The variance exists for $\mu > 2$, and for μ an odd integer, one can derive closed expressions for the hedging functions $\Phi_k^\star$ above. Figure 7.1 shows the price C of a European call option at seven days from maturity, in units of the standard deviation, as a function of the price of the underlying, using the optimal hedge derived from the formalism of this chapter (crosses). It also shows the residual risk which cannot be hedged away, as the

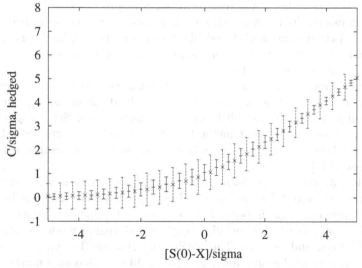

Fig. 7.1. Price of a European call option sevend days from maturity, determined from the optimal hedging strategy discussed in this chapter, for IID random variables drawn from a Student-t distribution (*crosses*) together with residual risk (*dashed error bars*). For comparison, the price and residual risk of the same call is shown when the return process is Gaussian in discrete time (*solid error bars*). Due to discreteness of time, a finite residual risk remains even for a Gaussian return process, unlike in the continuous-time Black–Scholes theory. Both the call price C and the initial difference between the price of the underlying and the strike price, $S(0) - X$, are measured in units of the standard deviation σ of the daily returns. By courtesy of K. Pinn. Reprinted with permission from Elsevier Science from K. Pinn: Physica A **276**, 581 (2000). ©2000 Elsevier Science

dashed error bars. A Student-t distribution with $\mu = 3$ has been assumed. For comparison, the solid error bars show the call price and residual risk of a Gaussian return process *in discrete time*. While for a continuous-time Gaussian return process, the risk can be hedged away completely by following the Black–Scholes Δ-hedging strategy (cf. Chap. 4), for a discrete-time process, a residual risk always remains [165]. The figure nicely demonstrates both the effects of the fat-tailed distribution, and of discrete trading time.

What about real markets? Figure 7.2 compares the market price of an option on the BUND German government bond, traded at the London futures exchange, to the Black–Scholes price. The inset shows the deviations from a correctly specified theory, represented by the straight line with slope of unity in the main figure. There is a systematic deviation between the Black–Scholes and the market price so that the market price is higher. Black–Scholes therefore underestimates the option prices, because it underestimates the risk of an option position. The market corrects for this. On the other hand, the comparison between the theoretical price calculated from (7.17) using the optimal strategy (7.27) and the market price is much better, as shown in Fig. 7.3. The inset again shows the deviations from a correctly specified theory. These deviations are symmetric with respect to the line with slope unity, and essentially random. Also, their amplitude is a factor of five smaller than those between the market and Black–Scholes prices. The theory exposed in this chapter therefore allows for a significant improvement over the Black–Scholes pricing framework [17].

Notice, however, that the market did not have this theory at hand, to calculate the option prices. The prices were fixed empirically, presumably by applying empirically established corrections to Black–Scholes prices and prices calculated by different methods. This has led to speculations that financial markets would behave as adaptive systems, in a manner similar to ecosystems [115].

Earlier, arbitrage was defined as simultaneous transactions on several markets which allow riskless profits. This requires that risk can be eliminated completely. This is possible in the case of a forward contract quite generally. For options, it is possible only in a Gaussian world, as shown by Black, Merton, and Scholes. The notion of arbitrage becomes much more fuzzy in more general situations (e.g., options in non-Gaussian markets, etc.) where riskless hedging strategies are no longer feasible. Then, it will depend explicitly on factors such as the measurement of risk, risk premiums, etc., and is no longer riskless in itself.

7.5 Monte Carlo Simulations

Monte Carlo simulations are an important tool for option pricing. Starting from the ideas of Black, Merton, and Scholes and requiring that no arbitrage opportunities exist in a market, the important input for a calculation

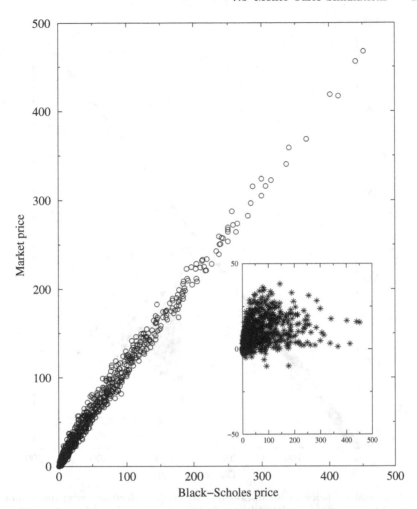

Fig. 7.2. Market price of an option on the BUND German government bond, compared to the Black–Scholes price. The inset shows the deviations from the ideal line with slope unity. The Black–Scholes price systematically underestimates the market price of the option. Reprinted from J.-P. Bouchaud and M. Potters: *Théorie des Risques Financiers,* by courtesy of J.-P. Bouchaud. ©1997 Diffusion Eyrolles (Aléa-Saclay)

of option prices by numerical simulation is the risk-neutral probability distribution of returns which, in real-world markets, is different from the normal distribution assumed in the Black–Scholes theory. One can either assume a distribution consistent with the empirical facts, or try to reconstruct the risk-neutral distribution from quoted option prices. Price charts then are generated from this risk-neutral distribution, the payoff of the option for each particular trajectory is evaluated, and finally the option price is calculated as

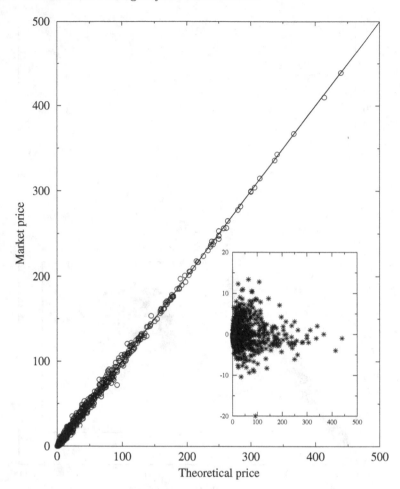

Fig. 7.3. Market price of an option on the BUND German government bond, compared to the price calculated by minimizing the risk of an integral wealth balance, explained in the text. The inset shows the deviations from the ideal line with slope unity. Deviations from the market price are distributed approximately symmetrically around zero. Reprinted from J.-P. Bouchaud and M. Potters: *Théorie des Risques Financiers,* by courtesy of J.-P. Bouchaud. ©1997 Diffusion Eyrolles (Aléa-Saclay)

the expectation value of the payoffs over the various trajectories. This basic procedure works for simple options, such as European plain-vanilla calls and puts. For American-style options or path-dependent options, efficient extensions have been developed [166]. One major drawback of these approaches is that the variance of the option price is rather important. More importantly, though, its derivatives such as $\Delta = \partial f / \partial S$, which are important for hedging purposes and trading strategies, come out to be extremely inaccurate.

One can, however, also use the theory exposed in the preceding section to develop an efficient Monte-Carlo approach to option pricing [167]. The following features of the approach of Sect. 7.4 directly carry over to the Monte Carlo variant:

- At the same time, the option price, the optimal hedge, and the residual risk are calculated.
- No assumption is made on a risk-neutral measure, or on the nature of the stochastic process, except the absence of linear correlations on the time scale of an elementary Monte Carlo step. One can use complex return processes, and even do a historical simulation where the historically observed price increments of the underlying are used.
- In addition, one obtains an important reduction of the variance of the option price and hedge. When the residual risk is minimized by finding the optimal hedging strategy, the variance of the option prices is automatically minimized.

Being optimally hedged by construction, the method has been called Hedged Monte Carlo.

For simplicity, we only consider a European call option. Numerical option pricing always works backward in time because at maturity $T = N\tau$ the option price C_N is known exactly and is equal to the payoff of the option. At time t_k, the price of the underlying is S_k, the option price is C_k, and the hedge is $\Phi_k(S_k)$, as above. In the absence of linear temporal correlations in the prices, the wealth balance ΔW becomes the sum of local changes ΔW_k between steps k and $k+1$. The same applies to its variance, the residual risk, which can be minimized locally. The analog to (7.21) at time step k is

$$R_k^2 = \left\langle \left(e^{-r\tau} C_{k+1}(S_{k+1}) - C_k(S_k) + \Phi_k(S_k) \left[S_k - e^{-r\tau} S_{k+1} \right] \right)^2 \right\rangle , \quad (7.29)$$

where the expectation value is taken with the historical probability distribution of the underlying [167]. $C_k(S_k)$ and $\Phi_k(S_k)$ must be chosen so as to minimize R_k^2 given $C_{k+1}(S_{k+1})$ and S_{k+1}. In order to implement this minimization numerically, one decomposes C_k and Φ_k over a set of suitable basis functions

$$C_k(S) = \sum_{\alpha=1}^{M} \gamma_k^\alpha C^\alpha(S) , \quad \Phi_k(S) = \sum_{\alpha=1}^{M} \phi_k^\alpha F^\alpha(S) . \quad (7.30)$$

In actual applications, the basis functions $F^\alpha(S)$ and $C^\alpha(S)$ have been chosen piecewise linear and piecewise quadratic, respectively. In this way, the problem has been reduced to a variational search for the coefficients γ_k^α and ϕ_k^α, and one is left with an ordinary least-squares minimization of

$$\sum_{\ell=1}^{N_{MC}} \left(e^{-r\tau} C_{k+1}(S_{k+1}) - \sum_{\alpha=1}^{M} \gamma_k^\alpha C^\alpha(S_k^\ell) + \sum_{\alpha=1}^{M} \phi_k^\alpha F^\alpha(S_k^\ell) \left[S_k^\ell - e^{-r\tau} S_{k+1}^\ell \right] \right)^2 .$$
$$(7.31)$$

Using a delta hedge

$$\phi_k^\alpha = \gamma_k^\alpha \, , \quad F^\alpha = \frac{dC^\alpha(S)}{dS} \tag{7.32}$$

simplifies the problem even further and often produces very good results [167].

In order to assess its accuracy, this method is tested on a standard Black–Scholes problem [167]. The asset price $S(t)$ follows geometrical Brownian motion with a drift rate $\mu = r = 5\%/y$ and a volatility $\sigma = 30\%/\sqrt{y}$. A three-month European call option is priced with $X = S(0) = 100$, and the Black–Scholes price is $C_0^{BS} = 6.58$. For 500 simulations containing 500 paths each, $N = 20$ time intervals and $M = 8$ basis functions have been used. Hedged Monte Carlo gives a call price $C_0^{HMC} = 6.55 \pm 0.06$, a very good approximation to the Black–Scholes price indeed. The unhedged risk-neutral Monte Carlo scheme [166] would yield a price $C_0^{RNMC} = 6.68 \pm 0.44$. A reduction in the standard deviation of the call price by a factor of seven has been achieved.

The example of a European call option has been chosen for pedagogical reasons and, of course, Hedged Monte Carlo is not restricted to it. For example, American-style options with early exercise features have been successfully priced and hedged, with results superior to established approaches [167]. The simulations can also be performed for exotic, path-dependent options. As an example of historical simulation, a series of one-month options on Microsoft has been priced using the price chart of eight years of daily quotes. As explained in Chap. 4, one can invert the Black–Scholes equation to calculate an implied volatility σ_{imp} from an option price. Performing this inversion for the series of simulated prices, a volatility smile not unlike those observed in real-world option markets is found.

7.6 Option Pricing in a Tsallis World

In Sect. 5.5.7 we showed that power-law distribution functions for random variables obeying special, non-linear Langevin equations with a rather peculiar feedback between macroscopic and microscopic variables could be obtained from an extension of statistical mechanics. In that approach, an entropy somewhat different from the usual definition was maximized, and the corresponding statistical mechanics was not extensive. To be specific, the distributions with entropic indices of 3/2 or 5/3 produce tail indices for the power-law distributions $\mu = 3$ resp. 2, and would thus be able to describe financial time series [168].

The correspondence between Tsallis statistics and financial markets is made by postulating that the return of an asset over an infinitesimal time scale (i.e., in continuous-time finance) follows the Tsallis version of Brownian motion

$$d \ln S = \mu dt + \sigma d\Omega \quad \text{with} \quad d\Omega = P(\Omega)^{(1-q)/2} dz . \tag{7.33}$$

dz describes ordinary Brownian motion. The probability density $P(\Omega)$ both makes the differential equation non-linear and mediates the peculiar macroscopic–microscopic feedback effects discussed in Sect. 5.5.7. It can be determined self-consistently from a Fokker–Planck equation and behaves as a power law with exponent $2/(2-q)$ in Ω, as does the distribution of $\ln S$ in that variable. Equation (7.33) describes an Itô process, cf. (4.40). Using Itô calculus, a differential equation for the price can be derived [169]:

$$dS = \left[\mu + \frac{\sigma^2}{2} P^{1-q}(\Omega) \right] S dt + \sigma S d\Omega, \tag{7.34}$$

which might be dubbed geometric Tsallis motion.

From this point on, one would like to compose a portfolio of a suitable quantity of the underlying and a European call or put option (worth f) on it which, by a magic trick, is described by a Black–Scholes-type equation

$$\frac{\partial f}{\partial t} + rS \frac{\partial f}{\partial S} + \frac{1}{2} \frac{\partial^2}{\partial S^2} \sigma^2 S^2 P^{1-q}(\Omega) = rf . \tag{7.35}$$

Again, the $P(\Omega)$ term induces a non-linear dependence on S but vanishes as $q \to 1$ (geometric Brownian motion). Itô's lemma has been applied and, as with the Black–Scholes problem, a Delta hedge apparently makes the portfolio riskless.

The basic procedure now follows the standard Black–Scholes scheme, although there are a few subtleties to be considered due to the different statistics. In particular, the non-linearity $P^{(1-q)/2}(\Omega)$ in the stochastic differential equations (7.33) and (7.34) requires a particular treatment of the martingale property of the stochastic process. Transforming explicitly to an equivalent martingale measure introduces an alternative noise term into the integration of dS, namely

$$d\tilde{z} = \frac{\mu + \frac{\sigma^2}{2} P^{1-q}(\Omega) - r}{\sigma P^{(1-q)/2}(\Omega)} dt + dz . \tag{7.36}$$

Following (4.95), the derivative price can be written as

$$f = e^{-rT} \langle h[S(T)] \rangle_Q , \tag{7.37}$$

where $h[S(T)]$ is the payoff function of the derivative, Q is the equivalent martingale measure, and the price of the underlying at maturity T is

$$S(T) = S(0) \exp \left(\int_0^T \sigma P^{(1-q)/2}(\Omega) d\tilde{z}_s + \int_0^T \left[r - \frac{\sigma^2}{2} P^{1-q}(\Omega) \right] ds \right) . \tag{7.38}$$

The expectation value in (7.37) is taken over a Tsallis distribution, and the final result – the price of a European call option – is the difference of two lengthy integrals. Apparently, this theory gives rather realistic option prices. When represented in terms of an implied volatility (invert the Black–Scholes equation for the true option price and solve for σ), one nicely reproduces the main features of the characteristic skewed volatility smile observed in real option markets [169].

7.7 Path Integrals: Integrating the Fat Tails into Option Pricing

When deriving the Black–Scholes solution of option pricing and hedging in Sect. 4.5.1, we mentioned that the Black–Scholes equation could be solved by path-integral methods, defining a Black–Scholes "Hamiltonian", (4.93), on the way [51]. Also, the integral framework based on a global wealth balance and the minimal variance hedging strategy of Sect. 7.4 is very reminiscent of the path integrals used in physics [50]. This is not accidental: one can in fact systematically derive path-integral representations for the conditional probability distributions encountered in finance, introducing on the way "Hamiltonians" [170]. The door to quantum finance has been opened.

To make the notation easier, define the log-price of an asset as $x \equiv \ln S$, and assume that its evolution equation is determined by a stochastic differential equation

$$\frac{dx}{dt} = \mu_x + \sigma\varepsilon(t) \, . \tag{7.39}$$

The relative rate of return of the asset price follows

$$\frac{1}{S}\frac{dS}{dt} = \mu + \sigma\varepsilon(t) \, . \tag{7.40}$$

Geometric Brownian motion follows these equations, cf. (4.53) and (4.62), with independent $\varepsilon(t)$ drawn from a Gaussian, and with $\mu_x = \mu - \sigma^2/2$. The difference between both growth rates is the *noise-induced drift*. Here, however, we allow the independent $\varepsilon(t)$ to be taken from some general distribution

$$p(x) = \int \frac{dz}{2\pi} e^{izx} \hat{p}(z) \equiv \int \frac{dz}{2\pi} e^{izx - H(z)} \, . \tag{7.41}$$

For non-Gaussian probability distributions, the relation between μ_x and μ is not fixed, and depends on the specific distribution considered. The characteristic function $\hat{p}(z)$ was introduced in (5.16), and the last identity defines a *Hamiltonian* associated with this distribution. To keep consistency with Sect. 5.4, we use the variable z in the characteristic function. Analogy with physics would suggest using p instead, but this would conflict with the use of p for probabilities. For a Gaussian distribution with zero mean, the Hamiltonian is $H_G = \sigma^2 z^2/2$. The Gaussian Hamiltonian describes a free particle with mass $m = 1/\sigma^2$ and momentum z. For a symmetric, stable Lévy distribution with zero mean, (5.42), the Hamiltonian is $H_L = a|z|^\mu$, with no obvious interpretation in terms of a physical system. The definition of the cumulants c_n in (5.17) immediately suggests the following power-series expansion of the Hamiltonian:

$$H(z) = \sum_{n=0}^{\infty} \frac{c_n (iz)^n}{n!} , \tag{7.42}$$

the equivalent to the cumulant expansion of the characteristic function. Two Hamiltonians related to $H(z)$ are useful:

$$\bar{H}(z) = H(z) - ic_1 z , \tag{7.43}$$
$$H_r(z) = H(z) - ic_1 z + irz . \tag{7.44}$$

The conditional probability distribution for finding x_b at time t_b, conditioned on x_a at t_a, then is given by the path integral [170]

$$p(x_b, t_b | x_a, t_a) = \int \mathcal{D}\varepsilon \int \mathcal{D}x \exp\left\{ -\int_{t_a}^{t_b} dt \tilde{H}[\varepsilon(t)] \right\} \delta\left(\frac{dx}{dt} - \varepsilon \right) . \tag{7.45}$$

The function $\tilde{H}(x)$ is defined by

$$\tilde{H}(x) \equiv -\ln p(x) . \tag{7.46}$$

The path integral in (7.45) is evaluated by cutting the time interval into N slices of length τ each, integrating over all $x(t_n)$, and taking the limit $N \to \infty$, $\tau \to 0$ with $t_b - t_a = N\tau = \text{const.}$ [50]. In complete analogy to physics, one can calculate a partition function, a generating function, and all moments and correlation functions in the path-integral formulation. The path integrals also satisfy a Chapman–Kolmogorov–Smoluchowski equation (3.10), implying that they describe Markov processes. This is not surprising, though, as we required independent increments in (7.39) from the outset.

Also, a general Fokker–Planck-type equation

$$\frac{\partial}{\partial t} p(x_b, t_b | x_a, t_a) = -H\left(-i\frac{\partial}{\partial x} \right) p(x_b, t_b | x_a, t_a) \tag{7.47}$$

can be derived, where the canonical substitution $z \to -i\partial_x$ has been performed in the Hamiltonian. Clearly, $H(z)$ in general is not a quadratic function of z, and (7.47) therefore contains higher-order terms beyond the drift and diffusion terms present in the canonical Fokker–Planck equation. In that sense, (7.47) is more correctly termed a Kramers–Moyal equation, And, from what has been said in Chap. 6, there is no equivalent Langevin equation in this case [37, 146]. The unconditional probability distribution $p(x, t)$ also satisfies (7.47),

$$\frac{\partial}{\partial t} p(x, t) = -H\left(-i\frac{\partial}{\partial x} \right) p(x, t) , \tag{7.48}$$

a Schrödinger equation in imaginary time (with a substitution $p \to \psi$, the illusion is complete) [170].

The stochastic processes considered here, in general, are not Itô processes of the form (4.40). Therefore Itô's lemma, (4.57), does not apply in the form given there. However, one can use the Schrödinger equation above to derive a generalized Itô relation [170]. The evolution of a function f of a stochastic variable obeying (7.39) with increments drawn from an arbitrary probability distribution is given by

$$\frac{\mathrm{d}f[x(t)]}{\mathrm{d}t} = \frac{\partial f[x(t)]}{\partial x}\frac{\mathrm{d}x}{\mathrm{d}t} - \bar{H}\left(i\frac{\partial}{\partial x}\right) f[x(t)] \, . \tag{7.49}$$

$\bar{H}$ appears instead of H because the first derivative of f has been taken out of H, cf. (7.43), to emphasize the similarity with the equivalent Gaussian expression (4.57).

From (7.39), the relation between the stochastic variable $x(t)$ and the asset price is $S(t) = \exp[x(t)]$. Using the generalized Itô relation, we can relate μ_x to μ by

$$\mu_x = \mu + \bar{H}(i) = \mu + H(i) - iH'(0) \tag{7.50}$$

and relate the log-return rate $\mathrm{d}x = \mathrm{d}\ln S$ to the relative return rate [170]:

$$\frac{1}{S}\frac{\mathrm{d}S}{\mathrm{d}t} = \frac{\mathrm{d}x}{\mathrm{d}t} - \bar{H}(i) = \frac{\mathrm{d}x}{\mathrm{d}t} - H(i) + iH'(0) \, . \tag{7.51}$$

Integrating the expectation value of this equation from zero to t gives the expected asset price at t,

$$\langle S(t) \rangle = S(0)\mathrm{e}^{\mu t} = S(0)\exp\left[\mu_x t - H(i) + iH'(0)\right] \, . \tag{7.52}$$

Path integrals are useful for calculating expectation values of stochastic variables, or functions thereof. As explained in Sect. 4.5.3, in an option-pricing context, this implies that one has to use the equivalent martingale process of the underlying, rather than the historical price process. What is the equivalent martingale process to (7.39)? The simplest solution is

$$\mathrm{e}^{-\mu t}S(t) = \mathrm{e}^{-\mu t}\mathrm{e}^{x(t)} = \mathrm{e}^{-\mu t}\mathrm{e}^{\mu_x t + \sigma \int_0^t \mathrm{d}t'\varepsilon(t')} \, . \tag{7.53}$$

A martingale distribution which gives such a process is

$$p^M(x_b, t_b | x_a, t_a) = \mathrm{e}^{-\mu t}\int \mathcal{D}\varepsilon \int \mathcal{D}x \exp\left\{-\int_{t_a}^{t_b} \mathrm{d}t \tilde{H}_{\mu_x}[\varepsilon(t)]\right\} \delta\left(\frac{\mathrm{d}x}{\mathrm{d}t} - \varepsilon\right) \, . \tag{7.54}$$

This, however, is not the only distribution with a time-independent expectation value. There is an entire family of equivalent martingale distributions with this property, among them

$$p^{Mr}(x_b, t_b | x_a, t_a) = e^{-rt} \int \mathcal{D}\varepsilon \int \mathcal{D}x \exp\left\{ -\int_{t_a}^{t_b} dt \tilde{H}_r[\varepsilon(t)] \right\} \delta\left(\frac{dx}{dt} - \varepsilon \right).$$

$$(7.55)$$

This distribution is also called the natural martingale [170].

The application to option pricing now uses a differential equation for the wealth of a portfolio consisting of $N_S(t)$ assets of price $S(t)$, $N_f(t)$ options of price $f(t)$, and $N_B(t)$ units of a risk-free bond (or cash) of price $B(t)$. The aim is to determine a hedging strategy $\{N_S(t), N_f(t), N_B(t)\}$ which makes the portfolio

$$\Pi(t) = N_S(t)S(t) + N_f(t)f(t) + N_B(t)B(t) \tag{7.56}$$

grow exponentially without fluctuations:

$$\frac{d\Pi(t)}{dt} = r_\Pi \Pi(t). \tag{7.57}$$

The risk-free position $B(t)$ grows with the risk-free interest rate r. The absence of arbitrage then implies that

$$r_\Pi = r. \tag{7.58}$$

As in the Black–Scholes theory, in the absence of transaction costs, the trading strategy is self-financing, i.e., there is no net cash flow into or out of the portfolio. This is expressed by

$$\frac{dN_S(t)}{dt}S(t) + \frac{dN_f(t)}{dt}f(t) + \frac{dN_B(t)}{dt}B(t) = 0. \tag{7.59}$$

Injecting this equation into (7.57) cancels the terms involving the bond (which is why cash or bonds did not appear in our discussion of the Black–Scholes theory). Rewriting all remaining contributions in terms of the option price $f(t)$ and its derivatives, and in terms of the log-price $x(t)$, the Δ-hedge of Black and Scholes is found by requiring that the fluctuating variable dx/dt must disappear from the equations. At the same time, the option price satisfies the Fokker–Planck-type equation

$$\frac{\partial f}{\partial t} = rf - [r + \bar{H}(\mathrm{i})]\frac{\partial f}{\partial x} + \bar{H}\left(\mathrm{i}\frac{\partial}{\partial x}f\right). \tag{7.60}$$

This is a straightforward generalization of the Black–Scholes equation (4.75), as can be checked by using the quadratic Hamiltonian $H_{\mathrm{gBm}} = \sigma^2 z^2/2$ of geometric Brownian motion. The general solution of this equation is

$$p(x_b, t_b | x_a, t_a) = e^{-r(t_b - t_a)} \int_{-\infty}^{\infty} \frac{dz}{2\pi} e^{\mathrm{i}z(x_b - x_a) - [\bar{H}(z) + \mathrm{i}\{r + \bar{H}(\mathrm{i})\}z][t_b - t_a]}.$$

$$(7.61)$$

This equation, however, must be solved numerically. Unfortunately, no examples have been worked out to date which would demonstrate the potential power of the method [170].

Path-integral techniques can also be useful when path-dependent options are priced and hedged [171]. Examples of path-dependent options are Asian, barrier, or lookback options. The payoff of an Asian option usually is determined by the average of the price of the underlying during a certain period [10]. The payoff of a barrier option is triggered by the underlying passing above or below a certain threshold price. The payoff of a lookback option depends on the maximal or minimal stock price realized during the lifetime of the option. For a European call it is the difference between the maximum and the minimum of the price of the underlying while, for a European put, it is the difference between the maximal price and the price at maturity of the underlying stock. The general ideas are somewhat similar to the preceding presentation, which is why we will be rather brief here.

Assume that we know the risk-neutral stochastic process of the underlying, and assume further (for simplicity, this is not a requirement) that it follows geometric Brownian motion. Then the price f of a path-dependent option at maturity is

$$f[S(T), \mathcal{I}, T] = h[S(T), \mathcal{I}] = h[e^{x(t)}, \mathcal{I}] , \qquad (7.62)$$

where $h[\ldots]$ is the payoff profile of the option, and the path-dependent random variable

$$\mathcal{I} = \int_t^T ds\, w(s)\, g[x(s), s] \qquad (7.63)$$

is written as an integral over an arbitrary function g with a sampling function $w(s)$. For continuous sampling $w(s) = 1$ while, for discrete sampling, $w(s)$ is a series of delta functions. In a risk-neutral world, the option price is the discounted expectation value of the payoff

$$f[S(t), t] = e^{-r(T-t)} \langle h[e^{x(T)}, \mathcal{I}]\rangle_t \qquad (7.64)$$

$$= e^{-r(T-t)} \int_{-\infty}^{\infty} dx(T) \int_{-\infty}^{\infty} d\mathcal{I}\, p[x(T), \mathcal{I}|x(t)]\, h[e^{x(T)}, \mathcal{I}]. \qquad (7.65)$$

From the preceding discussion, it is obvious that the conditional probability distribution can be represented as a path integral (given here for the special case of geometric Brownian motion)

$$p[x(T), \mathcal{I}|x(t)] = \frac{1}{2\pi} \exp\left[\frac{\mu\{x(T) - x(t) - \mu(T - t)\}}{\sigma^2}\right]$$

$$\times \int_{-\infty}^{\infty} dk e^{-ik\mathcal{I}} K[x(T), x(t); T - t] , \qquad (7.66)$$

$$K[x(T), x(t); T-t] = \int_{x(t)}^{x(T)} \mathcal{D}x(s) \exp\left\{ -\frac{1}{2\sigma^2} \int_t^T ds \left[\frac{dx(s)}{ds} + V[x(s), s] \right] \right\},$$

$$(7.67)$$

$$V[x(s), s] = -2 i k \sigma^2 w(s) g[x(s), s]. \tag{7.68}$$

$K[x(T), x(t); T-t]$ is a propagator, and the path integral in (7.67) integrates over all paths connecting $x(t)$ at the initial time t with $x(T)$ at maturity T. $V[x(s), s]$ is a potential whose shape is determined by the path-dependent random variable $\mathcal{I}$.

These path integrals, and the corresponding option prices, cannot be evaluated analytically in general. Matacz [171] has shown, however, that a partial average can be performed systematically based on the path-integral representation, which considerably reduces the numerical effort compared to standard numerical methods such as Monte Carlo. The path integral in $K[x(T), x(t); T-t]$ is evaluated by discretizing time and deriving a cumulant expansion for the propagator ($s = t + n\epsilon$, $\epsilon = (T-t)/N$):

$$K[x(T), x(t); T-t] = \int_{-\infty}^{\infty} dx_{N-1} \ldots dx_1 \prod_{n=1}^{N} K[x_n, x_{n-1}; \epsilon], \quad (7.69)$$

$$K[x_n, x_{n-1}; \epsilon] = \sqrt{\frac{1}{2\pi\sigma^2\epsilon}} \exp\left[-\frac{(x_n - x_{n-1})^2}{2\sigma^2\epsilon} \right.$$

$$\left. + \sum_{m=1}^{\infty} \frac{1}{m!} \left(-\frac{\epsilon}{2\sigma^2} \right)^m C_m(x_n, x_{n-1}; \epsilon) \right]. \quad (7.70)$$

Notice that the path dependence of the option has entirely been transformed into the details of the cumulants. The cumulant expansion at the same time is a power-series expansion in ϵ, the length of a time slice. A partial averaging of the short-time propagator $K[x_n, x_{n-1}; \epsilon]$ can then be performed by simply truncating the cumulant expansion at some order. For example, a propagator correct to second order in ϵ is obtained by dropping all cumulants beyond the first. The first cumulant is given by

$$C_1[x_n, x_{n-1}; \epsilon] = -i2\sigma^2 k \int_0^1 d\tau w(\tau) \int_{-\infty}^{\infty} dp_\tau \frac{e^{-p_\tau^2/2\nu_\tau^2}}{\sqrt{2\pi\nu_\tau^2}} g[\bar{x}_\tau + p_\tau, \tau] \quad (7.71)$$

with the abbreviations $\nu_\tau^2 = \sigma^2\epsilon(1-\tau)\tau$, $\bar{x}_\tau = \tau(x_n - x_{n-1}) + x_{n-1}$, and $\tau = (s - s_{n-1})/\epsilon$. If required, higher-order terms can also be calculated.

The option price finally becomes in this first-order cumulant approximation

$$f[S(t), t] = e^{-r(T-t)} \int_{-\infty}^{\infty} dx_N \ldots dx_1 \, p[x_N, \ldots, x_1|x_0]$$

$$\times h\left[e^{x_N}, 2\sigma^2\epsilon \sum_{n=1}^{N} C_1(x_n, x_{n-1}; \epsilon) \right] + \ldots . \quad (7.72)$$

Often, the first cumulant can be calculated analytically. Within our approximation of geometric Brownian motion, it is simply the Gaussian transform of the function g containing the path dependence of the option. An important practical advantage is that the size ϵ of the time slices entering the partially averaged cumulants can be chosen much bigger than the sampling scale of the options, which determines the structure of the sampling function w. This is an important simplification in the evaluation of the multidimensional integral in (7.72), which can be evaluated by standard Monte Carlo methods [171]. Again, however, no benchmark examples are provided which would allow for a critical assessment of the virtues and drawbacks of this method.

Another perspective is opened up by applying directly numerical methods to the Lagrangian (or Hamiltonian) which is generated by a path-integral formulation of the conditional probability distribution functions [172]. One such method is simulated annealing, which is an extension of a Monte Carlo importance sampling method. The aim is to find the global minimum of a ragged energy landscape. To this end, simulated annealing works at finite temperature. The process is started at high temperature, and the temperature then is lowered in order to trap the system in an energy minimum. Normally, this minimum will not be the global but rather a local minimum. In order to find the global minimum, the system is reheated and recooled in cycles. In finance, the equivalent of the ragged energy landscape would be stochastic volatility. The global minimum dominates the evolution of the conditional probability density with time. Once it has been calculated from the path-integral representation, one again we can use it for expectation-value derivative pricing in a risk-neutral world.

7.8 Path Integrals: Integrating Path Dependence into Option Pricing

It is surprising that less work has been done on the use of path integrals to incorporate path dependences into option theory. The most prominent example of path-dependent options are the plain vanilla American-style options. Depending on the actual path followed by the price $S(t)$ of the underlying, early exercise may or may not be advantageous [10]. In Sect. 4.5.4, we have discussed that the correct pricing of American options requires approximate valuation procedures even when the price of the underlying follows geometric Brownian motion. Some of them certainly can be improved. Exotic options with path-dependent payoff profiles are other examples where the methods described in the following can be useful.

The central problem in pricing path-dependent options is the evaluation of conditional expectation values such as those used on the right-hand sides of (4.100) and (4.101). We can write them in the general form

$$\langle h[S(t)] \mid S(t') \rangle = \int_{-\infty}^{\infty} dx \, h \left[e^{x(t)} \right] q(x, t \mid x't') \,. \tag{7.73}$$

The notation $x = \ln S$ has been kept from above. $q(x \mid x')$, where explicit time variables are dropped from now on, is the transition probability of the log-price between times t' and t. $h(S)$ is the payoff function of the option taken at price S of the underlying. From our earlier discussions, there are a few indications that path integrals could be useful in evaluating the transition probability $q(x \mid x')$, and expectation values involving this quantity: (i) In Sect. 4.5.2, we saw that path integrals led to a quantum Black–Scholes Hamiltonian (4.93) for the European options with the standard solution [51]. (ii) The Fokker–Planck equation employed in Chap. 6 also admits a path-integral representation [37]. The transition probabilities $q(x \mid x')$ only depend on the stochastic process involved and not on the specific option considered. Option properties only enter through the expectation values (7.73).

By iterating the Chapman–Kolmogorov–Smoluchowski equation (3.10), and discretizing time between t' and t into $n + 1$ slices of length $\Delta t = (t - t')/(n + 1)$, we write the transition probability as [173]

$$q(x \mid x') = \int_{-\infty}^{\infty} \dots \int_{-\infty}^{\infty} dx_1 \dots dx_n \frac{1}{\sqrt{(2\pi\sigma^2 \Delta t)^{n+1}}}$$

$$\times \exp \left\{ -\frac{1}{2\sigma^2 \Delta t} \sum_{k=1}^{n+1} \left[x_k - x_{k-1} - \left(r - \frac{\sigma^2}{2} \right) \Delta t \right]^2 \right\} \,. \tag{7.74}$$

Formally, we set $x = x_{n+1}$ and $x' = x_0$. A direct evaluation of $q(x \mid x')$ by Monte Carlo simulation requires very long simulation times when good accuracy is sought. On the one hand, by taking the continuum limit, one can derive a path integral representation [173]

$$q(x \mid x') = \int \mathcal{D} \left[\sigma^{-1} \tilde{x} \right] \exp \left\{ -\int_{t'}^{t} d\tau \, L \left[\tilde{x}(\tau), \frac{dx}{d\tau}; \tau \right] \right\} \tag{7.75}$$

with a Lagrangian

$$L \left[\tilde{x}(\tau), \frac{dx}{d\tau}; \tau \right] = \frac{1}{2\sigma} \left[\frac{dx}{d\tau} - \left(r - \frac{\sigma^2}{2} \right) \right]^2 \tag{7.76}$$

equivalent to the Black–Scholes Hamiltonian (4.93).

On the other hand, one can use substitutions common in the evaluation of path integrals to transform (7.74) into a form which allows a fast and accurate Monte Carlo evaluation. Two steps are necessary to overcome two important obstacles in the evaluation of (7.74) or (7.75). Firstly, the integral kernels are nonlocal in time: (7.74) depends on the difference $x_k - x_{k-1}$, and (7.75) on $dx/d\tau$. An expression local in time would allow to separate the multi-dimensional resp. path integral into a product of independent one-dimensional integrals. Secondly, the integral should be brought into a form which allows a Monte Carlo evaluation which is fast and accurate at the same time. As discussed in Sect. 4.5.4, the convergence of a direct Monte Carlo evaluation is rather slow. One therefore seeks a representation of the integral where Monte Carlo simulations give a good convergence.

The first goal is achieved by the substitution

$$y_k = x_k - k\left(r - \frac{\sigma^2}{2}\right)\Delta t \tag{7.77}$$

which eliminates the drift term in (7.74). The argument of the exponential in (7.74) is transformed

$$\sum_{k=1}^{n+1}\left[x_k - x_{k-1} - \left(r - \frac{\sigma^2}{2}\right)\Delta t\right]^2$$

$$= \sum_{k=1}^{n+1}[y_k - y_{k-1}]^2$$

$$= \boldsymbol{y}^T \cdot \mathbf{M} \cdot \boldsymbol{y} + [y_0^2 - 2y_0 y_1 + y_{n+1}^2 - 2y_n y_{n+1}] \; . \tag{7.78}$$

$\boldsymbol{y}^T = (y_1, \ldots, y_n)$ is the transpose of $\boldsymbol{y}$, and $\mathbf{M}$ is a tridiagonal matrix which can be diagonalized by an orthogonal matrix $\mathbf{O}$ with eigenvalues m_i and eigenvectors w_i. We obtain for the transition probability

$$q(x \mid x') = \frac{e^{-(y_0^2 + y_{n+1}^2)}}{\sqrt{(2\pi\sigma^2\Delta t)^{n+1}}} \prod_{i=1}^{n} \int_{-\infty}^{\infty} dw_i \exp\left\{-\frac{1}{2\sigma^2\Delta t}\right.$$

$$\left. \times \sum_{i=1}^{n}\left(m_i\left[w_i - \frac{(y_0 O_{1i} + y_{n+1}O_{ni})}{m_i}\right]^2 - \frac{(y_0 O_{1i} + y_{n+1}O_{ni})^2}{m_i}\right)\right\} . \tag{7.79}$$

The coupled, multi-dimensional integral over the x_k, resp. y_k, now is decoupled into a product of one-dimensional integrals over the w_i-variables with a Gaussian kernel.

A naive Monte Carlo integration of the integral over w_i uses uniformly distributed random numbers w_i, and determines the value of the integral as the product of the average of the kernel at the positions w_i times the area sampled [174]. The error depends on the standard deviation of the kernel at the positions w_i, and decreases as the inverse square root of number of w_i. This is the problem of slow convergence. The problem will be computationally more efficient if we can transform to a structure where the kernel is constant (or almost constant), and the random numbers are no longer distributed uniformly. This technique is known an importance sampling [174] and, for our kernel, is achieved by the substitution

$$dh_i = \sqrt{\frac{m_i}{2\pi\sigma^2\Delta t}} \exp\left\{-\frac{m_i}{2\pi\sigma^2\Delta t}\left[w_i - \frac{(y_0 O_{1i} + y_{n+1}O_{ni})}{m_i}\right]^2\right\} dw_i \; . \tag{7.80}$$

The resulting transition probability

$$q(x \mid x') = \frac{e^{-(y_0^2+y_{n+1}^2)}}{\sqrt{(2\pi\sigma^2\Delta t)^{n+1}}} \prod_{i=1}^{n} \int_{-\infty}^{\infty} dh_i \tag{7.81}$$

$$\times \exp\left\{-\frac{1}{2\sigma^2\Delta t}\frac{(y_0 O_{1i} + y_{n+1} O_{ni})^2}{m_i}\right\}$$

possesses the desired features: A constant kernel the integral which – in Monte Carlo – is sampled by random numbers h_i drawn from a Gaussian with mean $(y_0 O_{1i} + y_{n+1} O_{ni})^2/m_i$ and variance $\sigma^2\Delta t/m_i$ [173]. With these transformations, the asset price is given by

$$S_i = \exp\left\{\sum_{k=1}^{n} O_{ik} h_k + i\left(r - \frac{\sigma^2}{2}\right)\Delta t\right\}. \tag{7.82}$$

The normal distribution of the h_i implies the log-normal distribution of the prices S_i, as required for geometric Brownian motion. This simple representation of the price in terms of the random sampling variables h_i makes this method well suitable for evaluating path-dependent options.

This path intergral representation and its discretized counterpart transformed as above, are useful for several purposes.

- Firstly, it is a competitive alternative, both in terms of accuracy and calculation speed, to finite difference methods, to binomial trees, and to Green function methods [173].
- Secondly, for American options, the continuum limit in time, $\Delta t \to 0$, can be combined with an "infinitesimal trinomial tree" in the space of the log-prices, x_k, to allow a seminanalytical evaluation of the fundamental integral (7.73). Specifically, noting that the transition probability, for small Δt, is an almost δ-function peak in $x_k - x_{k-1}$, the payoff function $h(e^x)$ is expanded to second order on the trinomial tree $x_k^j = x_{k-1} + r\Delta t + j\sigma\sqrt{\Delta t}$ with $j = -1, 0, 1$. The integral (7.73) is evaluated analytically with the second-order expanded payoff function, and the second-order coefficient is determined by numerical differentiation on the trinomial tree. The implementation of this scheme makes almost negligible the numerical effort of calculated prices and hedges for American options (for geometric Brownian motion) [173].
- The path integral representation can be generalized to path-dependent options on assets following multidimensional, correlated geometric Brownian motion. Examples include options dependent on baskets of stocks, or baskets containing stocks, bonds, and currencies. Often, the statistical properties of the basket price are less well known than those of the constituent assets. Path-dependent exotic options on such baskets can be evaluated by generalizing the techniques described above [175].

8. Microscopic Market Models

In the preceding chapters, we described the price fluctuations of financial assets in statistical terms. We did not ask questions about their origin, and how they are related to individual investment decisions. In the language of physics, our approach was macroscopic and phenomenological. We considered macrovariables (prices, returns, volatilities) and checked the internal consistency of the phenomena observed. In this chapter, we wish to discuss how these macroscopic observables are possibly related to the microscopic structure and rules governing capital markets. We inquire about the relation of microscopic function and macroscopic expression.

8.1 Important Questions

Hence we face the following open problems:

- Where do price fluctuations come from? Are they caused by events external to the market, or by the trading activity itself?
- What is the origin of the non-Gaussian statistics of asset returns?
- How do the expected profits of a company influence the price of its stock?
- Are markets efficient?
- Are there speculative bubbles?
- What is the reference when we qualify market behavior as normal or anomalous?
- Can computer simulations be helpful to answer these questions?
- Can simulations of simplified models give information on real market phenomena?
- Is there a set of necessary conditions which a microscopic market model must satisfy, in order to produce a realistic picture of real markets?
- What is the role of heterogeneity of market operators?
- Is there something like a "representative investor"?
- What is the role of imitation, or herding in financial markets? Is such behavior important, if ever, only in exceptional situations such as crashes, or also in normal market activity?
- Can realistic price histories be obtained if all market operators rely, in their investment decisions, on past price histories alone (chartists) or on company information alone (fundamentalists)?

• Are game-theoretic approaches useful in understanding financial markets?

8.2 Are Markets Efficient?

The efficient market hypothesis states that prices of securities fully reflect all available information about a security, and that prices react instantaneously to the arrival of new information. In such a perspective, the origin of price fluctuations in financial markets is the influx of new information. Here, the origin of price fluctuations would be exogeneous. The information could be, e.g., the expected profit of a company, interest rate or dividend expectations, future investments or expansion plans of a company, etc., which constitute the "fundamental data" of the asset. Traders who hold such an opinion are "fundamentalists" who therefore search/wait for important new information and adjust their positions accordingly.

Opposite to this opinion is the idea that the fluctuations and price statistics are caused by the trading activity on the markets itself, rather independently of the arrival of new information. Here, the origin of the fluctuations is endogenous. Related to this picture is the hypothesis that past price histories carry information about future price developments. This is the basis of "technical analysis". Its practitioners are the "chartists", who attempt to predict future price trends based on historic data. They base their investment decisions on the signals they receive from their analysis tools.

Concerning the crash on October 27, 1997 (the "Asian crisis"), one might ask if and to what extent the cause of the price movements was indeed the collapse of major banks in Asian countries, if they were caused to a large extent by the traders themselves who reacted – perhaps in exaggerated manner – to the news about the bank collapse, or if there was just an accidental coincidence. In a similar way, there are conflicting views about the origins of the crash on Wall Street on October 19, 1987, which cannot be linked unambiguously to a specific information flow.

Unfortunately, it is difficult in practice to make a clear case for one or the other paradigms. One reason is that most traders do not base their investment decisions on one method alone but rather use a variety of tools with both fundamental and technical input. However, the point can also be seen checking one or both paradigms empirically. As an example, Fig. 8.1 shows the expected profit per share of three German blue chip companies: Siemens, Hoechst (now Aventis), and Henkel. The expectation is for the business year 1997, and its evolution from mid-1996 through 1997 is plotted. If the fundamentalist attitude is correct, the evolution of the stock prices should somehow reflect these evolving profit expectations. Figure 8.2 shows the evolution of the Henkel stock price over a similar interval of time. The expected profits of this company increased monotonically from DM4.00/share to DM4.50/share. With the exception of the period July–October 1997, culminating in the crash on October 27, 1997, the stock price by and large followed an upward trend,

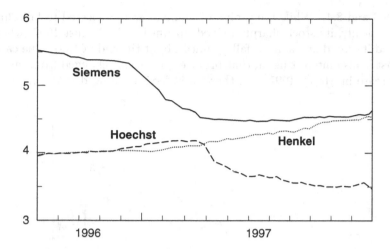

Fig. 8.1. Expected profit per share of three German blue chip companies for 1997, in DM, as a function of time from mid-1996 through 1997: Siemens (*solid line*), Hoechst (*dashed line*), and Henkel (*dotted line*). Adapted from Capital 2/1998 courtesy of R.-D. Brunowski, based on data provided by Bloomberg

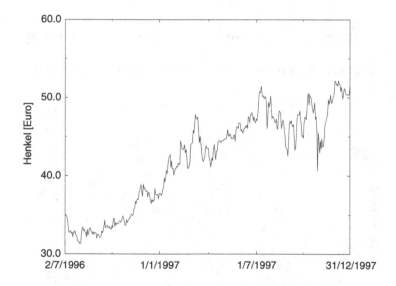

Fig. 8.2. Share price of Henkel from 2/7/1996 to 31/12/1997

too, in agreement with what fundamentalists would claim. If a moving average with a time window of more than 100 days is taken, the drawdowns in summer 1997 are averaged out, and the parallels are even more striking. The situation is, however, much less clear for Siemens and Hoechst, shown in

Figs. 8.3 and 8.4. While the profits of Siemens were expected to fall almost monotonically, its stock sharply moved up until early August 1997, when it reversed its trend and started falling until about the end of 1997. The case of Hoechst is also interesting, in that profit expectations changed from increase to decrease in March 1997, and there is indeed a strong drawdown in their

Fig. 8.3. Share price of Siemens from 2/7/1996 to 31/12/1997

Fig. 8.4. Share price of Hoechst (now Aventis) from 2/7/1996 to 31/12/1997

stock price in that period. However, the further evolution does not appear to be strongly correlated with the expected profits per share. These three examples show that, while there is some evidence for the influence of fundamental data on stock price evolution, this evidence is not so systematic as to rule out other, possible endogenous, influences.

Another issue of market efficiency, often discussed in conjunction with crashes, is about speculative bubbles. In such a bubble, prices deviate significantly from fundamental data, and increasingly so in time. They are believed to be caused by some positive feedback mechanism, such as imitation, or herding behavior, and self-fulfilling prophecies are often involved. An important issue in economics is whether such bubbles can be detected, controlled, and avoided. One explanation forwarded for Black Monday on Wall Street, the crash on October 19, 1987, is related to a hypothetical speculative dollar bubble. It is not universally shared, however.

Currency markets, e.g., are very speculative with only a small fraction of the transaction being executed for real trading purposes (paying a bill in foreign currency). Most transactions are due to speculation. The sheer amount of trading volume raises doubts about market efficiency. Tobin therefore proposed raising a small tax on currency transactions, in order to raise the threshold of speculative profits, in order to prevent the formation of bubbles. The question, of course, is whether such a Tobin tax would be successful, or whether it would adversely affect currency markets.

The big problem with speculative bubbles, however is their timely diagnosis. To this end, one must know the fundamental data, and they must be translated into asset prices with the correct market model. Any misspecification of the model will inevitably lead to incorrect diagnoses about bubbles. As a recent example, take the internet, or "New Economy" bubble 1996–2000. During this period, the DAX returned about 30% per year, cf. Fig. 1.2. While from about 2001, this period has been recognized as a speculative bubble, essentially nobody voiced such an interpretation during the period in question.

Unlike in physics, where controlled laboratory experiments are usually carried out to answer similar questions, economics does not allow for such experiments. Computer simulation of models for artificial markets is therefore the only possibility of clarifying some aspects of these problems. The situation is rather similar to climate research where large-scale experiments are also impossible, but there is an obvious need for (at least approximate) answers to a variety of questions ranging from weather forecasting, to the greenhouse effect, to the ozone hole, etc. For a physicist, a market is basically a complex system away from equilibrium, and such systems have been simulated in physics with success in the past.

8.3 Computer Simulation of Market Models

Computer simulations of markets have a long history in economics. Two early examples were concerned precisely with market efficieny [5], and aspects of the 1987 October crash on Wall Street [176].

8.3.1 Two Classical Examples

Stigler challenged both the statements, and the assumptions underlying a report of a committee of the US congress on the regulations of the securities markets in the US [5]. This report tested market efficiency by two methods which, in an essential way, relied on continuous stochastic processes for the prices, in order to be significant. In the course of his arguments, Stigler devised a simple random model of trading at an exchange. Starting from a hypothetical order book with 10 buy orders at subsequent prices on one side (labeled 0, ..., 9), and no sell orders on the other side, prices are generated from two-digit random numbers. The parity of the first digit (even or odd) indicates if the price is bid (buy) or ask (sell). There are rules when transactions take place (bid > ask), how to treat unfulfilled orders, etc. This simple model creates a strongly fluctuating transaction price, certainly not the smooth price histories assumed in the tests conducted in the report.

Another market model was developed by Kim and Markowitz in response to speculations about the role of portfolio insurance programs during the 1987 October crash on Wall Street [176]. A published report ascribed a large part of the crash to computerized selling of stock by portfolio insurance programs run by large institutional investors. This view, however, was disputed by others, and no consensus could be reached. The aim of Kim and Markowitz' work was to study if a small fraction of portfolio insurance sell orders could sufficently destabilize the market, to lead to the crash.

Portfolio insurance is a trading strategy designed to protect a portfolio against falling stock prices. The specific scheme, constant proportion portfolio insurance, implemented in Kim and Markowitz' model illustrates the general ideas. At the beginning, a "floor" is defined as a fraction of the asset value, say 0.9. The "cushion" is the difference between the value of the assets (including riskless assets such as bonds or cash), and the floor. At finite times, one ideally leaves the floor unchanged, and the cushion changes with time as the stock price varies. (In practice, however, the floor must be adjusted both for deposits and withdrawals of money, and for changes in interest rates of the riskless assets.) One now defines a target value of the stock in the portfolio as a multiple of the cushion. As an example, suppose that a portfolio worth $ 100,000 consists of $ 50,000 in stock and $ 50,000 in cash, and that the target value of stock is five times the cushion. The floor is then $ 90,000 and the cushion is $ 10,000. Now assume that the value of the stock falls to $ 48,000. The cushion reduces to $ 8,000, and the target value of stock falls to $ 40,000. The portfolio manager (or his computer program) will sell stock worth $ 8,000.

In addition to portfolio insurers, the model contains two populations of "rebalancers". These agents will attempt to maintain a fixed stock/cash ratio in their portfolio, and give buy and sell orders accordingly. The two groups of rebalancers have different preferred stock/cash ratios, i.e., different risk aversion. At the beginning, all three populations receive the same starting capital, half in stock and half in cash. One rebalancer group has a preferred stock/cash ratio larger than $1/2$, the other one smaller. This element of heterogeneity is important for the simulation. There is a set of rules which determine the course of trading. The stock price changes because the two rebalancer groups will place orders to reach their preferred stock/cash ratio. This will generate orders from the portfolio insurance population, etc.

With rebalancers alone, important trading activity takes place at the beginning but quickly dies out because they can reach their preferred stock/cash ratios. As the fraction of portfolio insurance population on the market increases from zero to two-thirds, the volatility of the prices increases by orders of magnitude. Returns and losses of 10–20 % in a single day are not uncommon. It is therefore very conceivable that portfolio insurance schemes contributed to the crash on October 19, 1987. Interestingly, when the simulations allowed margins for the dealers, i.e., the possibility of short selling or buying on credit, the market exploded even with a 50% portfolio insurer population and 33% margin. Prices would then diverge, and the simulation had to be stopped.

This work gives a good impression of the sensitivity of such models in general, and the need to specify them correctly in terms of both rules and initial conditions. Meanwhile, computer simulations in economics have become more complex and have greater performance, and some scientists have attempted to model a stock market under rather realistic conditions. In the most advanced simulations, agents can evaluate their performance and change their trading rules in the course of the simulation [177]. In some cases, however, the models have become so complex that a correct calibration is difficult, to say the least.

8.3.2 Recent Models

The physicist's approach, on the other hand, is usually to formulate a minimal model which depends only on very few factors. Such a simple model may not be particularly realistic, but the hope is that it will be controllable, and allow for definite statements on the relation between observables, such as prices or trading volumes, and the microscopic rules. Once such a simple model is well understood, one might make it more realistic by gradually including additional mechanisms. Such models will be presented in Sect. 8.3.2.

The first model was chosen rather arbitrarily to introduce the general principle. Other models, in part historically older, will be discussed later.

Space, however, only allows us to discuss the most general principles, and we refer the reader to a more specialized book for more details [20].

A Minimal Market Model

One minimal model for an artificial market was proposed and simulated by Caldarelli et al. [178]. It consists of a number of agents who start out with some cash and some units of one stock, an assumption common to most models. The agents' aim is to maximize their wealth by trading. The only information at their disposal is the past price history, i.e., an endogenous quantity. There is no exogenous information. The agents therefore behave as pure chartists. Therefore, this model addresses the interesting question of whether realisitic price histories can be obtained even in the complete absence of external, fundamental information.

Structure of the Model There are N agents (e.g., $N = 1000$), labeled by an integer $i = 1, \ldots, N$. Their aim is to maximize their wealth $W_i(t)$ at any instant of time t:

$$W_i(t) = B_i(t) + \Phi_i(t) S(t) \ . \tag{8.1}$$

Here, $B_i(t)$ is the amount of cash owned by agent i at time t, and it is assumed that no interest is paid on cash ($r = 0$ in the language of the preceding chapters). $S(t)$ is the spot price of the stock, and $\Phi_i(t)$ is the number of shares that agent i possesses at t. It is also assumed that there is no long-term return from the stock, i.e., the drift of its stochastic process vanishes: $\mu = 0$. Agents change their wealth (i) by trading, i.e., simultaneous changes of $\Phi_i(t)$ and $B_i(t)$, and (ii) by changes in the stock price $S(t)$.

The trading strategies of the agents are "random" in a sense to be specified, across the ensemble of agents, but constant in time for each agent. In order to "refresh" the trader population, at any time step, the worst trader $[\min_i W_i(t)]$ is replaced by a new one with a new strategy. This, of course, is to simulate what happens in a real market where unsuccessful traders disappear quickly. Apart from this replacement, there are no external influences, and the system is closed.

Trading Strategies The agents place orders, i.e., want to change their $\Phi_i(t)$ by $\Delta\Phi_i(t)$ with

$$\Delta\Phi_i(t) = X_i(t)\Phi_i(t) + \frac{\gamma_i B_i(t) - \Phi_i(t) S(t)}{2\tau_i}. \tag{8.2}$$

There are two components implemented here.

The first term is purely speculative: $X_i(t)$ is the fraction of the number of shares currently held by agent i, which he wants to buy ($X_i > 0$) or sell ($X_i < 0$) in the next time step. Each agent evaluates this quantity from the price history based on the rules that define his trading strategy, i.e., from a set of technical indicators. One may determine $X_i(t)$ from a utility function $f_i(t)$ through

$$X_i(t) = f_i [S(t), S(t-1), S(t-2), \ldots,] \quad , \tag{8.3}$$

i.e., the agent follows technical analysis to reach his investment decision. Each agent's utility function f_i is now parametrized by a set of ℓ indicators I_k. These indicators are available to all agents. Possible indicators are

$$
\begin{aligned}
I_1 &= \langle \partial_t \ln S(t) \rangle_T \sim \left\langle \ln \frac{S(t)}{S(t-1)} \right\rangle_T \\
I_2 &= \langle \partial_t^2 \ln S(t) \rangle_T \\
I_3 &= \langle [\partial_t \ln S(t)]^2 \rangle_T
\end{aligned}
\tag{8.4}
$$

$$\ldots$$

The symbols $\langle \ldots \rangle_T$ denote moving averages over a time window T, i.e., from $t-T$ to t. An exponential kernel is claimed to be chosen [178] but this is not clear from the explicit expressions given.

The individual trading strategies are then defined through the set of weights η_{ik} which the agents i use on the indicator I_k, to compose their utility function. Each agent forms his or her global indicator according to

$$x_i = \sum_{k=1}^{\ell} \eta_{ik} I_k (\{S\}) \ . \tag{8.5}$$

The utility function is then implemented as a simple function of the global indicator

$$X_i(t) = f(x_i) \tag{8.6}$$

and should have the following properties: (i) $|f(x)| \le 1$ since $X_i(t)$ is the *fraction* of stock to be sold (bought) at time t, and short selling is not permitted; (ii) $\text{sign}(f) = \text{sign}(x)$, i.e., negative indicators trigger sell orders and positive indicators lead to buying; (iii) $f(x) \to 0$ for $|x| \to \infty$, implementing a cautious attitude when fluctuations become large. This may be unrealistic, especially when $x \to -\infty$, so exceptional situations in practice may not be covered by this model. The function chosen in [178] is

$$f(x) = \frac{x}{1 + (x/2)^4} \ . \tag{8.7}$$

Notice, however, that this $f(x)$ violates condition (i)!

The second term in (8.2) represents consolidation. The idea is that every trader, according to his attitude towards risk, has a favorite balance between riskless and risky assets. In a quiet period, he or she will therefore try to rebalance his portfolio towards his personal optimal ratio, in order to be in the best position to react to future price movements. This is exactly the strategy of the "rebalancers" of Kim and Markowitz [176]. The optimal stock/cash ratio is given by

$$\gamma_i = \frac{\Phi_i S}{B_i} \tag{8.8}$$

and is reached with a time constant τ_i. This interpretation of the second term is reached from (8.2) by putting the first term to zero (in a quiet period, the indicators should be small or zero, making X_i vanish). An important difference between this model and the one by Kim and Markowitz lies in the implementation of heterogeneity: here, there is a single population of traders with heterogeneous strategies (random numbers) whereas in the earlier work, there were three populations with homogeneous strategies.

To simulate random trading strategies, the variables η_{ik} and γ_i, τ_i are chosen randomly (although it is not specified from what distribution). These numbers completely characterize an agent.

Order Execution, Price Determination, Market Activity Price fixing and order execution are then determined by offer and demand. Agents submit their orders calculated from (8.2) as *market orders*. The total demand $D(t)$ and offer $O(t)$ at time t are simply sums of the individual order decisions

$$D(t) = \sum_{i=1}^{N} \Delta\Phi_i(t)\Theta\left[\Delta\Phi_i(t)\right] ,$$

$$O(t) = -\sum_{i=1}^{N} \Delta\Phi_i(t)\Theta\left[-\Delta\Phi_i(t)\right] , \tag{8.9}$$

where $\Theta(x)$ is the Heavyside step function. Usually, demand and offer are not balanced. If $D(t) > O(t)$, the shares are alloted as

$$\begin{cases} \overline{\Delta\Phi}_i(t) = \Delta\Phi_i(t)\frac{O(t)}{D(t)} & \text{if } \Delta\Phi_i(t) > 0 \\ \overline{\Delta\Phi}_i(t) = \Delta\Phi_i(t) & \text{if } \Delta\Phi_i(t) < 0 . \end{cases} \tag{8.10}$$

Each agent who wanted to buy gets a fraction of shares $\overline{\Delta\Phi}_i(t) < \Delta\Phi_i(t)$ of his buy order, while the sell orders are all executed completely. The reverse holds if $O(t) > D(t)$. The new price is then fixed as

$$S(t+1) = S(t)\frac{\langle D(t)\rangle_T}{\langle O(t)\rangle_T} . \tag{8.11}$$

Apparently [178], the moving averages here extend over the same time horizon as the indicators underlying the investment decisions. One may have a critical opinion on this fact. The order execution and price fixing is somewhat different from real markets, discussed in Sect. 2.6. It is not clear to what extent the outcome depends on these details.

The model is then run as follows: (i) initialize the market by defining all agents through their random numbers $\eta_{ik}, \gamma_i, \tau_i$, and by giving all dealers their starting capital $B_i(0), \Phi_i(0)$ while the initial stock price is $S_i(0)$. Few specifications are found in the literature [178] on how this is done precisely. In our own simulations, we gave all dealers the same amount of cash $B_i(0) = B$ and shares $\Phi_i(0) = \Phi$ so that the initial value of cash and shares was equal $\Phi S(0) = B$. Trading and price fluctuations then initially arise just because this equipartition does not correspond to the preferred consolidation level of the agents. Following this, the different indicators acquire nonzero values, and will take their influence on the operators' investment decisions. After a finite transient, the results should become independent of these starting details. (This statement has, however, not been checked extensively.)

At $t = 1$, finite $\Delta\Phi_i(t), D(t), O(t)$ are found, and the dealers who had issued buy orders change the number of stocks in their portfolio, and their amount of cash as

$$\Phi_i(t+1) = \Phi_i(t) + \overline{\Delta\Phi}_i(t) , \tag{8.12}$$
$$B_i(t+1) = [1 + \varepsilon_i(t)]B_i(t) - S(t)(1 + \pi)\Delta\Phi_i(t) , \tag{8.13}$$

and likewise for the dealers with sell orders (but $\pi = 0$). $\varepsilon_i(t)$ is a small random number of order 10^{-3} whose origin and importance have remained rather obscure (looking like a random interest rate), and π are transactions costs. The second term in (8.13) is just the price of the shares acquired. The new price $S(t + 1)$ is then fixed according to (8.11), and the wealth balance $W_i(t)$ is evaluated for each operator. Finally, the worst operator is replaced by a new one, the indicators are updated, and new orders are placed.

Results Price fluctuations are clearly the first issue one is interested in. Figure 8.5 shows price histories obtained in a simulation. At least to the eye, they look rather realistic, indicating that many of the essentials of real markets might have been captured by this simple model. More importantly, the data show scaling behavior within the limits of their accuracy. The upper panel of Fig. 8.6 shows raw data for the probability distribution $p(\Delta S_\tau, \tau)$, of price changes $\Delta S_\tau = S(t + \tau) - S(t)$ over a time horizon τ, for various $\tau = 4, \ldots, 4096$. Again these probability distributions look rather similar to those obtained on real markets, especially concerning the fat tails for large changes. The pronounced peak for small price changes is not usually observed on real markets. This peak may be due to the use of an exponential memory kernel in the indicators [86]. If the price changes and probability distributions are rescaled as

$$\Delta S_\tau \quad \rightarrow \Delta S_\tau / \tau^H$$
$$p(\Delta S_\tau, \tau) = \tau^{-H} p(\Delta S_\tau \tau^{-H}, 1) \tag{8.14}$$

with a Hurst exponent $H = 0.62$, all data collapse onto a single universal curve, as shown in the lower panel of Fig. 8.6. Observation of scaling of this kind suggests that the different distributions observed are generated from a single master curve by variation of one parameter, the time horizon τ over

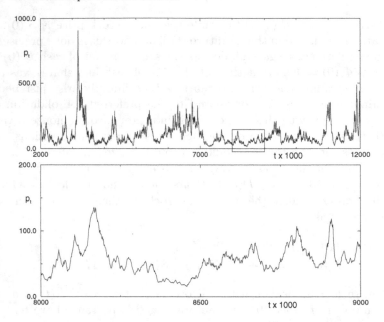

Fig. 8.5. Price history for a system of 1000 agents. Prices p_t in the figure correspond to $S(t)$ in the text. The parameters are $\epsilon = 0.01$ and $\pi = 10^{-3}$. The lower part is a zoom of the area in the upper rectangle. By courtesy of M. Marsili. Reprinted from G. Caldarelli, et al.: Europhys. Lett. **40**, 479 (1997), ©1997 EDP Sciences

which the returns are evaluated, and that the same underlying mechanism is responsible for the functional form of all probability distributions. The value of the Hurst exponent is derived from a power law found for the return probability to the origin over the horizon τ, $p(0, \tau) \sim \tau^{-H}$ [178]. For Lévy distributions, $\mu = 1/H = 1.61$, quite close, in fact, to the values $\mu \sim 1.4$ found in empirical studies, e.g., by Mantegna and Stanley [69] of the S&P500 index. If copying of successful strategies is allowed, e.g., when new traders replace the unsuccessful ones, similar data are obtained, but the exponent $H = 0.5$ now, i.e., scaling is like that for a random walk.

Extremal events, i.e., $|\Delta S_\tau| \to \infty$, obey slightly different statistics

$$p(\Delta S_\tau, \tau) \to |\Delta S_\tau|^{-2} \quad \text{for} \quad |\Delta S_\tau| \to \infty . \tag{8.15}$$

This is higher than both a Lévy flight ($p \sim |\Delta S_\tau|^{-(1+\mu)}$) and practice ($p \sim |\delta S_\tau|^{-4}$, cf. Sect. 5.6.1), and might indicate that traders act according to different rules in such extreme situations [178].

The distribution of wealth, after a sufficiently long run, is described by Zipf's law,

$$W_n \sim n^{-1.2} \tag{8.16}$$

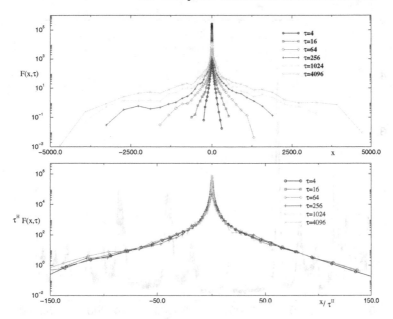

Fig. 8.6. Raw data for the probability distribution of price changes (*upper panel*), and rescaled probability distributions (*lower panel*). The scaling procedure is explained in the text. x is ΔS_τ, and $F(x, \tau)$ is $p(\Delta S_\tau, \tau)$ in the text. By courtesy of M. Marsili. Reprinted from G. Caldarelli, et al.: Europhys. Lett. **40**, 479 (1997), ©1997 EDP Sciences

where the traders have been reordered according to their wealth, i.e., $W_1 > W_2 > \cdots > W_{1000}$. Quite early, Zipf had found that the distribution of wealth of individuals in a society follows a power law [179].

Criticism Despite these encouraging results, there are a few problems with this work. Some of them have been mentioned above, e.g., the fact that the required bounds on $f(x)$, (8.6), are violated, or that the published kernel for the moving averages does not produce exponential decay in time.

Moreover, while the authors state that the results are rather independent of initial parameters and robust against variation, it appears that fine-tuning of parameters is necessary, indeed, at least into certain parameter ranges. Together with A. Rossberg (Bayreuth/Kyoto), we have written a program for this model and attempted to calibrate it against the published results. These attempts have failed so far. While for special values of the parameters, we indeed observed a rather dynamical price history over half a million time steps, this was rather the exception than the rule. Such an example is shown in Fig. 8.7. More typically, we have observed price histories such as that shown in Fig. 8.8 where a rapid "equilibration" of the system into a state with strongly bounded price variation occurs. Here, the price variations seem to be rather similar to those one would observe from a Gaussian random walk.

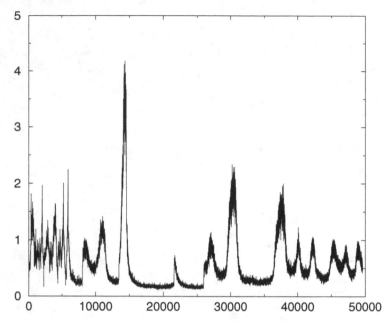

Fig. 8.7. Exceptional results of a simulation of the CMZ model by Rossberg and Voit. The price history $S(t)$ is shown in arbitrary units. Only every 10th data point of a 500 000 step simulation is shown

Looking at their microstructure, however, reveals that they are quasiperiodic and not random. The origin of this quasiperiodicity is not clear, at present. In the same way, the differences between these more typical results of the simulations by Roßberg and Voit, and those of Caldarelli, et al., have not yet been understood.

The Levy–Levy–Solomon (LLS) Model

An earlier model simulation by Levy, Levy, and Solomon [180] emphasizes the role of agent heterogeneity on the price dynamics of financial assets. Here, we only discuss the most elementary aspects of this model. There is much literature on this model and various extensions [20].

Structure of the Model Levy, Levy, and Solomon consider an ensemble of agents which can switch between a risky asset (stock) and a riskless bond [180]. The bond returns interest with rate r. There is a positive dividend return on the stock, and additional (positive or negative) returns arise from the variation of the stock price. Time steps in this model are taken as years. Unlike the previous model which focuses on short-term speculative trading, this model takes a long-term perspective and has a strong fundamentalist element.

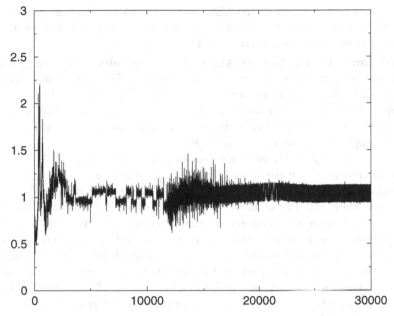

Fig. 8.8. Typical results of a simulation of the CMZ model by Rossberg and Voit. Shown is the price history $S(t)$ in arbitrary units. Only every 10th data point of a 300 000 step simulation is shown

The evaluation of order volumes and prices differs from the preceding model. The traders have a memory span of k time steps. The price, or return, which they expect for the next time step, is taken from the past k prices with equal probability $1/k$. From these expected prices, they determine their order volume by maximizing a utility function $f[\langle W(t+1)\rangle]$ of their expected wealth $\langle W(t+1)\rangle$ at the next time step. The utility function should be monotonically increasing and concave, e.g., $f(W) = \ln W$.

Prices are determined by demand and supply. To do this, LLS assume a series of hypothetical prices $S_h(t+1)$ for the next time step. The wealth of an investor at $t+1$ will then depend on this price, and on his order volume. The agent can now determine, for each hypothetical price $S_h(t+1)$, his corresponding order volume $X_h(t+1)$ from his utility function. Then, the hypothetical order volumes $X_h(S_h, t+1)$ are summed over all investors, to determine the aggregate demand and supply functions of the market. This is rather similar to our determination of the same functions in a stock exchange auction with limit orders. The stock price is then determined by the intersection of the demand and supply functions, as in Sect. 2.6. Up to this point, everything is deterministic. Randomness is now introduced by giving the $X(t)$ a random component which is drawn from a Gaussian.

An important element of the LLS work is that they simulate two different versions of the model, one with a homogeneous trader population, and another one with hetereogeneous traders.

Agent Homogeneity Versus Agent Heterogeneity The homogeneous model has been specified in the preceding section. The only trader-specific component is the random number added to the order volumes of the various traders. Interest rates were taken as 4% per year, and the initial dividend yields were 5% per year. Dividends were increased by 5% annually. Similar numbers apply to the S&P500 index [180].

Such a model goes through a series of booms and crashes [180]. After an initial transient, the stock price rises exponentially with the return rate of the dividends. This rapid rise makes the investors very bullish about the stock, and they will invest into the stock as much as possible. However, in such a homogeneous situation, a small change in return can lead to a discontinuous change of investment preferences, and trigger massive sales. The market crashes and reaches a bottom at a much lower level. Again, it will become more homogeneous, and a small increase of returns will trigger a boom: investors sell the bond and buy the stock, and the price increases sharply. This pattern reproduces periodically, with the period equal to the memory span of the investors.

Additional heterogeneity can be introduced in several ways. One can give the agents different memory spans, or different utility functions. In both cases, the return histories lose their periodicity. In the simplest case with two populations with different memory spans, the returns still oscillate between the two limiting values of the homogeneous model, but the oscillations are "less periodic" than before. Not surprisingly, they become more aperiodic when the memory spans of the traders are randomized, and when in addition, they get different utility functions. Finally, when another population is introduced which holds a constant investment proportion in the stock, price histories are simulated which compare favorably with the actual evolution of the S&P500.

This work shows, among other things, that heterogeneity is an important element in the financial market. A "representative investor" as assumed in many theoretical arguments of economics, is a construction which is not justified by the behavior of real markets. Moreover, it shows that several elements of heterogeneity must be present simultaneously, in order to produce apparently realistic time series, such as heterogeneity of memory, of expectations, and investment strategies. When the market becomes more homogeneous, crashes are inevitable. Notice finally that so much has been learned about real markets because of the extensive discussion of simulation results which *deviate significantly* from real market behavior [180].

Ising Models, Spin Glasses, and Percolation

In the previous models, the amount of stock bought or sold by the traders was a continuous variable. One can achieve a higher degree of simplification by replacing this continuous variable by a discrete three-state variable $\Delta\Phi_i(t)$ [181]: $\Delta\Phi_i(t) = +1, 0, -1$ according to whether the trader i wants to buy one unit of stock at time t, stay out of the market, or sell one unit of stock. The greater simplification allows one to introduce additional features of complexity into the model.

An article by Iori addresses three possibly important mechanism of price dynamics in financial markets: heterogeneity, threshold trading, and herding [181]. Heterogeneity will no longer be discussed here. We have seen in the preceding section that it is essential. When the possible order volumes are restricted to $1, 0, -1$, threshold trading is a necessity. However, it is also an important fact in reality. An investor will not enter a market whenever he receives a positive signal (e.g., from his utility functions discussed above, or from technical or fundamental analysis) however small. In the presence of transaction costs, the expected profit from the trade must at least provide for these costs. Moreover, investors usually buy or sell stock only when they are *sufficiently* bullish or bearish about it. Thus, orders are placed only when the signals received are beyond certain thresholds. In Iori's model, traders have heterogeneous thresholds $\xi_i^{\pm}(t)$ which vary with time, and the actions are taken as a function of a trading signal $Y_i(t)$ according to

$$\Delta\Phi_i(t) = \begin{cases} +1 \text{ if } & Y_i(t) \geq \xi_i^+(t) \,, \\ 0 \text{ if } & \xi_i^-(t) < Y_i(t) < \xi_i^+(t) \,, \\ -1 \text{ if } & Y_i(t) \leq \xi_i^-(t) \,. \end{cases} \tag{8.17}$$

The third important aspect is communication between the agents, leading to herd behavior in its extreme consequences. Direct communication has not been modeled in the previous sections. There, the traders "interacted" only through the common variable of the past price history. Here, communication is explicitly modeled in the trading signal which each agent receives at time t:

$$Y_i(t) = \sum_{\langle i,j \rangle} J_{ij} \Delta\Phi_j(t) + A\nu_i(t) + B\varepsilon(t) \,. \tag{8.18}$$

J_{ij} is the interaction, or communication, between agents i and j, and the symbol $\langle i, j \rangle$ restricts the sum to those j which are nearest neighbors to i. $\nu_i(t)$ represents idiosyncratic noise of the traders, and $\varepsilon(t)$ is a noise field common to all traders. This could be, e.g., the arrival of new information. The model assumes that the traders "live" on a two-dimensional square lattice. However, this assumption can probably be relaxed, and it would certainly be interesting to introduce a more realistic communication structure. The idea of "small-world networks" [182] could prove useful here.

Depending on the choice of the interaction parameters J_{ij}, one recovers variants of interesting physical problems. If all $J_{ij} = 1$, one has the random-field Ising model [183]. $\Delta\Phi_i(t)$ plays the role of the spins (for consistency with the remainder of this book, we avoid the symbol S for the Ising spins here), and the model has spin 1 (the inactive state $\Delta\Phi_i(t) = 0$ is not allowed in the standard spin-1/2 model). As a function of the noise level, this model has a transition from a paramagnetic to a ferromagnetic state. If $J_{ij} = 1$ with a certain probability p, and zero otherwise, one obtains a bond percolation problem [184]. Finally, with J_{ij} random, a spin-glass problem is generated [185]. In this case, as well as in the random-field Ising limit, the first term in (8.18) is the Weiss molecular field.

In this model, a price history is generated although apparently the stock prices do not influence the traders' decisions to buy or sell. The traders receive cash and stock as in the preceding sections. Before the first trade, a consultation round is opened. Traders whose idiosyncratic signals $A\nu_i(t)$ exceed the thresholds manifest their ordering decisions $\Delta\Phi_i(0)$. Then traders decide sequentially if they want to revise their decisions under the influence of the communication term $\sum J_{ij}\Delta\Phi_j(0)$, i.e., follow their neighbors. This process continues until convergence is reached. Then orders are placed, and the price $S(t)$ is changed according to demand and supply, (8.9), as

$$ S(t+1) = S(t) \left(\frac{D(t)}{O(t)} \right)^\alpha \quad \text{with} \quad \alpha = \frac{D(t) + O(t)}{N} . \qquad (8.19) $$

The numerator of the exponent α is the trading volume, and the denominator is the number of traders, i.e., the number of sites of the square lattice, $N = L^2$. At the same time, due to (8.17), it is the maximal number of stocks that can be traded at any single time step. The power-law dependence in the price law creates stronger price changes when there is a large imbalance between demand and supply, which is reasonable. The dependence of the exponent on the trading volume generates a correlation of price changes with trading volume. α reduces the influence of an imbalance of demand and supply if it is created only by very few traders. At the end, the thresholds of the traders are adjusted by multication by $S(t+1)/S(t)$. This is the only way the actual prices can influence the trading decisions of the agents.

When the model is simulated in the percolation mode, one can clearly observe the influence of communication. In the absence of thresholds, the price fluctuations increase by an order of magnitude when the probability of $J_{ij} = 1$ is increased from 0.4 to 0.8 [181]. Finite but fixed thresholds stabilize the system. Even for $p = 1$, i.e., in the random-field Ising limit, the price fluctuations are strongly bounded, and presumably give rise to Gaussian statistics of returns. Interactions between the agents increase the fluctuations but do not change them qualitatively. Occasional big fluctuation periods, i.e., volatility clustering, is observed only when interactions are combined with adjusting thresholds. The lower curve in Fig. 8.9 shows the results of such a simulation. Periods of quiescence and turbulence are observed in this market. Trading is hectic in turbulent times, as shown by the positive correlation of volatility and trading volume. This effect is also observed in real financial markets [186], and has been built into this model through the structure of the exponent α.

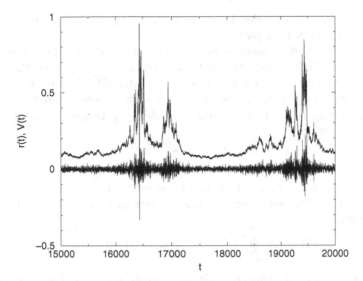

Fig. 8.9. Return of stock $r(t)$ (*lower curve*) and trading volume $V(t)$ (*upper curve*) in a simulation of a random-field Ising model for stock markets. Notice the correlation between volatility and trading volume. By courtesy of G. Iori. Reprinted from G. Iori: Int. J. Mod. Phys. C **10**, 1149 (1999), ©1999 by World Scientific

News arrival ($\varepsilon(t) \neq 0$) also leads to a synchronization of the traders even in the absence of interaction. Adjusting thresholds, and communication, however increase the volatility clustering in the time series [181]. Finally, with all the important factors present at the same time, the model reproduces the important features of financial time series discussed in Chap. 5, such as fat tailed probability distributions, a crossover from Lévy-like to more Gaussian statistics as the time scale of the returns increases, and the long-time correlations of the absolute returns and volatility.

A particularly simple model of threshold trading was introduced by Sato and Takayasu [187]. Here, all dealers (labelled by i) publish their bid and ask prices B_i and A_i, and they all have the same bid–ask spread $\Lambda = A_i - B_i$. A trade can be concluded between dealers i and j when $B_i > A_j$, and one chooses those traders who propose the maximal bid and minimal ask price. The transaction price S is fixed as the arithmetic mean of the bid and ask prices. In each time step, traders change their bid (and ask) prices as

$$B_i(t+1) = B_i(t) + a_i(t) + c\left[S(t) - S(t_{\text{prev}})\right],\qquad (8.20)$$

where $a_i(t)$ denotes the ith dealer's expectation of the bid price in the next time step (the idiosyncratic noise above) and c is the dealers' response to a change in market price since the last trade at t_{prev}, i.e., a trend-following attitude. Finally, it is assumed that the traders' resources are limited, and that therefore they want to become sellers after buying and buyers after selling. This can be included by changing the sign of their $a_i(t)$ after each trade in which they took part.

This simple model generates interesting price histories [187]. For $c \approx 0$, price changes follow exponential statistics. For larger c, however, they follow power laws, and for $c = 0.3$, e.g., a Lévy-like probability distribution function with an exponent $\mu \approx 1.5$ is found. Larger c gives even smaller exponents. More interesting is the fact that one can derive a Langevin-like stochastic difference equation

$$\Delta S(t_{k+1}) = c n_k \Delta S(t_k) + \phi_k \qquad (8.21)$$

in terms of three more elementary stochastic processes. t_k is the time of the kth trade, and $n_k = t_{k+1} - t_k$ is the time interval between two successive trades. n_k is a stochastic variable and drawn from a discrete exponential distribution

$$W(n) = \sum_{m=1}^{\infty} \frac{1 - e^{-\gamma}}{e^{-\gamma}} \exp(-\gamma m) \delta(n - m) . \qquad (8.22)$$

Even for $c = 0$ price fluctuations exist at the trading times. They are denoted by ϕ_k, and drawn from a Laplace distribution

$$U(\phi) = \frac{1}{2\sigma} \exp(-|\phi|/\sigma) . \qquad (8.23)$$

Finally, $\Delta S(t_k)$ is the price change at the last trade. From a detailed analysis of the individual stochastic processes in terms of the microscopic parameters c, Λ, the number of traders, and the width of the distribution of the a_i, one can derive conditions on these parameters to find, e.g., power-law scaling in the distribution of returns [187]. It is interesting that very recent empirical studies also seem to find evidence for a decomposition of the price or return process of a financial time series, into more elementary processes involving the waiting times between trades, etc. [114, 188].

Increasing communication led to stronger price fluctuations in the model by Iori, and communication was an essential ingredient to obtain a realistic volatility clustering. Herding must be an important factor in financial markets. On the one hand, economic studies produce evidence for herd behavior [189]. On the other hand, there are also mathematical and physical arguments showing that the independent agent hypothesis cannot be a good approximation under any circumstances [190]. The argument basically goes as follows. Assume that price changes of a stock are roughly proportional to excess demand. If then the probability of a certain demand by an individual agent has a finite variance, and agents act independently, the central limit theorem guarantees the convergence of the excess demand distribution to a Gaussian. By proportionality, price changes should then also obey Gaussian statistics. Even if we do not assume a finite variance of the individual demand distribution, the generalized central limit theorem would still require convergence to a stable distribution. The persistent observation of price changes being strongly non-Gaussian and nonstable, cf. Chap. 5, such as a truncated Lévy distributions, or a power-laws with non-stable exponents

$$p(\Delta S) \sim |\Delta S|^{-(1+\mu)} \exp(-a|\Delta S|) \quad \text{or} \quad p(\Delta S) \sim |\Delta S|^{-4} \tag{8.24}$$

is solid evidence against an independent agent approach.

The influence of communication and herding alone can best be studied by focusing on an even simpler model: percolation, as proposed by Cont and Bouchaud [190]. Agents again have three choices of market action: buy, sell, or inactive, as in Iori's model. They can form coalitions with other agents who share the same opinion, i.e., choice of action. N agents are assumed to be located at the vertices of a random graph, and agent i is linked to agent j with a probability p_{ij}. A coalition is simply the ensemble of connected agents (a cluster) with a given action $\Delta\Phi_i$. This, of course, precisely defines a percolation problem [184].

Agents in a cluster share the same opinion and do not trade among themselves. They issue buy and sell orders to the market with probabilities $P(\Delta\Phi_i = +1) = P(\Delta\Phi_i = -1) = a$, and remain out of the market with $P(\Delta\Phi_i = 0) = 1 - 2a$. a is the traders' activity, and for $a < 1/2$, a fraction of traders is inactive. If all $p_{ij} = p$, the average number of agents, to which one specific agent is connected, is $(N-1)p$. In order to solve the model, one is interested in the limit $N \to \infty$. In this limit, $(N-1)p$ should remain finite so that the probability of a link scales as $p = c/N$. Finally, price changes are assumed to be proportional to the excess demand

$$\Delta S \propto \sum_i \Delta\Phi_i . \tag{8.25}$$

Random graph theory now makes statements on the sizes W of clusters in the limit $N \to \infty$ [190]. When $c = 1$, there is a power-law distribution of cluster sizes in the large-size limit

$$p(W) \sim W^{-5/2} \quad \text{for} \quad W \to \infty . \tag{8.26}$$

For c slightly below unity $(0 < 1 - c \ll 1)$, the power law is truncated by an exponential

$$p(W) \sim W^{-5/2} \exp\left(\frac{(c-1)W}{W_0}\right) \quad \text{for} \quad W \to \infty . \tag{8.27}$$

For $c = 1$, variance and kurtosis are infinite. They become finite but large when $c < 1$. Notice the similarity of (8.27) to the truncated Lévy distributions with $\mu = 3/2$, discussed earlier. When c is close to unity, an agent forms a link with one other agent, on the average. Larger clusters can still form from many binary links.

The law of price variations can be calculated in closed analytical form [190]. In the limit where $2aN$ is small, i.e., most of the traders are inactive, it reduces to (8.26) or (8.27), depending on the value of c, with the replacement $W \rightarrow \Delta S$. From this model, one would therefore predict that the Lévy exponent $\mu = 3/2$, which is close to the empirical results discussed in Chap. 5, should be universal.

In terms of percolation theory, the model formulated by Cont and Bouchaud [190] is in the same universality class as Flory–Stockmayer percolation [191]. Numerical simulations, however, become easier when the random graph underlying the Cont–Bouchaud model is replaced by a regular hypercubic lattice. On such lattices, critical behavior with the functional forms of cluster sizes similar to (8.26) and (8.27) is obtained, but the exponent in the power laws generally differs from $5/2$. Only for lattices in more than six dimensions is the $5/2$-power law recovered [184].

Stauffer and Penna performed extensive Monte Carlo simulations of such percolation problems [191]. They verified the power-law scaling of the probability distribution function of price changes, and its exponential truncation, on hypercubic lattices from two to seven dimensions if the activity a of the traders was chosen sufficiently small. They were also able to show that a crossover to more Gaussian return statistics occurred in the Cont–Bouchaud model when the activity a was increased, at least on hypercubic lattices.

The Cont–Bouchaud model can be extended further to include interactions between the coalitions of traders (percolation clusters), and the influence of fundamental analysis [192]. To this end, one replaces the percolation clusters by superspins Σ_i. A superspin is a spin with variable magnitude. This magnitude is the size of the original percolation clusters, and can be drawn from an appropriate distribution function, such as (8.26), eventually with a free exponent μ. In spin language, the excess demand on the market is equivalent to the magnetization of the superspin model, and the price changes are proportional to it. If the spin magnitudes are drawn from a power-law distribution without truncation, one can show that the distribution of magnetization, and with it the distribution of returns, carries the same power laws as the spin size distribution, if its Lévy exponent was $\mu < 2$.

Ferromagnetic interactions between the superpsins then correspond to herding behavior of the coalitions of traders in the Cont–Bouchaud model [192]. In practical terms, one might think of the managers of mutual funds imitating their colleagues' behavior. This is modeled by an "exchange integral" J_{ij}, in the same way as in the first term of (8.18). The local energy

$$E_i = - \sum_{j \neq i} J_{ij} \Sigma_i \Sigma_j \qquad (8.28)$$

is a measure of the disagreement of trader i with the prevalent opinion. Conformism leads to energy minimization. If one assumes the same $J_{ij} = J$ between all spins, a ferromagnetic state results in the physics version, and a boom or crash in the finance version of the model. This is unrealistic. One way to avoid such a totally ordered state is through the introduction of a fictitious temperature T which, when sufficiently high, returns the system to its paramagnetic state. In this case, the expected return is zero. This still gives power-law price statistics, and can be mapped onto an equivalent zero-temperature, zero-interaction model.

The agents so far behaved as noise traders, i.e., their opinions were randomly chosen. In this superspin variant, one can also include opinions which might arise from fundamental analysis of a company [192]. This can be done by a (coalition dependent) random field $h_i(t)$ which introduces a bias into the spin energy. Equation (8.28) is then changed to

$$E_i = -\sum_{j \neq i} J_{ij} \Sigma_i \Sigma_j + \Sigma_i h_i \ . \tag{8.29}$$

Such a field must be time-dependent, too. Assume that there is a certain stock price justified from fundamental analysis. If the actual price is much higher than the fundamental price, there should be a bias towards selling. When the price falls sufficiently below the fundamental price, a buying bias must arise. If this scheme is implemented, however, in a model with interactions, finite temperature, and a 50% population of agents with a fundamental bias, the price changes become quasiperiodic, and a bimodal return distribution curve is found. As a consequence, either bubbles and crashes on real markets are caused by much less rational behavior than included in the model [192], or the herding effect between the coalitions of traders has been overemphasized.

Adaptive Trader Populations

In the models discussed above, there was either just one population of traders with heterogenous trading strategies (random number parameters), or there were two or more populations of traders with different strategies (rebalancers, portfolio insurers, noise traders, fundamentalists, etc.). In all cases, these populations were fixed from the outset, and traders were not allow to change camp when they saw that their competitors' strategies were more successful. Changing camp, however, is certainly an important feature of herding in real financial markets where the operators often use a variety of analysis tools to reach their investment decisions, and the influence of the various tools on the decisions may well change with time.

Strategy hopping is at the center of a model formulated by Lux and Marchesi [193]. It had also been included in a simulation by Coche, of a more realistic model with much higher complexity [177] which we do not discuss here in detail. In the Lux–Marchesi model, traders are divided into two groups: fundamentalists and noise traders. Fundamentalists use exogeneous news arrival, modelled by geometric Brownian motion for a "fundamental price" S_f (the returns are normally distributed and the prices are drawn from a log-normal distribution). An example of this process is shown in Fig. 8.10. Noise traders, on the other hand, rely on chart analysis techniques and the

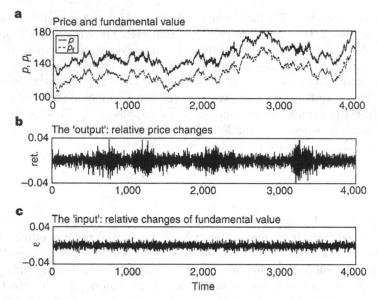

Fig. 8.10. Simulation of the Lux–Marchesi model. Panel (**a**) shows the history of the fundamental prices (S_f of our text is denoted p_f here), and of the actual share prices (S in the text is p here). Both price series have been offset for clarity. There is thus no long-term difference between both series, and the model market is efficient in the long run. Panel (**b**) displays the return on the share, while panel (**c**) is the return process of the fundamental price. Notice the very different return dynamics of both time series. By courtesy of T. Lux. Reprinted by permission from Nature **397**, 498 (1999) ©1999 Macmillan Magazines

behavior of other traders as information sources. Moreover, noise traders are divided into an optimistic and a pessimistic group. When the share price rises, optimistic noise traders will buy additional shares while pessimistic noise traders will start selling.

The important feature of this model is the possibility for strategy change by the traders. Noise traders change between optimistic (+) and pessimistic (−) with rates

$$\pi_{-\to+} = \nu_1 \frac{n_c}{N} e^{U_1} , \quad \pi_{+\to-} = \nu_1 \frac{n_c}{N} e^{-U_1}$$

$$\text{with} \quad U_1 = \alpha_1 \frac{n_+ - n_-}{n_c} + \frac{\alpha_2}{\nu_1} \frac{1}{S} \frac{dS}{dt} . \tag{8.30}$$

Here, $N = n_c + n_f$ is the total number of traders, and $n_c = n_+ + n_-$ is the number of noise traders of optimistic (n_+) or pessimistic (n_-) opinion. n_f is the number of fundamentalist traders. The first term in the utility function U_1 measures the majority opinion among the noise traders, and the second term measures the price trend. ν_1 and $\alpha_{1,2}$ are the frequencies of reevaluation of opinions and price movements. If both signals, majority opinion and price movement, go in the same direction, a strong population change will take place. If they point in opposite directions, the migration between the two noise trader subgroups will be much less pronounced.

Switching between the noise trader and fundamentalist group is driven by the difference in profits of both groups. Four rates are needed because of the two subgroups of noise traders

$$\pi_{f\to+} = \nu_2 \frac{n_+}{N} e^{U_{2,1}} \ , \quad \pi_{+\to f} = \nu_2 \frac{n_f}{N} e^{-U_{2,1}} \ ,$$
$$\pi_{f\to-} = \nu_2 \frac{n_-}{N} e^{U_{2,2}} \ , \quad \pi_{-\to f} = \nu_2 \frac{n_f}{N} e^{-U_{2,2}} \ . \tag{8.31}$$

n_f denotes the number of fundamentalists, and ν_2 the reevaluation frequency for group switching. The utility functions here are more complicated because the profits of both groups are different. The fundamentalists' profit is given by the deviation of the stock price from its fundamental price, but the profit is realized only in the future when the stock price returns to the fundamental value. They must be discounted therefore with a discounting factor $q < 1$, and are given by $q|S - S_f|/S$. The profit of the optimist chartists is given by the excess return of dividends (D) and share price changes $\nu_2^{-1} dS/dt$ per asset, over the average market return R. The profit of the pessimistic chartists is just its negative, and is realized when prices fall after assets have been sold. The utility functions $U_{2,1}$ and $U_{2,2}$ then become

$$U_{2,1} = \alpha_3 \left\{ \frac{D + \frac{1}{\nu_2} \frac{dS}{dt}}{S} - R - q \left| \frac{S - S_f}{S} \right| \right\} \ ,$$
$$U_{2,2} = \alpha_3 \left\{ R - \frac{D + \frac{1}{\nu_2} \frac{dS}{dt}}{S} - q \left| \frac{S - S_f}{S} \right| \right\} \ . \tag{8.32}$$

The order size of the noise traders is assumed to be the average transaction volume, and their contribution to the excess demand is then proportional to the difference between optimistic and pessimistic noise traders. The fundamentalists, on the other hand, will order in proportion to the perceived deviation of the actual stock price from its fundamental price, and the total excess demand is just the sum of both contributions. In the Lux–Marchesi model, the price changes are not deterministic but are given by probabilities which depend on the excess demand [193].

An interesting result of the simulations of this model is that, *on the average,* the market price equals the fundamental value [193]. This is shown in panel **a** of Fig. 8.10 where both price series have been offset for clarity. In the long run, the model market is efficient, and there are no persistent deviations of the share price from its fundamental value. As is apparent from panels **b** and **c**, however, the return processes of input and output are very different. The input process for the news arrival is geometric Brownian motion. The output process exhibits much stronger fluctuations, and volatility clustering. When the statistics of the return process of the share price is analyzed, one finds fat tails with power-law decay when the time scale of the returns is one time step. The exponent of the power law has been estimated as $\mu \approx 2.64 \pm 0.077$, but this may vary with parameters [193]. This value of μ compares favorably with the empirical studies discussed in Sect. 5.6.1. When the time scale is increased, the probability for large events decreases more steeply, and shows evidence for crossover to a Gaussian, again in agreement with what is found in real markets. Also, there are long-time correlations in the absolute returns of the shares, and volatility clustering.

The driving force of the model performs geometric Brownian motion. The peculiar scaling behavior found in the output stochastic process therefore must be the result of the interactions among the agents [193]. Analysis of the simulations shows that big changes of volatility are caused by the switching of traders between the various groups. The volatility is usually high when there are many noise traders. When the fraction of noise traders exceeds a critical value, the system becomes unstable. However, the action of the fundamentalists who can make above-average profits from such situations soon brings back the system into a stable regime. This mechanism is rather similar to the phenomenon of intermittency in turbulence.

The preceding discussion has emphasized the variety of mechanisms contributing to the price dynamics of real financial markets: herding, fundamental analysis, portfolio insurance, technical analysis, rebalancing portfolios, threshold trading, dynamics of trader opinions, etc. Every model discussed contained a unique mix of these factors, and emphasized different aspects of real markets. In the future, it will certainly be important to quantify more precisely the influence of the individual factors in specific markets. It is conceivable, e.g., that the role of fundamental analysis is different in stock, bond, and currency markets. A first step towards more quantitative investigation of markets by computer simulation of simplified models will be a careful calibration against a minimal set of market properties, such as those discussed in Chap. 5.

8.4 The Minority Game

Game theory deals with decision making and strategy selection under constraints. Game theory as applied by economists is built on one standard assumption of economics – that agents behave in a rational manner. Loosely speaking, agents know their aims, and what are the best actions to achieve them. The non-trivial problem comes from constraints, and conflicting though structurally similar behavior of the other agents. This assumption of rationality eliminates randomness from the games, and makes them essentially deterministic. In this perspective, games involve an optimization problem. The benefits of a player are often described by a utility function which, of course, depends on the strategies of all players. Under the assumption of complete information sharing between all players, the solution of the game is a Nash equilibrium. A Nash equilibrium is a state which is locally optimal simultaneously for each player, e.g., a local maximum of all utility functions.

In a physics perspective, such Nash equilibria in deterministic games might be viewed as zero-temperature solutions, where all possible (classical) fluctuations are frozen [194]. Introducing fluctuations, or randomness, then would correspond to finite-temperature properties. Depending on the importance of fluctuations, the properties of a finite-temperature system may or may not be close to those of its zero-temperature solution.

In the presence of randomness, games, in essence, will turn into scenario simulations. One may wonder to what extent game theory can improve our understanding of financial markets. Financial markets certainly provide the basic ingredients of game theory: a common goal and the necessity of strategy selection and decision making under constraints. However, uncertainty is an essential feature of capital markets, and while some information is available at high frequency and quality, the information on the strategies of other players is very limited, and can be guessed at best. In this section, we will explore a very simple game where agents have to make decisions using strategies chosen from a given set. They are selected based on their perceived historical performance using the available common information. Players however do not know their fellows' strategies. While the models grown from this seed have evolved some way towards the market models discussed above, the emphasis is different. Before, each agent either operated according to a random fixed strategy, or stochastically switched strategy according to an indicator function. Here, the question is how to select winning strategies for the agents in a market, possibly by simple deterministic rules despite the randomness present in the game, and how the stylized facts of financial markets may be generated by the interplay of agents with heterogeneous strategies. This process may be closer to real life where often strategies are selected and switched on a trial-and-error basis.

8.4.1 The Basic Minority Game

Take a population of an odd number N_p of players, each with a finite number of strategies, N_S. At every time step, every player must choose one of two alternatives, ± 1, buy or sell, attend a bar or stay at home, etc. without knowing the choices of the other players [195, 196]. To be specific, we take binary digits, and the decision of player i at time t is denoted by $a_i(t)$. The rule then is to reward those players on the minority side with a point. The winner is the player with the maximal number of points.

The time series of 0 or 1 is available to all players as common information. A strategy of length M is a mapping of the last M bits of the time series of results into a prediction for the next result, e.g., for $M = 3$ it maps the eight 3-bit signals into a set of eight predictions

$$\left\{ \begin{pmatrix} -1 \\ -1 \\ -1 \end{pmatrix}, \begin{pmatrix} -1 \\ -1 \\ 1 \end{pmatrix}, \begin{pmatrix} -1 \\ 1 \\ -1 \end{pmatrix}, \begin{pmatrix} -1 \\ 1 \\ 1 \end{pmatrix}, \begin{pmatrix} 1 \\ -1 \\ -1 \end{pmatrix}, \begin{pmatrix} 1 \\ -1 \\ 1 \end{pmatrix}, \begin{pmatrix} 1 \\ 1 \\ -1 \end{pmatrix}, \begin{pmatrix} 1 \\ 1 \\ 1 \end{pmatrix} \right\}$$

$$\rightarrow \{1, -1, -1, 1, 1, 1, -1, 1\} \ . \tag{8.33}$$

The "history" $h(t)$ is the signal broadcast to all players at a given instant of time, i.e., the last M outcomes of the game. Agents react to information, and they modify this information through their own actions. Different strategies are distinguished by the different predictions from the same signals. There are 2^M signals of M bits, and two possible predictions for each signal. The space of strategies of length M therefore is of size 2^{2^M} ($= 256$ for $M = 3$). M is an indicator of the memory capacity of the agents. When the number of strategies available to a player $N_S \ll 2^{2^M}$, very few strategies will be used by two or more players. On the other hand, when the inequality is violated, many players will have a common reservoir of strategies, and only very few strategies will not be available to another player.

At every turn of the game, the players evaluate the results of all their strategies on the outcome of the game, and assign a virtual point to all winning strategies, no matter if the strategy actually used in the game was among them (in which case the player won a real point) or not. At every time step, the player uses that strategy from his set which features the highest number of virtual points, i.e., which would have been his most successful strategy based on the historical record. The game is initialized with a random strategy selection [196]. All agents enter the game with the same weight, i.e., there are no rich agents who can invest much, and no poor agents who only can invest small sums.

To understand the outcome of the game, consider two extreme situations. One possible result is that only one player selects one side, and all the remaining $N_p - 1$ players take the other one. In this simple game, a single point is awarded in this turn of the game, to the winning player. The other extreme is an almost draw, when $(N_p - 1)/2$ players take the minority side and $(N_p + 1)/2$ players form the majority. In this case, $(N_p - 1)/2$ points are awarded. If one imagines the points to come from a reservoir, the second case would be interpreted as a very efficient use of resources by the player ensemble (they gather the maximal number of points to be gained in a single trial) whereas the first result would imply a huge waste. Clearly, this is opposite to a lottery where a fixed amount of money is distributed to the winners, and a lonely winner would gain much more than a winner in a large crowd.

The record of the game then is the time series of actions

$$A(t) = \sum_{i=1}^{N_p} a_i(t) \ . \tag{8.34}$$

Points are awarded to all players with $a_i(t) = -\text{sign } A(t)$. When the game is simulated, the time series $A(t)$ oscillates rather randomly around 0. The variance of the time series is high when M is small, and vice versa. In the interpretation suggested before, players with larger memories then would better use the available resources because, as an ensemble, they would score a higher number of points on the average. Remarkably, this behavior is achieved by selfish players who only search to optimize their own performance.

Is there an optimal strategy for an individual player in this game? For the ensemble of players, the optimal score is $(N_p - 1)/2$ per turn. The maximal average gain per trader and per game therefore is $1/2$. Can this gain be realized by an individual player with a simple strategy, systematically choosing one side, say $a_i(t) = 1$? If this were the case indeed, then other players would be attracted to make similar choices, too, because those of their strategies predicting an outcome 1 on a given signal would accumulate more virtual points. Then, however, the prediction 1 would quickly become a majority action, and not win points any longer. Notwithstanding these findings, in every game, there are players with success rates higher than $1/2$. When the number of strategies N_S of each player increases, the success rate decreases. Players more often switch strategies and face more difficulties in identifying outperforming strategies in their pool. Quite generally, the less players switch strategies, the higher their success rates.

However, strategies are good or bad only on a given time horizon. When the virtual points of all strategies are analyzed, the distribution at short times is rather wide, indicating that there is a big spread between good and bad strategies. As time increases, the distribution shrinks. This tells us that, on the long run, all strategies become equal. Success or failure then is linked to the good or bad timing in the use of specific strategies [196].

8.4.2 A Phase Transition in the Minority Game

The standard deviation of $A(t)$ (volatility) displays a very interesting behavior as the memory size of the agents is varied [197]. For small M, the volatility

$$\sigma_A = \sqrt{\text{var } A(t)} \tag{8.35}$$

decreases steeply from high values when the memory size M of the agents increases. σ_A increases gently with M from rather low values, beyond a critical memory size. Opposite behavior is found as a function of N_p, i.e., the effective parameter for the transition is $\alpha = 2^M/N_p$, the information complexity per player. The critical memory size increases with the number of strategies N_S available to each player. With reference to the extreme situations discussed above, the highly volatile low-α (low M, large N_p) regime describes a "symmetric" (the meaning will become clear in Sect. 8.4.4) information-efficient phase. This phase is named "crowded" because, due to the limited number of strategies available, the "crowding" of several players on one strategy is likely [197]. The more players present in the game at constant memory size, (i.e., size of the strategy space) or the smaller the agent memory, i.e., information, at constant number of players, the more likely this crowding effect is. Also, many of the strategies available are actually used, and information is processed efficiently. On the other hand, in the "dilute" large-α (large M, small N_p) phase, the strategy space is huge, and it is extremely unlikely that two agents will use the same strategy. This phase is termed "asymmetric" (cf. below), and information is not used efficiently: many strategies remain unexplored.

An interesting explanation can be given for these findings in terms of crowding effects [197]. Suppose that there is a specific strategy R used by N_R agents, who thus act as a crowd. For each strategy R, there is an anticorrelated strategy $\overline{R}$ where all predictions are reversed. The $N_{\overline{R}}$ agents using $\overline{R}$ form the anticrowd. R and $\overline{R}$ form a pair of anticorrelated strategies. Pairs of strategies are uncorrelated. When $N_R \approx N_{\overline{R}}$, the actions of the crowds and of the anticrowds almost cancel, and σ_A will be small. On the other hand, when $N_R \gg$ or $\ll N_{\overline{R}}$, herding dominates and generates a high volatility.

It turns out that the behavior of the volatility is almost unchanged when a reduced strategy space made up only of pairs of anticorrelated strategies is used [197]. Different pairs being uncorrelated, the signal $A(t)$ can be decomposed into the contributions of the various groups. Each of these groups essentially performs a random walk of step size $|N_R - N_{\overline{R}}|$. The variance of these walks then determines the standard deviation σ_A, and it turns out that the extent of crowd–anticrowd cancellation determines the non-monotonic variation of the volatility.

In terms of this crowd–anticrowd picture, the asymmetric large-α phase corresponds to N_R, $N_{\overline{R}} \in \{0, 1\}$. Strategies are either selected once, or not at all. The volatility is almost that of a discrete random walk with unit step size. When more agents play, the simultaneous use of R and $\overline{R}$ will become more likely, giving cancellations, i.e., zero step size in the random walk, and σ_A decreases. With even more players, the crowd sizes on R and $\overline{R}$ will become sizable but likely very different. The step size of the random walk will grow, as does the volatility. The behavior of σ_A can thus be interpreted in terms of repulsion, attraction, and incomplete screening of crowds and anticrowds [197].

8.4.3 Relation to Financial Markets

In the basic form described above, the minority game shares some features of financial markets. Agents have to take choices under constraints, uncertainty, and with limited information. The most fundamental decision in a financial market is binary: buy or sell. Speculators may find their strategies by trial and error, and their strategy pool may be limited. There is competition in the minority game as in markets, and agents cannot win all the time. Furthermore, there is no a priori definition of good behavior in markets. Good behavior is defined with respect to the behavior of the competitors, a posteriori, and based on success. As a corollary, the definition of good behavior may change when the reference behavior of the other agents changes. The choice to take in the minority game amounts to predict a future event – which depends only on the choices of all other players [198]. Of course, one might argue that, in a fundamental perspective, financial markets are also influenced by the arrival of external information. However, many readers will know from experience that on certain days, external information strongly moves markets, while on other days, it is completely ignored by the operators. Most likely, psychology is at the origin.

However, there are also important differences. Firstly, with an average expectation of a winning trade of ideally 50%, it is not clear why agents trade at all. One tentative answer which has been given to this question is the presence of non-speculative trades in a market, originated by "producers" or investors. In a commodity market, there will be producers who sell their goods, and buyers who need those goods for their utility, rather than for profit. An investor might buy shares in a stock market for gaining control over a company, rather than for speculative profits. One can then set up an argument that these producers would introduce predictable patterns into the markets which would be exploited by speculators who can adapt much more quickly to a market situation than producers [198]. It is not clear, however, to what extent such an argument could explain trading in FX markets, where more than 90% of the trading volume is speculative in origin, and which are extremely liquid.

Secondly, one suspects that the aim of the players taking part in the minority game corresponds to a contrarian trading strategy. "Be part of the minority!" implies to buy when everybody is selling and vice versa. However, in financial markets, one often finds extended trending periods where the most successful strategy would be to buy when the majority buys and sell when the majority sells (hopefully early enough, though). Is it more appropriate then to view financial markets as the playground for a majority game rather than a minority game? Remember that the presence of different trader populations and the switching of their trading philosophy in the Lux–Marchesi model (cf. the preceding section) was essential to produce realistic time series in that artificial market.

Thirdly, in a real-world market, operators may not trade in a given time interval. This is ignored in the minority game. A straightforward generalization to a "grand-canonical minority game" would open such an avenue. In order to decide whether to trade or not, an agent should compare her strategies to a benchmark. The basic minority game only compares the relative merits of all her strategies, and trading is done also when all strategies lose out. To remedy for that deficiency, a rule can be introduced that an agent only trades when at least one of her strategies has a positive score. Here, one still faces the problem that the success of the strategy at the origin of the decision to trade, is virtual while any loss incurred while being in the market, would be real. A more realistic benchmark would be to trade only when at least one strategy is available whose success *rate* is superior to a threshold [198].

The minority (and majority) games can be derived from a market mechanism [199], once price formation and market clearing are defined. We proceed as in Sect. 8.3.2, i.e., determine the price at which the market is cleared from the aggregated demand $D(t)$, the aggregated supply $O(t)$, and the price quoted in the last time step, according to

$$S(t) = S(t-1)\frac{D(t)}{O(t)} , \tag{8.36}$$

$$D(t) = \sum_{i=1}^{N_p} a_i(t)\Theta[a_i(t)] = \frac{N_p + A(t)}{2} , \tag{8.37}$$

$$O(t) = -\sum_{i=1}^{N_p} a_i(t)\Theta[-a_i(t)] = \frac{N_p - A(t)}{2} . \tag{8.38}$$

The return on an investment for one time step from t to $t+1$ [$a_i(t) = 1$, $a_i(t+1) = -1$] is

$$\delta S_1(t+1) = \ln\left(\frac{S(t+1)}{S(t)}\right) \approx \frac{S(t+1)}{S(t)} - 1 . \tag{8.39}$$

Of course, the information on $S(t+1)$ is not available to the players when they must place their orders. The best they can do is to base their decision on their expectation for the return on their investment. Assume that the expectation of player i at time t for the price of the asset at $t+1$ is

$$E_t^{(i)}[S(t+1)] = (1 - \psi_i)S(t) + \psi_i S(t-1) . \tag{8.40}$$

Let each player place an order at t according to that expectation, calculate the payoff on the investment over one time step, and compare the payoffs of the majority and the minority sides. It turns out that agents with $\psi_i > 0$ are on the winning side when they are in the minority, i.e., they follow a contrarian investment strategy. They expect that the future price movement is negatively correlated with the past move. On the contrary, agents with $\psi_i < 0$ are trend followers and play a majority game. Thus it appears that real markets may be described best as mixed minority–majority games.

8.4.4 Spin Glasses and an Exact Solution

A slightly modified, "soft minority game", can be solved exactly using methods from spin-glass physics in the limit $N_p \to \infty$ [201]. Agents do not simply choose the strategy with the highest virtual score, but proceed in a probabilistic manner: a strategy is chosen with a probability which depends exponentially on its virtual score in the game. Moreover, the binary payoff of one point when the strategy played was successful, is changed into a gain function linear in the population difference between minority and majority sides,

$$g_i(t) = -a_i(t)A(t), \qquad (8.41)$$

i.e., the minority wins points or money, and the majority loses them. By definition, this is a negative sum game. The total average loss in the system then is

$$-\sum_i \langle g_i \rangle = \sigma_A^2 . \qquad (8.42)$$

This equation reemphasizes the interpretation of σ_A as a measure of the waste in the system.

The dynamical equations of the minority game then suggest a description in terms of a Hamiltonian which is reminiscent of disordered spin systems [200, 201]. To see the essentials, we limit ourselves to $N_S = 2$ strategies which would correspond to spin $1/2$. To distinguish strategies $s_i \in \{\uparrow, \downarrow\}$ from actions $a_{i,s}$ (the subscript emphasizes that the action a_i depends on a strategy s_i), decompose $a_{i,s}(t)$ as

$$a_i^{h(t)}(t) = \omega_i^{h(t)} + s_i(t)\xi_i^{h(t)}, \quad \omega_i^h = \frac{a_{i,\uparrow}^h + a_{i,\downarrow}^h}{2}, \quad \xi_i^h = \frac{a_{i,\uparrow}^h - a_{i,\downarrow}^h}{2} . \qquad (8.43)$$

ω_i^h represents a fixed bias in the strategies of agent i, whereas ξ_i^h represents the flexible part. Of course, they depend on the history $h(t)$ of the game. The time dependence of $a_i(t)$ now is attributed to two time-dependent factors: one is the particular history $h(t)$ realized in the game during the M rounds preceding t. This is why ω_i^h and ξ_i^h depend on t only through $h(t)$. The second factor is the time dependence of $s_i(t)$, which reflects the choice of strategy made by agent i at time t based on the available history and his strategy selection rules (probabilistic or deterministic).

Introducing $\Omega = \sum_i \omega_i$, $A(t)$ can be rewritten as

$$A^h(t) = \Omega^h + \sum_{i=1}^{N_P} \xi_i^h s_i(t) , \qquad (8.44)$$

and its variance becomes

$$\sigma_A^2 = \overline{\Omega^2} + \sum_i \left[\overline{\xi_i^2} + 2\overline{\Omega \xi_i} \langle s_i \rangle \right] + \sum_{i \neq j} \overline{\xi_i \xi_j} \langle s_i \rangle \langle s_j \rangle . \qquad (8.45)$$

Here, $\langle x \rangle$ denotes the temporal average of a quantity x while $\overline{x}$ is the average over histories. Unless necessary, the history superscript h is dropped under the history averages. All $2M$ histories are explored for long enough times.

This allows us to decompose a temporal average into one conditioned on history $\langle x^h \rangle$, followed by one over histories, i.e., $\langle x \rangle = \overline{\langle x^h \rangle}$. By symmetry, $\langle A \rangle = 0$. However, for particular histories, there may be a finite expectation value $\langle A^h \rangle \neq 0$. One may then calculate the average over the histories of the history-dependent expectation values of A,

$$\overline{\langle A^h \rangle^2} = \overline{\Omega^2} + 2 \sum_i \overline{\Omega \xi_i} \langle s_i \rangle + \sum_{i,j} \overline{\xi_i \xi_j} \langle s_i \rangle \langle s_j \rangle \equiv H . \tag{8.46}$$

When the scores of the strategies are updated using a reliability index

$$U_{s,i}(t+1) = U_{s,i}(t) - 2^{-M} a_{s,i}(t) A(t) \tag{8.47}$$

and a probabilistic strategy selection rule $P[s_i(t) = s] \sim \exp[\Gamma U_{s,i}(t)]$ is adopted, the evolution of $\langle s_i \rangle$ with time scale $\tau = 2^{-M} \delta t$ can be cast in the form

$$\frac{\mathrm{d}\langle s_i \rangle}{\mathrm{d}\tau} = -\Gamma \left(1 - \langle s_i \rangle^2\right) \left(\frac{\partial H}{\partial \langle s_i \rangle}\right) . \tag{8.48}$$

Formally, these are the equations of motion for magnetic moments $m_i = \langle s_i \rangle$ in local magnetic fields $\overline{\Omega \xi_i}$ interacting with each other through exchange integrals $\overline{\xi_i \xi_j}$. H in (8.46) then is a spin-glass Hamiltonian [200].

Such Hamiltonians can be studied using the replica trick familiar from the theory of spin glasses [185] and it turns out that, under the standard assumptions, the ground state of the Hamiltonian which describes the stationary state always is in the replica-symmetric phase [200, 201]. Within the replica-symmetric phase, there is a transition, however, as a function of the ratio between the information complexity 2^M and the number of players. When this number is small, the probability distribution of the strategies used in the game is continuous while, for a large ratio, it contains two delta functions at the positions of static strategies $a_i = \pm 1$ in addition to a Gaussian distribution. Agents contributing to the delta functions do not switch strategies while those under the continuous distributions stochastically change strategies. This is the phase transition seen in the dependence of the volatility on the memory length/agent number discussed in Sect. 8.4.1.

The small-α phase is called "symmetric" because both $\langle A \rangle = 0$ and $\langle A^h \rangle = 0$. In the "asymmetric" large-α phase, we have $\langle A \rangle = 0$ but $\langle A^h \rangle \neq 0$ at least for some histories. $\langle A^h \rangle$ therefore is akin to an order parameter in a symmetry-breaking phase transition. Here, it is the symmetry between the histories which is lost at the critical α_c. In the asymmetric phase, for those histories with $\langle A^h \rangle \neq 0$, there is a best strategy

$$a_{\mathrm{best}}^h(t) \equiv a_{\mathrm{best}}^h = -\mathrm{sign} \langle A^h \rangle \tag{8.49}$$

which allows for a positive gain $\overline{|\langle A^h \rangle|} - 1$. In this phase, the market is predictable. The measure of predictability is $H \equiv \overline{\langle A^h \rangle^2}$, (8.46). Using (8.36)–(8.38), we have

$$A(t) = D(t) - O(t) \,, \tag{8.50}$$

i.e., $A(t)$ also is the excess demand in the market. When $\langle A^h \rangle \neq 0$, there are persistent periods of excess demand/supply where price will move in one direction. The volatility is somewhat better than coin tossing but not dramatically so, because of the crowd–anticrowd repulsion. That information is used not very efficiently is evidenced by σ_A, which is significantly above its minimum at α_c. The game becomes more information-efficient when players are added who more evenly cover the strategy space.

In the symmetric small-α phase, $H = 0$, i.e., the market is unpredictable. Moreover, $\langle A(t) \rangle = 0$, i.e., there is no excess demand on the average, and prices are stable. When there are very many players at moderate information complexity, herding takes place due to the incomplete crowd–anticrowd screening, and the volatility increases again. The waste of resources/total loss of the population is minimal at the transition $\alpha = \alpha_c$.

When the agents include a term into their strategy selection probability which rewards the strategy actually used by them in the game with respect to virtual strategies, a replica-symmetry broken solution can be found. The interesting point is that the replica-symmetry broken solution describes a Nash equilibrium. Nash equilibria in the minority game correspond to pure (static) strategies $a_i = \pm 1$ independent of t. The replica-symmetric solution, on the other hand, does not correspond to a Nash equilibrium. However, the trimodal solution for the strategy probability including the delta-function peaks at pure strategies contain some of its ingredients.

8.4.5 Extensions of the Minority Game

A variety of extensions can be formulated in order to bridge the gap between the basic game and a model for financial markets. The agent population can be made heterogeneous in various dimensions such as memory size, strategy diversification, evolutionary strategies, etc., and agents may choose to stay out of the market. When the game is played with mixed memory sizes, players with longer memories perform better than those with shorter memories [196]. When the payoff function is changed to lottery-type, i.e., the payoff (both in real and virtual points) increases with decreasing number of winners, the probability distribution of $A(t)$ becomes bimodal – it is monomodal in the standard game. This is a remarkable example of self-organization because the most likely configurations are avoided by the players at the expense of somewhat less likely ones.

One can introduce explicitly hedgers who only possess one strategy. They do not enter the marketplace for speculation but for "fundamental" (exogeneous) reasons, cf. Sect. 2.5. They might as well be producers who use the market for selling or buying goods. In the game, their role is to introduce information through their trading activity which is supposed to be due to drivers external to the game [200]. Also, noise traders, who take random decisions, can be included. Further extensions could include insiders and spies.

It is particularly interesting that the minority game can be extended to allow for predictions of moves in actual markets [202]. It is based on the "grand canonical" extension of the minority game where agents trade or stay out of the market depending on the comparison of their scores (virtual or real) with a threshold value. Thus the number of active traders has become variable. Also, the threshold can be made a dynamic quantity. One restriction is that the threshold should be positive, i.e., a trader should only use strategies which have won more often than lost. As a second restriction, the threshold should increase when the player's scores decrease, i.e., one should take less risk after losing for some period of time. These rules generate quite diverse populations of traders. One may further diversify the trader population in terms of wealth (initial capital), investment size (wealthy investors will place big orders), and investment strategy (trend following versus contrarian, or minority versus majority games). The mechanism of price formation is assumed to be similar to (8.36).

This extended mixed minority–majority game is trained on a financial time series, converted into a binary sequence, e.g., by just recording the signs of market moves. In other words, the game is fed with a signal where Zipf analysis, discussed in Sect. 5.6.3, has demonstrated that non-trivial correlations exist [112, 113]. Such correlations have been uncovered specifically in the USD/JPY exchange rates [113] which have been used in this experiment. Players then take their actions based on that signal history $h(t)$. The sign history is an external signal whereas $A(t)$ in the minority game was generated internally to the game. The feedback effect included in $A(t)$ has been removed. However, the game and the time series of aggregated actions $A(t)$ are used to carry the game forward into the future. When using hourly quotes of ten years of USD/JPY exchange rates, the game performs much better than random, and the accumulated wealth of the total agent population is increasing steadily. The actual increase, however, depends on the pooling of the agents' predictions which is not specified for the best performances [202]. The trading strategy certainly is somewhat oversimplified: depending on the minority game prediction, put the investment on the USD or JPY side and, after one hour, withdraw it. Neglect transaction costs, slippage, etc. Despite these simplifications, the game apparently produces many of the stylized facts of financial markets: fat-tailed return distributions, price–volume correlations, volatility clustering, ... [202, 203].

More importantly, when run into the future for several time steps, the game also generates prediction corridors for future prices of the asset [204]. In many cases, large changes can be predicted accurately in the sense that the probability density function of the returns possesses a large mean and a narrow variance. In other cases, the prediction of a sign change comes out correctly although the prediction corridors are rather wide. Large price movements such as crashes or booms apparently can be predicted with some degree of reliability based on the minority game. Johnson et al. have filed a patent application on these algorithms [204].

As a final remark, it has been shown that a winning strategy can be set up by playing two different losing games one after the other (Parrondo's paradox) [205]. It would certainly be interesting to include such effects into the minority game.

9. Theory of Stock Exchange Crashes

Crashes of stock exchanges, and speculative markets more generally, have occurred ever since trading securities and commodities has become an important activity. A historical example is the "tulipmania", the rise and subsequent crash of prices for tulip bulbs on Dutch commodity markets in 1637 [206, 207], or the South Sea bubble in England, where Newton lost much of his fortune, cf. Chap. 1. Modern financial crashes are discussed below. Since in such events enormous fortunes are at stake, efforts towards an improved understanding are mandatory.

9.1 Important Questions

In this chapter, we will attempt answers to the following important questions concerning financial crashes:

- What are the origins of stock exchange crashes?
- Are crashes compatible with rational behavior of investors?
- Are they endpoints of "speculative bubbles" and signal the return of market prices to their "fundamental values"?
- Do crashes signal phase transitions in markets?
- Are there parallels to earthquakes or avalanches?
- Are earthquakes predictable?
- Are crashes part of the normal statistics of asset price fluctuations, or are they outliers?
- Can crashes be predicted? Are there crashes which have been predicted successfully in the past?
- Are there examples of anticrashes, i.e., trend reversals from falling to rising prices which follow patterns established for crashes?
- Can one measure the strength of crashes in the same way as the Gutenberg–Richter scale measures the strength of earthquakes?
- Are there signals for the end of a crash?

9.2 Examples

Here is a list of the more recent examples of financial crahes, some of which readers may well remember.

1. The "Asian crisis" on October 27, 1997 and the "Russian debt crisis" starting in summer 1998, have been discussed briefly in Chap. 1. Figure 1.1 shows these two events in the variation of the DAX, the German stock index, from October 1996 to October 1998. The Asian crisis is a drawdown of about -10% on the German stock market on October 27, 1997, with a very quick recovery. Interestingly, the aggregate drawdown over scales even as short as a week was rather small. Notice, however, that the DAX stopped its upward trend in July 1997, and one question we wish to discuss here is to what extent this can be viewed as a kind of precursor of the crash. Indeed, there have been predictions of this crash [208].

   On the contrary, the drawdowns of the stock markets in Asia was much stronger. The Hang Seng index of the Hong Kong stock exchange, e.g., lost 24% in a week. The index is shown as the dotted line in Fig. 9.1. The solid line shows the variation of the US S&P500 index in the four years prior to the 1997 crash. The long-term upward trend is stopped by

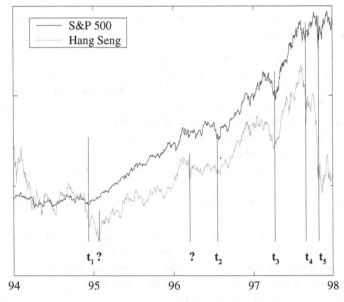

Fig. 9.1. Extrema of variation of the S&P500 and the Hong Kong Hang Seng index, prior to the 1997 crash. Notice that the Hang Seng index has two pronounced minima not lying on the log-periodic sequence marked by the vertical lines. By courtesy of J.-P. Bouchaud. Reprinted from L. Laloux, et al.: Europhys. Lett. **45**, 1 (1999), ©1999 EDP Sciences

the drawdown in late October 1997. Similar to the European markets, its amplitude was much smaller than on the Asian markets. However, unlike the German market, the index continued to increase throughout summer 1997 although there have been certain periods of local short-time decrease, marked by the vertical lines. The labels "t_i" and "?" on these lines will be explained in Sect. 9.4.

The impact of the Russian debt crisis on the German stock market was very different from the Asian crisis. The decrease was much less abrupt – though much more persistent and of much larger amplitude, at the end. Over four months, the DAX lost 39% which corresponds to an average loss of 2.7% per week. It is obvious from Fig. 1.1 that losses of this order of magnitude occurred regularly, almost every week, between July and October 1998.

With reference to the discussion in Sect. 5.3.3, notice that stop-loss orders would not have protected investors from sizable losses in the Asian crisis while they would have offered protection throughout most of the Russian debt crisis. On the other hand, due to the quick recovery of the markets after the Asian crisis, an investor simply holding his assets for a few more weeks would have wiped out most, if not all of his losses.

2. Figure 9.2 shows the variation of the Dow Jones Industrial Average in the Wall Street crash of October 1987 [209]. The index lost about 30% in one day. To put that into perspective, the loss in a single day is comparable

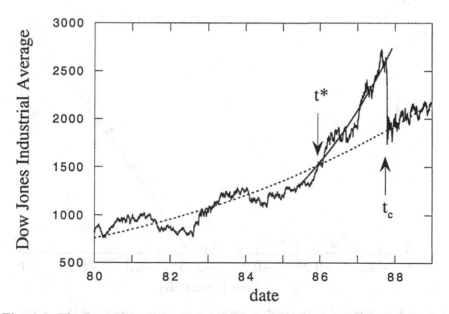

Fig. 9.2. The Dow Jones Industrial Average during the October crash 1987. By courtesy of N. Vandewalle. Reprinted with permission from Elsevier Science from N. Vandewalle, et al.: Physica A **255**, 201 (1998). ©1998 Elsevier Science

to the decline of the DAX over the entire four month Russian debt crisis period in autumn 1998! This was the largest crash of the century.

3. Other important crashes took place in 1929, and at the outbreak of World War 1. Figure 9.3 shows the largest weekly drawdowns of the Dow Jones in this century. The biggest crash was 1987, followed by World War 1, and the 1929 crash. Notice that on this scale, the Asian and Russian crisis are completely negligible, and contribute to the leftmost points in this figure. (Of course, they are no longer negligible when the variations of the Asian or Moscow stock exchanges are plotted.)

Figure 9.3 uses an exponential distribution to fit the weekly drawdowns of the Dow Jones index. *If* this procedure is endorsed, crashes would appear as outliers: they would not be subject to the same rules as "ordinary" large drawdowns and be governed by separate mechanisms. Indeed, this point of view has been defended in the recent research literature by several groups, and we will discuss it in the present chapter.

Notice, however, that the assumption of an exponential distribution is arbitrary, to some extent, and that statistics is difficult on singular events such as a major crash. In the framework of stable Lévy distributions, discussed in Chap. 5, crashes would be part of the statistical analysis, and not be generated by exceptional mechanisms. This may also apply to power-law statistics

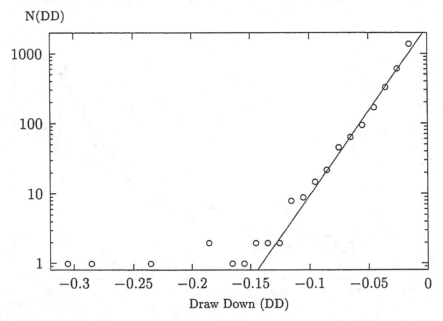

Fig. 9.3. Number of large negative weekly price variations of the Dow Jones in the 20th century. By courtesy of D. Sornette. Reprinted from D. Sornette and A. Johansen: Eur. Phys. J. B **1**, 141 (1998), ©1998 EDP Sciences

with nonstable tail exponents. Most likely, in such frameworks, crashes will not be predictable.

Theories based on exceptional mechanisms underlying crashes therefore can only be tested on their predictive power.

For all crashes, various economic "causes" have been discussed in the literature. Hull [10] lists a variety of such possibilities. For the 1987 crash, e.g., it was observed that investors moved from stocks to bonds, as the return of bonds increased to almost 10% in summer 1987. Another cause may have been the increasing portfolio hedging, using index options and futures, combined with the implementation on computers which generated automatic sell orders once the index fell below a certain limit. This effect has been modelled explicitly in the computer simulation by Kim and Markowitz [176], cf. Sect. 8.3.1. Changes in the US tax legislation may have contributed. Rising inflation and trade deficits weakened the US dollar throughout 1987, and this may have pushed overseas investors to sell US stocks. Finally, one may think about imitation and herd behavior. However, it seems to be a common feature of the major crashes that no single economic factor can be identified reliably as the triggering event.

Looking at the behavior of the market operators, a crash occurs when a synchronization of the individual actions takes place. In normal market activity, the individual buy and sell orders are not strongly correlated, and rather weak price or index variations result. In a crash, on the other hand, all operators decide to sell, and there are no compensating buy orders which would maintain market equilibrium. The market seems to behave collectively.

An increasing synchronization, or correlation, is observed in physics when a phase transition, especially a critical point, is approached. Examples are the transition from a paramagnet to a ferromagnet, or from an ordinary metal to superconductivity. Certainly, there are important differences, in that crashes take place as a function of time while the critical points in physics usually are reached by careful fine-tuning of an external control parameter. The idea of critical points has been generalized to self-organized critical points in open nonequilibrium systems [79], and the question is if stock exchange crashes can be considered as critical points, or self-organized critical points, as they occur in physics.

There are other nonequilibrium situations in nature whose phenomenology seems to be similar to market crashes, and where ideas and models about phase transitions and critical points have been formalized, too: earthquakes and material failure. We shall discuss them in the following section, before returning to the (admittedly phenomenological) description of stock exchange crashes.

9.3 Earthquakes and Material Failure

Earthquakes and material failure are both characterized by a slow building up of strains, and a sudden discharge. The idea of these phenomena being critical points in time has been discussed in the literature for some time. There is some evidence for this view, although it is still controversial.

1. Figure 9.4 shows the cumulative Benioff strain prior to the earthquake occurring on October 18, 1989 near Loma Prieta (northern California). The cumulative Benioff strain $\varepsilon(t)$ is defined as

$$\varepsilon(t) = \sum_{n=1}^{N(t)} \sqrt{E_n} \ . \tag{9.1}$$

n is the number of small earthquakes from some starting date $t = 0$ until t, and E_n is the energy liberated in quake n. The appearance of energy under the square-root can simply be understood in terms of a spring obeying Hooke's law: at a given strain ε, the energy stored in the spring is $E = (f/2)\varepsilon^2$ where f is the spring constant. Fig. 9.4 also shows a fit of $\varepsilon(t)$ to a power law in time-to-failure

$$\varepsilon(t) = A + B|t_\mathrm{f} - t|^\mu \ , \qquad A > 0 \ , \ B < 0 \ , \ 0 < \mu < 1 \ . \tag{9.2}$$

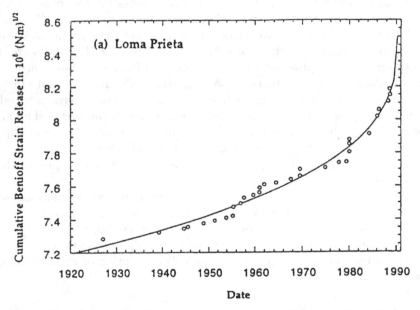

Fig. 9.4. Cumulative Benioff strain before the Loma Prieta earthquake in 1989 (*dots*) and fit to a power law (*solid line*). By courtesy of D. Sornette. Reprinted from D. Sornette and C. G. Sammis: J. Phys. I (France) **5**, 607 (1995), ©1995 EDP Sciences

Power laws are the hallmarks of critical points, and the fit apparently supports the idea of a critical point occurring in time. Notice that $\varepsilon(t)$ stays finite at t_f but $d\varepsilon(t)/dt \to \infty$ as $t \to t_f$. Notice that the deviations between the measured points and the power-law fit do not look exactly random. There are hints of oscillatory behavior.

2. Both the cumulative Benioff strain, and the concentration of Cl^- ions, before the earthquake in Kobe (Japan) on January 17, 1995, show a similar increase [210]. Again, oscillations seem to be superposed on the smooth power-law variation of (9.2).

3. On a laboratory scale, acoustic emissions recorded before the failure of materials under increasing load show similar variations.

For earthquakes and material failure, models have been developed which substantiate power-law behavior, and thus the critical point hypothesis, and even additional oscillations as the critical point is approached. Their most important ingredient is their hierarchical structure.

An important model for the description of earthquakes is due to Allègre et al. [211], and pictured schematically in Fig. 9.5. One starts from a cube formed by joining eight bars by bolts in the corners of the cube. On the next level, eight bigger bars form a bigger cube, and eight of the small cubes of

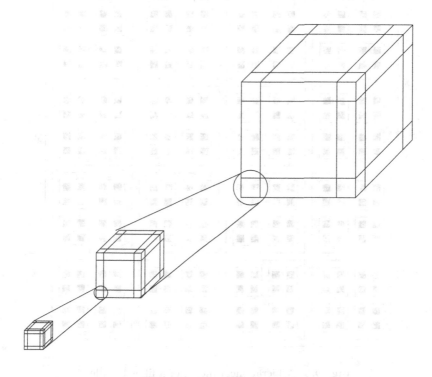

Fig. 9.5. The Allègre model

the preceding level are used as bolts to join the bars. This rule is continued to ever larger scales. The load on the bars and bolts of the biggest cube is distributed over all levels of the hierarchy. If this load is increased, the weakest bolt which is on the lowest level may break. Eventually, more than one bolt will break. This will lead to a redistribution of the load on the next level of hierarchy, and bolts may fail there, too, either immediately, or once the load is increased further, and so on. Finally, the highest levels of the hierarchy will break, resulting in a catastrophic event.

Similar ideas may be invoked for the failure of materials, e.g., composed of fibers. Figure 9.6 illustrates a hierarchical model for a fiber bundle. The cross-section of the bundle is shown, and the fibers are oriented perpendicular to the figure. The mechanism for failure of such a bundle under increasing load is rather similar to that of the cubic structures of the Allègre model.

Both models show some kind of critical behavior, and power laws, as the load on the structure is increased. Their criticality is different, however, from the ordinary critical points of physics in one important aspect. Power laws are related to scale invariance. Critical points associated with phase transitions in standard physical systems (magnetism, superconductivity, etc.) exhibit

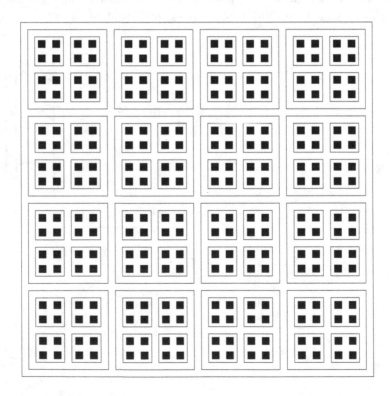

Fig. 9.6. A hierarchical model of a fiber bundle

continuous scale invariance. Under a change of scale $x \rightarrow x' = \lambda x$, scale invariance of a system implies that a function $f(x)$ reproduces itself, perhaps up to some prefactor, i.e.,

$$f(x) = \mu f(x') = \mu f(\lambda x) ,\qquad (9.3)$$

with real λ, μ. This equation is solved by power laws,

$$f(x) = C x^{\alpha} ,\qquad (9.4)$$

which lead to the condition

$$\lambda^{\alpha} \mu = 1 , \text{ i.e., } \alpha = -\ln \mu / \ln \lambda .\qquad (9.5)$$

Physically, continuous scale invariance comes out because the properties at the phase transition are determined completely by a diverging correlation length (the "synchronization" mentioned above), which is much larger than typical lattice constants, or nearest-neighbor distances. Notice that the underlying structures or Hamiltonians of such systems are not scale invariant, and that scale invariance only results from the spontaneous collective behavior.

As is obvious from Figs. 9.5 and 9.6, there can be no continuous scale invariance in hierarchical models. If they are continued to infinity, there will be no scale on which the "microscopic" structural details can become negligible because collective behavior would set in on much longer length scales. Unlike the models of statistical mechanics, however, hierarchical systems have a built-in *discrete scale invariance.* Under a discrete rescaling, $x \rightarrow x' = \lambda_n x$ with $\lambda_n = \lambda_0^n$, they reproduce themselves. For example, we have $\lambda_0 = 2$ for the structure in Fig. 9.6. An important consequence of discrete scale invariance is that critical exponents can become complex [212]. These complex exponents naturally come out of (9.5) when rewritten as

$$\lambda^{\alpha} \mu = \exp(2\pi i n) , \text{ i.e., } \alpha_n = -\frac{\ln \mu}{\ln \lambda} + \frac{2\pi i n}{\ln \lambda} .\qquad (9.6)$$

A priori, any n is permissible in (9.6). However, for the usual critical phenomena, solutions with $n \neq 0$ can be discarded because they would imply the existence of typical scales in the problem, which contradicts the scale invariance postulated to be at the origin of the power-law behavior. On a hierarchical structure, such an objection is not possible, and complex exponents must be allowed. As a consequence, when finite n are kept, a series of log-periodic oscillations is superposed on the power-law behavior

$$(t_{\mathrm{f}} - t)^{\alpha} \rightarrow (t_{\mathrm{f}} - t)^{\alpha} \left(1 + \sum_{n=1}^{\infty} c_n \cos \left[\frac{2\pi n}{\ln \lambda} \ln |t_{\mathrm{f}} - t| \right] \right)\qquad (9.7)$$

Such oscillations have indeed been observed both in earthquakes and financial data. An important practical advantage of the modified scaling law

(9.7) is that the determination, and in particular, a possible prediction of t_f, i.e., the time to failure or to an earthquake, become much more accurate if log-periodic oscillations lock in on the data in a fit. The disadvantage is that the number of fit parameters to be used on a noisy data set increases significantly, at least from four (pure power law) to seven [including the first log-periodic oscillation, cf. (9.8) below]. Under these circumstances, there may be many apparently equally good fits, and their interpretation as well as the selection of a "best" fit, become a nontrivial problem [213]. Analyzing the data in Fig. 9.4 a posteriori by fitting them to a pure power law such as (9.2), one would "predict" the Loma Prieta earthquake to have occurred at $t_f = 1990.3 \pm 4.1$. Using the first log-periodic oscillations, the prediction becomes $t_f = 1989.9 \pm 0.8$, i.e., is both significantly closer to the actual date of the earthquake, and carries a much smaller error bar. Figure 9.7 shows a fit to the same data as in Fig. 9.4 but using log-periodic corrections, showing the kind of agreement that can be reached. Similar fits can also be done on Kobe data [210].

This analysis has been done after the actual earthquake occurred. What about using the method to predict a quake? This has also been attempted by Sornette and Sammis [213]. Figure 9.8 shows data taken up to 1995 in the Komandorski islands, a part of the Aleutian islands in Alaska. Also shown is a fit to (9.7) which produces a (true) prediction of a major earthquake at

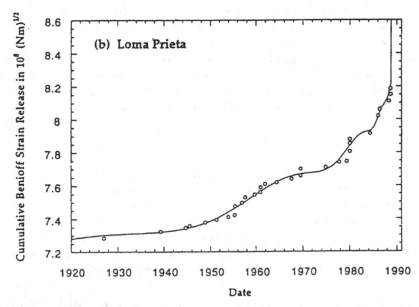

Fig. 9.7. Cumulated Benioff strain prior to the Loma Prieta earthquake (*dots*), fitted to a power law with log-periodic corrections (*solid line*). By courtesy of D. Sornette. Reprinted from D. Sornette and C. G. Sammis: J. Phys. I (France) **5**, 607 (1995), ©1995 EDP Sciences

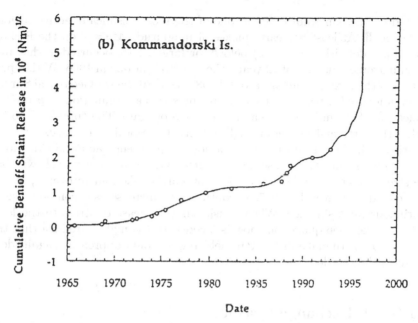

Fig. 9.8. Cumulated Benioff strain released by earthquakes of magnitude 5.2 or greater, in the Komandorski segment of the Aleutian islands (*dots*), and a fit to a power law with log-periodic corrections (*solid line*). By courtesy of D. Sornette. Reprinted from D. Sornette and C. G. Sammis: J. Phys. I (France) **5**, 607 (1995), ©1995 EDP Sciences

$t_f = 1996.3 \pm 1.1$, i.e., after the submission (January 1995) and publication (May 1995) of the paper. This prediction is to be compared to one based on a pure power law, (9.2), giving $t_f = 1998.8 \pm 19.7$, certainly too inaccurate to be of any use. Apparently the earthquake did not happen. However, as communicated to me by D. Sornette, from a considerably refined analysis method, the authors of the original prediction understand that it was an artifact of approximations used.

Earthquake predictions have also been attempted using different models, closer to the standard lines of geophysical research [214]. One model is based on the hypothesis that an earthquake occurs when a fault has been reloaded with the stress which was relieved in the most recent earthquake. The time from one earthquake to the next is the stress drop in the most recent earthquake divided by the fault stressing rate. It incorporates directly some of the physical processes which are believed to be at the origin of earthquakes. This would convey some degree of predictability to this "recurrence model". If, on the other hand, earthquakes occurred completely randomly, their timings would follow a Poisson distribution.

A test of this recurrence model has been performed in one of the supposedly ideal locations, Parkfield, California [215]. The town of Parkfield is located on the San Andreas fault, one of the most seismically active regions of the earth. At least five earthquakes of magnitude $M_S = 6$ on the Richter scale [for a definition, cf. (9.16) below], or larger, have occurred in this area with an average interval of 22 years, the most recent one in 1966. With a prediction of the next earthquake around 1988, in 1986 the US Geological Survey set up a focused experiment to measure the stress accumulation, capture the nucleation of the next rupture and watch it propagate. The "problem" today is that the earthquake never arrived. In fact, the recordings of the experiment constitute the longest documented period of quiescence at Parkfield. Moreover, using one in-situ data set and one from GPS signals, it was shown that the stress which was released in the 1966 earthquake had recovered, at the 95% confidence level, by 1987. It continues to increase as a consequence of continuous fault slippage. When considering a release of stress to the level just after the 1966 quake, one now is faced with the nightmare idea that the next major earthquake in the Parkfield region could approach magnitude 7 on the Richter scale [215].

9.4 Stock Exchange Crashes

In the initial phase of research, the basic postulate of all groups trying to predict crashes on stock exchanges was that they work according to the same principles as those of earthquakes, or overarching generalizations thereof. They would view financial crashes as phase transitions in a hierarchical system, characterized by discrete scale invariance, and being increasingly loaded with time. However, there is no evidence for mean-reversion in stock prices unlike the assumptions of, e.g., the recurrence model for earthquakes. More recently, research on financial crashes has gained a momentum of its own, and the relation to models of earthquakes has loosened somewhat [21].

Following the earthquake analogy, a stock price or index rising in time would build up some stress in the market. It would be released in a singular failure event, the crash, which would mark a critical point. If this hypothesis is endorsed, the variation $S(t)$ of a stock price, resp. index, prior to a crash should obey

$$S(t) = A + B(t_f - t)^\alpha \left\{ 1 + C \cos\left[\omega \ln(t_f - t) - \phi\right] \right\} \qquad (9.8)$$

or more complicated generalizations thereof. ϕ is the phase of the oscillations. An alternative or complement to fitting this expression is to analyze the times of occurrence t_n of pronounced minima in the price variation which are predicted to follow a geometric progression

$$\frac{t_{n+1} - t_n}{t_n - t_{n-1}} = \exp\left(\frac{2\pi}{\omega}\right) < 1 \ . \qquad (9.9)$$

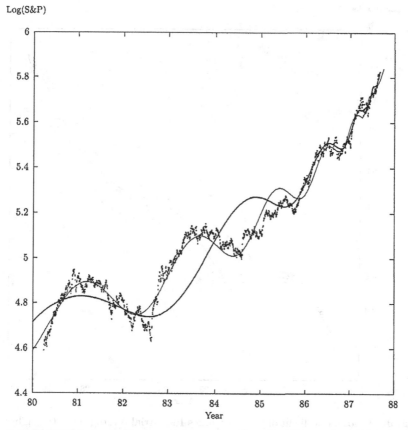

Fig. 9.9. S&P500 index in the seven years preceding the 1987 crash on Wall Street, and a fit to a power law with log-periodic oscillations. By courtesy of D. Sornette. Reprinted with permission from Elsevier Science from D. Sornette and A. Johansen: Physica A **245**, 411 (1997) ©1997 Elsevier Science

Price histories conforming approximately to (9.8) are called *log-periodic power laws*. With a positive power-law prefactor B, they correspond to bubbles. The understanding then is that a crash is the sudden collapse of a speculative bubble which has built up over a long time. Imitation and herding among market participants have pushed market prices of assets significantly above their fundamental values. The accelerating oscillations in (9.8) then reflect the competition between the instabilities of the inflating bubble due to sell orders on the one side, and the synchronization due to herding on the other side.

Figure 9.9 shows a fit of (9.8) to the S&P500 index in the years preceding the 1987 crash [216], showing clear signs of log-periodic oscillations. A similar fit is shown in Fig. 9.10 for the 1929 crash, using the Dow Jones index [216]. While these fits apparently describe the large-scale evolution of the data quite

Log(Dow Jones)

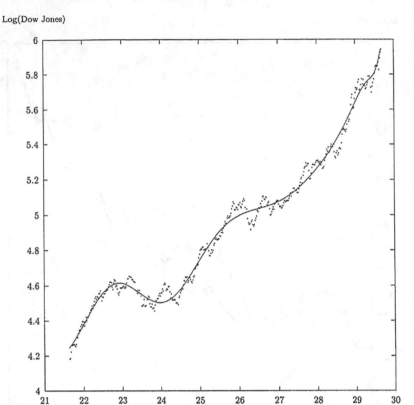

Year

Fig. 9.10. Log-periodic fit of the Dow Jones Industrial Average over the eight years preceding the 1929 crash. By courtesy of D. Sornette. Reprinted with permission from Elsevier Science from D. Sornette and A. Johansen: Physica A **245**, 411 (1997) ©1997 Elsevier Science

well, there are numerous additional oscillations in the data which are not accounted for by (9.8), and some subjective judgment certainly is required when using these methods to predict a crash.

The data shown in Fig. 9.2 can be analyzed in a similar way [209]. Vandewalle et al. first subtracted an exponential background corresponding to a long-term average growth rate of 0.1 per year, shown as a dotted line. An accelerated growth, corresponding to about 0.3 per year, sets in about two years before the crash (solid line). These departures from the long-term trend in the two years preceding the crash then are fitted to a variant of (9.8) where α is put to zero, i.e.,

$$|t_f - t|^\alpha \to \ln |t_f - t| \text{ as } \alpha \to 0 , \tag{9.10}$$

producing a rather successful description of the data. This is shown in Fig. 9.11. An advantage is that, the "exponent" being fixed now, there is

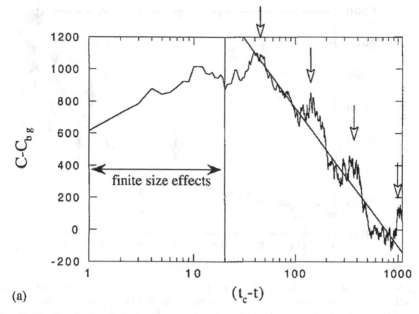

Fig. 9.11. Analysis of the excess evolution of the Dow Jones index over its long-term trend, in the two years prior to the 1987 crash, in terms of log-periodic oscillations. By courtesy of N. Vandewalle. Reprinted with permission from Elsevier Science from N. Vandewalle, et al.: Physica A **255**, 201 (1998). ©1998 Elsevier Science

one less fit parameter. If time t was taken to be temperature T, the law would correspond to the specific heat variation close to the critical point of the 2D Ising model [209]. Why this is the relevant quantity on which to model the evolution of a stock index, remains unclear. The claim of the authors that this variant would fit better than (9.8) with a power law [209] has, however, been disputed in the literature [217].

Based on these ideas, the crash in October 1997 ("Asian crisis") has been predicted by two groups. The prediction of Vandewalle, Bouveroux, Minguet, and Ausloos appeared in the popular press [208] first, and then in the scientific literature [218]. The analysis was performed both on the basis of (9.8) with $\alpha = 0$, and on the geometric progression of the extrema of the log-periodic oscillations, (9.9). The crash times predicted by both methods deviated from each other by less than the error bars. The corresponding data are shown in Fig. 9.12. An independent prediction of the 1997 crash by Didier Sornette is discussed in footnote 12 of [219]. Another group has given an analysis of this crash, using a rather similar theory, immediately after the event [220].

There have also been critical opinions on the predictability of financial crashes [221]. One problem is that the prediction based on log-periodic oscillations does not always work. Figure 9.13 shows a crash that did not take

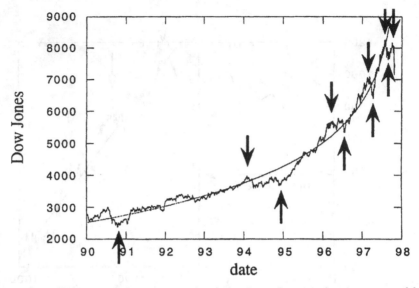

Fig. 9.12. Analysis of the 1997 crash of the Dow Jones index in terms of log-periodic oscillations. By courtesy of N. Vandewalle. Reprinted from N. Vandewalle, et al.: Eur. Phys. J. B **4**, 139 (1998), ©1998 EDP Sciences

place: the price variation of Japanese Government Bonds during 1993–1995 could be fit to a log-periodic variation, suggesting a crash by September 1995. This crash did not take place although the apparent quality of the fitted was best just during the year 1995! Based on log-periodic oscillations, there were also warnings of a crash possibly occurring in late 1998 on the web sites of Phynance technology [222], a young Belgian company marketing anticrash software based on the ideas discussed in this chapter, throughout the second half of 1998 and early 1999. No crash occurred, although the markets were extremely volatile, as readers may remember.

Moreover, there may be technical problems involved in the analysis which might make a prediction somewhat unreliable (consider your investment which depends on the quality of your prediction!) [221]. One, of course, is the rather large number of fit parameters implying that there will likely be several sets of equally good fit parameters yielding different crash times. In the analysis of the extrema of the putative log-periodic oscillations, one often encounters extrema which do not follow a hypothetical log-periodic sequence, but which are more pronounced than those which lie on the sequence. A decision then has to be made to either discard them or look for a different sequence. The latter will most likely yield a different prediction. This problem is illustrated in Fig. 9.1 [221]. The most prominent minima of the S&P500 index indeed lie on a log-periodic progression marked by t_i. However, the Hong Kong Hang Seng index has two additional minima labeled by question

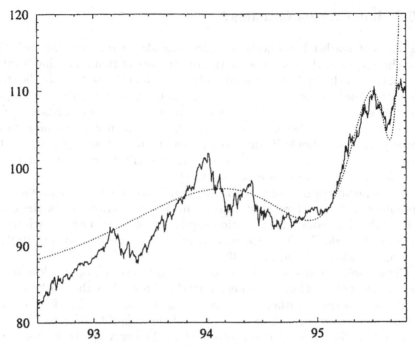

Fig. 9.13. Price variation of Japanese Government Bonds 1993–1995, and fit to a log-periodic variation. Note that the crash suggested by the fit did not take place. By courtesy of J.-P. Bouchaud. Reprinted from L. Laloux, et al.: Europhys. Lett. **45**, 1 (1999), ©1999 EDP Sciences

marks which do not fall into the log-periodic sequence. Nevertheless it was on the Asian markets that the crash started.

Finally, the predicted crash time is often not reached, but the crash can occur before – or not at all. Even with an accurate crash warning, an investor then has to decide how much time ahead he has to change his investment from risky assets to riskless ones, to protect it. Of course, if many investors do so well in advance, the crash might be avoided simply by the reaction of investors to a crash warning. Alternatively, investors might panic at a crash warning, and trigger the crash immediately. The warning then has become a self-fulfilling prophecy.

Despite these reservations, evidence for log-periodic oscillations in financial time series continues to accumulate. An analysis of both the Nasdaq composite index and of individual US stocks has shown that the crash in april 2000 was accompanied by significant log-periodic oscillations [223].

9.5 What Causes Crashes?

The efficient market hypothesis does not provide for crashes – at least not with the frequency they occur with. Its core statement that all available information on a stock is reflected immediately and in an unbiased way in the stock price, would only allow for a financial crash in the case of a truly catastrophic event. There is no systematic evidence in favor of such a mechanism.

Confronting the efficient market hypothesis to reality, one ecounters essentially three situations: (i) there is a crash due to a catastrophic event, (ii) there is a catastrophic event but no crash occurs, (iii) there is a crash but no catastrophic trigger event can be identified.

A prominent example of a crash triggered by a catastrophy is provided by September 11, 2001, where most markets in the world crashed. For the DAX, cf. Fig. 5.12. The cause–effect relationship is obvious here. Others include the outbreak of World War 1, or the coup against Gorbachev in August 1991, or the Nazi invasion of France in 1940.

Catastrophic events sometimes do not lead to crashes on stock markets. The outbreak of the Gulf War in early 1991 did not affect the stock prices in the Western world, or rather gave them a positive impetus. The fear of a war in Iraq, and its outbreak in 2003, increased the volatility of many financial markets but did not send them into decline. The earthquake in Taiwan in fall 1999 did not lead to a collapse of stock markets in Asia. The Kobe earthquake in Japan 1995 had a strong influence on *some* stocks but much less on the Japanese stock market as a whole. Also the South Asian tsunami on December 26, 2004, affected some stocks but it did not affect the financial markets as a whole, neither in Southern Asia nor worldwide.

On the other hand, often an entire market, or many world markets crash, and no single triggering event can be identified. According to Sect. 9.2, no single cause could be identified for the 1987 crash on Wall Street, resp. in the world markets [10, 224]. The situation is similar for many other crashes, e.g., the Black Monday in 1929, the Asian crisis in 1997 or the burst of the "dot.com" bubble on Nasdaq in 2000. There is a common feature in these cases, though: inflated expectations about the future evolution of economies. In 1929, the focus was on utilities, in 1987 on the effects of financial deregulation, in 1997 on the growth of the South-East Asian "Tiger States", in 2000 on the succes of telecommunication and computer industries.

Apparently, there are two classes of crashes in financial markets: crashes caused by catastrophic events ("exogeneous crashes") and crashes whose root and trigger must have been in the financial markets themselves ("endogeneous crashes") [225]. In the financial markets, do they show up with the same signatures? In other words, if we only possess the time series of a financial asset containing a crash event, could we unambiguously attribute the event to one of the two classes?

Indeed, one can. A systematic investigation of about fifty events from many different markets shows that the presence of a log-periodic power law of the type (9.8) or generalizations thereof is the discriminating factor [225]. Endogeneous crashes happen more or less close to the culmination point of a log-periodic sequence. The log-periodic precursor sequence therefore allows, with the reservations made above, a prediction of the event. Clear examples include in 1929 Black Monday (Fig. 9.10), the crash of October 1987 in many world markets (Figs. 9.2 and 9.9 for Wall Street), the Asian crash in 1997 (Figs. 9.1 and 9.12 for the Hang Seng and Dow Jones indices, respectively), and the 2000 crash on Nasdaq, among others. Common to these endogeneous crashes is that one cannot identify a single underlying cause or triggering event, and that the systematically happen after long bullish rallies [225].

Exogeneous crashes happen out of the blue, and are not preceded by a log-periodic power-law time series, as can be verified with the examples cited above and many more [225]. They are intrinsically unpredictable. An exogeneous crash in a specific market can, however, be due to the crash of another market. This can be seen on the time series of the DAX in 1987 which does not carry the log-periodic power-law signatures of Wall Street. Apparently, the endogeneous crash of Wall Street was perceived by german investors as an exogeneous, catastrophic event, and they reacted in panic.

Within a model of multifractal random walks [226, 227], building on the concepts discussed in Chap. 6, exogeneous and endogeneous crashes relate to different quantities and therefore produce, e.g., different decay of the volatility in the markets. The basic idea is as follows. Independently of its origin, the crash produces a volatility shock. Unlike in the simple models, volatility in real markets is a long-time correlated variable, cf. Chap. 5.6.3 and Figs. 5.24 and 5.25. The temporal decay of the excess volatility now depends on the nature of the perturbation, and the state of the market at the time of the perturbation [228].

For the exogeneous crash, the volatility decay is determined by the response of the market to a single piece of very bad news, i.e. to a delta function-like perturbation $\delta(t)$. Based on the linear response functions of the multifractal random walk model, a decay of the excess volatility $\sim 1/\sqrt{t - t_f}$ is found. The excess volatility after an exogeneous crashes indeed decays in this way while after an endogenous crashes, it does not [228].

For an endogenous crash, the volatility response conditional on a major volatility burst *within* the system is relevant. Evaluating the appropriate conditional response function, one finds that the excess volatility can formally be written as a power law of time, $\sim (t - t_f)^{-\alpha - \beta}$ [228]. The exponent α depends on the strength of the volatility perturbation. β contains a logarithmic time dependence itself. Unless $\alpha \gg \beta$, the volatility after an endogenous crash therefore does not decay as a pure power law.

Prior to an endogenous crash, a description of the market behavior in terms of incorporation of information into prices can only be given if it is assumed that there has been a particular sequence of small pieces of information which brought the market into an unstable state. The endogeneous crash itself finally is due only to an additional small piece of information. This is in line with the systematic failure of attempts to identify a trigger event in such a case [228].

9.6 Are Crashes Rational?

The consistency of the efficient market hypothesis with financial crashes is doubtful. A crash due to a catastrophic external event apparently is consistent. When a catastrophy occurs and no crash happens, one may argue that investors very quickly understand the limited impact on the economy, or that they see positive impacts counterbalancing the negative ones. E.g. after an earthquake or tsunami, tourism may decline temporarily, but at the same time construction certainly increases. However, the market reports often point out that a certain moderate or violent response to external events seems to depend on the assurance or fright of the markets as a whole. However, for the case of an endogeneous crash, the efficient market hypothesis has a problem: the crash simply should not happen.

One may turn around this argument and use it against the efficient market hypothesis. If crashes then occur as often as they do, this must be due to deviations from market efficiency. Expectations of future earnings may, in periods of general euphoria, create speculative bubbles which end in crashes. Such arguments have been invoked for the bubbles preceding historic crashes such as the tulipmania in the Netherlands of the 17th century [207], but also for those before the major crashes of this century in 1929 (driven by unrealisitc expectations from the utilities sector), 1987 (driven by general market deregulation), 1998 (driven by investment opportunities in Russia), and 2000 (driven by the euphoria about the "New Economy" of high-technology stocks) [223]. As a corrolary, crashes would be the consequence of irrational behavior of investors, of their "mad frenzy" [223]. In the preceding chapter, we have discussed some models which attempt to shed light on such irrational behavior as herding and imitation of agents.

However, despite the apparent failure of the efficient market hypothesis, and despite the wording often used to describe investor behavior during speculative bubbles (cf. preceding paragraph) "abnormal" price increases and crashes can occur, *with rational investors,* when a finite *exogeneous* probability of a crash is allowed for [229]. In other words, when exogeneous crashes can happen, endogeneous crashes may be the consequence.

When interest rates, transaction costs, etc. are neglected and a risk-neutral world is assumed, the efficient market hypothesis requires share prices to follow a martingale stochastic process

$$\langle S(t' > t) \rangle = S(t) . \tag{9.11}$$

Now assume that there is a nonzero probability of a crash. This can be modeled as a jump process $j(t) = \Theta(t - t_{\text{c}})$ which is zero before and unity after the crash occurring at an unknown t_c. t_c itself is now a stochastic variable, with a probability density function $q(t)$, a cumulative distrbution function $Q(t) = \int_{-\infty}^{t} dt' q(t')$, and a hazard rate $h(t) = q(t)/[1 - Q(t)]$. The hazard rate is the probability per unit time that the crash happens in the next time step if it has not happened yet. With such an exogeneous crash probability, the dynamics of the share price becomes [229]

$$dS = \mu(t)S(t)dt - \kappa S(t)dj . \tag{9.12}$$

In this equation, κ is the fraction of drawdown in the crash, and $\mu(t)$ is the return of the stock, treated as an open parameter at present. Apart from the crash probability, other sources of exogeneous noise have been neglected. In this case, if the crash probability was zero, the share price would stay constant. With a finite crash probability, however, the martingale condition for the share price becomes $\langle dS \rangle = 0$, and therefore requires a return on the stock before the crash

$$\mu(t) = \kappa h(t) . \tag{9.13}$$

This leads to a price dynamics before the crash

$$S(t) = S(t_0) \exp\left(\kappa \int_{t_0}^{t} dt' \, h(t') \right) . \tag{9.14}$$

The surprising result of this argument is that with a finite probability of a crash, even in a world of rational investors, there must be a boom period before the crash. The price increase before a crash is necessary to compensate for the losses during the crash [229]. However, in this simple model, the crash time follows a stochastic jump process and cannot be anticipated. Therefore, despite the booms preceding the crashes, abnormal profits cannot be earned. The situation may be better in real markets if the precursor signals discussed in the previous section consistently have predictive power.

Observe also that, despite much discussion to the contrary, there have been occasional reports discussing the most prominent features of the Dutch tulipmania in terms of market fundamentals [206].

9.7 What Happens After a Crash?

Despite some universality, we have also seen major differences between crashes. One example is provided by Fig. 1.1 containing the crashes of October 1997 and fall 1998. They are different in their shapes in the DAX time series but also in the duration of the "depression" they generated. The consequences of the 1997 event are no longer visible in the DAX quotes a few days after the crash. The 1998 drawdown lasted much longer: only one year after the event, the DAX again reached its precrash level. After the 1987 crash, the Dow Jones Industrial Average reached its precrash high after about two years. Figure 9.2 shows, however, that it resumed the long-term rise with a rate of about 0.1 per year which it had followed until about two years before the crash, almost immediately after. Finally, the consequences of a crash of the Japanese Nikkei 225 index in 1990 (not discussed above), have persisted for at least 10 years. At the time of writing the first edition of this book, the Nikkei index was at about 16,000 points, compared to about 40,000 at the beginning of 1990. In November 2002, when this book was updated for its second edition, the Nikkei traded below 9,000 points. On November 18, 2002, it closed at 8,346. In April 2005, it had risen back to about 11,000 points. How long do crashes persist?

Investors would like to have a signal identifying the trend reversal after a crash. In particular, one would like to have an exogenous variable, independent of the stock market. On a purely empirical basis, the interest rate spread on the bond market has been identified as such a variable recently [230].

A trend reversal after a crash should correspond to a change in the trader attitude from bearish to bullish. Bear markets are characterized by fear of the future evolution, bull markets rather by optimism about the future. The idea therefore is to search for a measure of the uncertainty which the market actors have about the future evolution. One possibility is to look at interest rates. In principle, the more uncertain the future, the more one expects high interest rates. The default risk of an debtor, which must be compensated by the interest payment, is the higher the more uncertain the repayment of the credit. The uncertainty on the repayment of a credit clearly is correlated with the future evolution of the economy. However, in practice, there is no strong and systematic correlation of interest rates with stock price evolution during and after a crash.

A different picture emerges, however, when one considers the spread in interest rates for credits extended to borrowers of different quality. If one takes as a measure of the interest rate spread the difference of the interest rates of bonds of the lowest credit rating with the rates of highly rated bonds, a strong correlation emerges [230]. Roehner has investigated this correlation for various crashes in the 19th and 20th centuries. He found significant correlations between the bottom line after a crash and a maximum in the interest rate spread after the crash, for all of the crashes in the last two centuries [230]. Figure 9.14 shows as an example the 1929 crash on Wall Street. The solid line is the stock index normalized to 100 at the beginning of the crash as a function of the number of months after the crash, the thick dotted line is the interest rate spread, and the thin dashed line is the interest rate. Throughout the series of crashes studied, similarly good correlations are found between stock price and interest rate spread (correlation -0.86 in 1929), but normally less good correlations between the stock prices and interest rates (the correlation coefficient of -0.72 is exceptionally high in 1929 compared to other dates). One can also establish parallels between the interest rate spread and

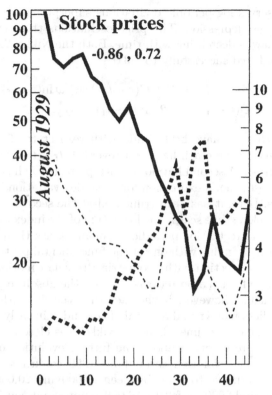

Fig. 9.14. Normalized stock prices (*solid line, left scale*), interest rate spread (*thick dotted line, right scale*), and interest rate (*thin dashed line*) at the New York Stock Exchange after August 1929. The horizontal axis numbers the months after the crash. The correlation stock price/spread is −0.86, and the stock price/interest rate correlation is −0.72. By courtesy of B. M. Roehner. Reprinted from B. M. Roehner: Int. J. Mod. Phys. C **11**, 91 (2000), ©2000 by World Scientific

a lack of consumer confidence in the market. Apparently both measure the uncertainty perceived by the market actors, about the future evolution of the stock markets, and of the economy more generally.

In all of our discussion, the crash seen as a phase transition occurred after prices rose with time. This corresponds to lowering the temperature towards a critical temperature in physics. However, critical phenomena in physics are also observed when one raises the temperature and approaches the critical temperature from below. Can we observe "reverse crashes" on financial markets?

With some caveats, one can, indeed. In the past, financial markets often entered severe depression after long bullish periods. However, these bull markets did not end in a crash but more gently crossed over into depression. Two examples are the Japanese Nikkei 225 stock index and the gold market [231].

One indeed observes log-periodic oscillations superposed on a power law as Japan entered the depression. The price oscillations then are decelerating, and the power law is decreasing with time. Both the Nikkei 225 and the gold price have been fitted successfully to [231]

$$\ln S(t) = A + B(t - t_f)^\alpha + C(t - t_f)^\alpha \cos\left[\omega \ln(t - t_f) - \phi_1\right]$$
$$+ D(t - t_f)^\alpha \cos\left[2\omega \ln(t - t_f) - \phi_2\right] . \tag{9.15}$$

The following changes have been made with respect to (9.8) describing a bubble. The time to the crash has been reversed, $t_f - t \rightarrow t - t_f$ to become time after the crash. A second harmonic with prefactor D has been added. In principle, the most general expression for the index variation is a log-periodic harmonic series. Here, it has been truncated at the second order. Finally, it turns out that on long time scales, the logarithm of the index variations $\ln S(t)$ provides better fits to the log-periodic harmonic series than the index $S(t)$ itself. Moreover, it is in line with one of the fundamental postulates discussed in Sect. 4.4.2 in connection with geometric Brownian motion, namely that investors are more focussed on returns than on the absolute prices.

Most remarkable, however, is the fact that the fit of the Nikkei index allowed the prediction of a trend reversal of this index in early 1999 [232]. The prediction was made at a time when the Nikkei was close to its 14-year low, and economists were skeptical about the further evolution of the Japanese markets. The further evolution throughout 1999 confirmed the prediction: the Nikkei index returned to levels between 19,000 and 20,000 points by the end of 1999. By mid-2000, it fell to 16,000 points, and continued falling to below 8,500 points in late 2002. It has recovered to levels of 10,000 ... 12,000 points, by early 2005.

In Chap. 8.2, bubbles have been defined as an overvaluation of market prices with respect to fundamental prices. Imitation and herding on the buy-side of the market fuelled by an optimistic outlook on the future evolution of the economy, was suspected to be the main driving mechanism behind a bubble. When a pessimistic outlook is predominant, exactly the same mechanisms, imitation and herding, on the sell-side of the market may lead to increasing synchronization and to decreasing prices. In such a situation, an *anti-bubble* may build up, again following some log-periodic power law price history. An anti-bubble corresponds to falling prices with log-periodic oscillations expanding in time. More specifically, prices during an anti-bubble will approximately follow (9.15). It is characterized by a power-law prefactor $B < 0$ [233]–[235]. t_f is the starting date of the anti-bubble.

Based on this theoretical framework, strong predictions have been published on the future bearish behavior of many of the world's financial markets [233]–[236]. For the US S&P500 index, based on data up to August 2002, a prediction was issued in September 2002 that (i) the index would reach its minimum at that time, (ii) reverse its trend to increase to a level of about 1,000 index points in late 2002 or early 2003, (iii) to slowly and slightly decrease until the second semester of 2003, and (iv) to sharply fall to below 700 index points in the first semester of 2004, always following (9.15) [233]. Underlying this predicted price variation is an anti-bubble which formed around August 2000, about four months after the collapse of the "new economy bubble" (or "dot.com bubble") on April 14, 2000 [223].

This bubble can be seen in Fig. 1.2 as the anomalous increase in the DAX from about 1996 to 2000. It was fuelled by collective beliefs that new communication technologies, more powerful computers, more intelligent software, the spreading use of the internet, etc. would give birth to a "new economy" with high growth rates where many traditional products and trading structures would be replaced by data and communication paths. Prices of companies like Cisco, Global Crossing, etc. were high because investors expected enormous future earnings – the current earnings per share of the companies at that time were actually rather low. Established blue chips like car makers traded at much lower prices or returns although their earning per share were rather high. The expectation of future earnings made the whole difference! The collapse of the bubble started on April 14, 2000 on the Nasdaq which lost about 37% until April 17, 2000 [223]. Other high-technology market segments in the world crashed in a similar way. The decline of these indices was not finished at the end of the crash, though, as investors were sent into depression after the end of the bubble, and negative sentiments prevailed on almost all markets. The consequences of the bubble collapse on the blue chip indices or very broad market indices such as the S&P500 were much milder, and could qualify for a crossover between a bubble and an anti-bubble.

Actually, many markets worldwide are well described by anti-bubble theory between mid-2000 and summer 2002 [233, 234]. The prediction made in summer 2002 about the future behavior of the S&P500 index based on the anti-bubble [233] was extended to the major stock indices of other countries [234], i.e. the anti-bubble went global. The prediction for the US market (sharp decline in 2004) was reemphasized in 2003 with a time scale set for validation by summer 2004 [235]. There are also reports of modifications with a slight shift in the dates of plunge and recovery. The year 2004 was held up, though, as the time of the decline with some recovery, perhaps, in 2005 [236].

It turned out, however, that only a small part of every prediction materialized! Summer 2002 indeed formed the bottom of many stock indices, and the predicitions of rising quotations through the second semester of 2002 generally were realized. However, the more spectacular part of the predictions ("Bear markets to return with a vengeance" [236]), namely that the trend reversal would be followed by another decline – first gentle, then steep – from early 2003 at least until 2004 did not happen on the world markets. After another, often deeper minimum in spring 2003, most market indices rose until the end of 2004, at least.

Here, we discuss the behavior of the DAX German blue chip index in more detail. Figure 9.15 displays the index (ragged solid line) together with a fit (smooth solid line) to (9.15) [234]. In the best fit based on the data up to September 30, 2002 (left vertical line in Fig. 9.15), t_f = October 6, 2000, i.e. the anti-bubble started almost half a year after the burst of the new economy bubble. Quite generally, there need not be a coincidence between the date of a crash (if one occurs) and the starting date of an antibubble. Similarly, one does not expect a symmetry between bubble and antibubble [235]. The other parameters are α = 0.94, ω = 8.47, ϕ_1 = 3.61, ϕ_2 = 4.58, A = 4.58, B = -0.0012, C = 0.00041, D = 0.00012. The negative value of B identifies the anti-bubble. The time t is measured in calendar days, unlike many other statistical analyses which refer to trading days.

Figure 9.15 shows that these expressions indeed give a good ex-post fit of the variation of the DAX, i.e. for the time period where data were available. The date where the prediciton was issued is marked by the left vertical line in Fig. 9.15. On the other hand, (9.15) does not give a reliable ex-ante description of the DAX. While prediction and actual realization still are consistent during the last quarter of 2002, they vary in completely different ways thereafter. After an intermediate high at about 3200 points in late 2003, the DAX falls to its nine-year low at 2202.96 points on March 12, 2003, while the prediction rises to about 3500 points. The DAX then rises gradually to about 4000 points until early 2004 to stay in this range for the rest of that year. The prediction, on the other hand levels off at 3500 points to enter the

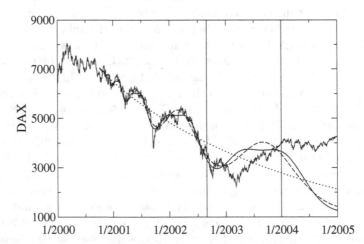

Fig. 9.15. Variation of the DAX from January 3, 2000 until December 30, 2004, and comparison to the anti-bubble prediction of Zhou and Sornette [234]. The DAX is the ragged *solid line*. The *dotted line* is the pure power-law component in (9.15). The *dashed line* includes the first log-periodic harmonic as well. The smooth *solid line* in addition includes the second log-periodic harmonic, i.e. describes (9.15) with the parameters given in the text. The left vertical bar labels the date where the prediction was issued. The right vertical bar is the shortest of the dates of validity of the prediction

bear market in early 2004. During 2004, the DAX was predicted to fall to almost 1000 index points, making up for a twenty-year low.

Several limits of validity have been attached to these predictions. One is at the end of 2003, marked by the right vertical line in Fig. 9.15 [234]. Others are in 2004, between the right vertical line and the right end of the figure [235, 236]. It is clear, though, that the prediction did not materialize in either of these time spans, and that significant deviations started as early as the beginning of 2003. The prediction also failed for all other indices investigated.

It therefore appears that log-periodic power-law behavior is a universal feature of speculative markets, no matter whether they are stock indices, individual stocks, commodities or currencies. They represent a kind of correlation very different from those discussed in Chap. 5. Apparently, log-periodic power-law price variations are common in financial markets and can both be associated with bubbles (bull markets) and anti-bubbles (bear markets). Likely, both are due to self-reinforcement of expectations and beliefs at the origin of trading decisions. Apparently, they are less stable and the problem of competing fits with different parameter sets is more serious, though, than advertised by their proponents. The fact that predictions are not systematically followed by markets, and sometimes fail, does not necessarily invalidate the concept as such. It indicates, however, that more research is mandatory before we can claim to understand crises and crashes in financial markets, and before reliable predictions can be made systematically.

9.8 A Richter Scale for Financial Markets

This chapter has drawn heavily on potential analogies between earthquakes and captial markets. For most of our discussion, we concentrated on the idea that these extreme events are related to the critical points discussed in physics, and on deterministic precursor signals. However, we have done little to quantify the magnitude of financial crashes. It is not even clear what features make up a "crash", or a "crisis" in a capital market. Should the "second black monday" on October 27, 1997, be called a "crash" in Germany or the US, where the stock indices lost about 7% in one day and recovered quickly, or only in Asia with, e.g., a 24% drawdown in Hong Kong (cf. Fig. 9.1)? Moreover, both in seismology, and in finance, the extreme events we call crashes are relatively rare, but there is much continuous seismic activity in the earth as well as much persistent turmoil on capital markets on smaller levels. We therefore need an accurate, quantitative measure of the state of financial markets.

In seismology, the Richter scale provides such an indicator. It is a logarithmic scale of the total seismic energy E_{tot} released in an earthquake. The magnitude M_S on the Richter scale is related to the total energy release by [237]

$$M_S = \frac{2}{3} \left(\ln E_{\text{tot}} - 11.8 \right) . \tag{9.16}$$

Moreover, the Gutenberg–Richter law

$$P(E_{\text{tot}}) \sim E_{\text{tot}}^{-1.5} \tag{9.17}$$

relates the probability per unit time, i.e., frequency, of an earthquake to its energy release, and thereby to its magnitude on the Richter scale. In other words, the Richter scale also measures the inverse frequency of earthquakes of a certain magnitude

$$M_S \approx \frac{2}{3} \ln \left(\frac{E_{\text{tot}}}{E_0} \right) = \frac{4}{9} \ln \left(\frac{1}{P(E_{\text{tot}})} \right) . \tag{9.18}$$

A group at Olsen & Associates, Zürich, has recently constructed an analogous scale for financial markets [108]. In fact, two such "scales of market shocks" (SMS) are needed: One is an absolute, universal scale which allows one to compare the influence of one specific event on a variety of assets. The other scale is an adaptive one which compares the relative importance of various events on a single asset.

An indicator measuring market shocks can be constructed in analogy with mechanics [108]. The kinetic energy is $E_{\text{kin}} = (m/2)v^2$ with (1D) velocity $v = \mathrm{d}x/\mathrm{d}t$ the derivative of position x. If we identify position in space with the logarithmic price $\ln S(t)$ of an asset, velocity is equivalent to time-scaled returns

$$v(t) \rightarrow r[\tau, S; t] = \frac{\ln S(t) - \ln S(t - \tau)}{\sqrt{\tau}} \equiv \frac{\delta S_\tau(t)}{\sqrt{\tau}} . \tag{9.19}$$

$\sqrt{\tau}$ appears in the denominator because of the stochastic nature of the price process. Unlike mechanics where the limit $\mathrm{d}t \rightarrow 0$ is well defined and usually finite, it is not obvious that a limit $\tau \rightarrow 0$ can be taken in (9.19). The $\sqrt{\tau}$-scaling in (9.19) removes the time scaling of the volatility of returns of geometric Brownian motion $\langle (\delta S_\tau)^2 \rangle \sim \tau$. In all other cases, the volatility of rescaled returns will continue to depend on the time scale τ, and may vanish or diverge as $\tau \rightarrow 0$.

A scaled volatility is then defined on an N-point grid in the time scale τ as the standard deviation of the scaled returns

$$v[\tau, S; t] = \sqrt{\frac{1}{N-1} \sum_{i=1}^{N} r^2 \left[\frac{\tau}{N}, S; t - \tau \frac{(i-1)}{N} \right]} . \tag{9.20}$$

The equivalent of the kinetic energy is then time-scaled variance, i.e.

$$E_{\text{kin}} \sim v^2 \rightarrow v^2[\tau, S; t] . \tag{9.21}$$

An indicator can be built on the expectation value of this quantity which, of course, is scale dependent. Big earthquakes are usually well separated from the background seismic activity. The integration of the energy release therefore poses no problems. In financial markets, the background signal is much stronger, and events cannot be clearly separated from their background. Therefore, time-rescaled variance $v^2[\tau, S; t]$ may be a better quantity to use in financial markets than bare variance $\langle(\delta S_\tau)^2\rangle$.

Now remember that volatilities are distributed log-normally, to a good approximation (cf. Chap. 5, [107, 108])

$$p(v) = \frac{1}{\sqrt{2\pi}\sigma_v v} \exp\left[-\frac{1}{2\sigma_v^2}\left(\ln\frac{v}{v_0}\right)^2\right] , \tag{9.22}$$

with maximum and mean at

$$v_{\max} = v_0 \exp(-\sigma_v^2) , \quad \text{and} \quad \bar{v} = v_{\max} \exp(3\sigma_v^2/2) , \tag{9.23}$$

respectively. v_0, and consequently $v_{\max}$ and $\bar{v}$ are τ-dependent when unscaled returns are used [107] and almost τ-independent with scaled returns [108] for smaller volatilities. A τ-dependence persists, however, for large $v \gg v_{\max}$. Analogy with (9.18) then suggests the following function for mapping volatility into the SMS indicator:

$$f_{\text{adap}}(v) = \frac{\text{sign}(v - v_{\max})}{2\sigma_v^2}\left(\ln\frac{v}{v_{\max}}\right)^2 . \tag{9.24}$$

By superposing this function on a log-normal distribution, one notices that $f_{\text{adap}}(v)$ is sensitive to large and small volatilities, but almost vanishes in the range of the normal background signals $v \approx v_{\max}$.

The adaptive scale of market shocks is finally defined as an integral of this indicator function over time scales

$$\text{SMS}_{\text{adap}} = \int d\ln\tau\, \mu(\ln\tau) f_{\text{adap}}(v[\tau]) \tag{9.25}$$

with a weight function

$$\mu(\ln\tau) = ce^{-x}\left(1 + x + \frac{x^2}{2}\right) \quad \text{with} \quad x = 2\left|\ln\frac{\tau}{\tau_{\text{center}}}\right| . \tag{9.26}$$

c is a normalization constant, and τ_{center} sets the time scale of maximum sensisitivity of the indicator. In practical applications $\tau_{\text{center}} = 1$ day has been used successfully. The universal scale of markets shocks SMS_{uni} is defined in the same way, except that a mapping function

$$f_{\text{uni}}(v) = v \left.\frac{df_{\text{adap}}}{dv}\right|_{v=3v_{\max}} \tag{9.27}$$

is taken. It is proportional to $v/v_{\max}$. $v_{\max}$ itself is strongly asset-dependent, and therefore ensures the normalization of the universal scale of market

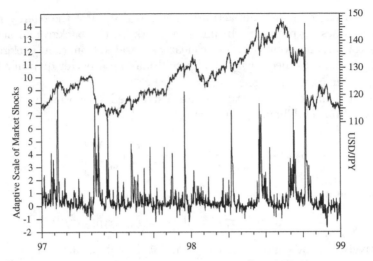

Fig. 9.16. Adaptive scale of markets shock for the USD/JPY markets in 1997/98 (*left scale*), and the corresponding price (*right scale*). By courtesy of G. O. Zumbach. Reprinted from *Introducing a Scale of Market Shocks*, Olsen & Associates preprint

shocks. Events in markets with different background volatilities thereby become comparable.

Figure 9.16 shows the exchange rate USD/JPY on the right scale, and the adaptive scale of market shocks on the left scale for the years 1997 and 1998. These years have been discussed throughout this book, as they were full of events. The scale of market shocks apparently works very well, and provides a much better distinction of exceptional from normal events than the price chart itself. Some strong peaks on the SMS are almost invisible in the price evolution. Conversely, strong price variations produce strong SMS signals. The reason for the high signal/noise ratio is the shape of the mapping function $f_{\text{adap}}(v)$ used in the SMS and its sensitivity to big events, and the use of $\tau_{\text{center}} = 1$ day which gives a good sensitivity to intraday fluctuations. More importantly, perhaps, most though not all of the major market shocks can be correlated with news headlines, be they on actual events or rumors.

10. Risk Management

In this chapter, we will describe the basic principles and methods of risk management. We define risk and various measures of risk. We discuss the types of risk which banks face, and how they actually manage them.

10.1 Important Questions

There are many important questions on risk.

- How is risk defined?
- How is risk measured quantitatively?
- What types of risk does a bank face?
- Are they independent of each other, or correlated?
- Are extreme risks and typical risks related in a simple manner, or do we need separate theories or methods for each?
- Why do people resp. institutions accept risk?
- What is the reward of accepting risk?
- What is the purpose of risk management?
- What are the tools for controlling, i.e., minimizing risks?
- Are there additional tools for complex portfolios of assets, compared with the hedging of a single security?
- Do the measures of risk, and the methods to control it, rely on Gaussian markets, or can they be adapted to the more general properties of asset prices discussed in Chaps. 5 and 6?
- How can we optimize the relation between risk and return?

Although measures of risk have been available and risk management functions in financial institutions have existed for a long time, the problem of correctly quantifying risk and prudently managing risk again has become very important recently. There have been unexpectedly big losses, e.g., at Barings Bank, Daiwa, Yamaichi, Hokkaido Takushoku Bank, Sanyo Securities, Allied Irish Bank, or Long Term Capital Management during the last couple years, or so. Rules have been established for financial institutions to control their risks, and banking has become one of the most heavily regulated businesses today.

However, many models used for risk management in banks and in the regulatory framework to which banks are subject, in one way or another rely on the Gaussian distribution for asset returns. Extreme risks are absent there!

10.2 What is Risk?

Future is uncertain. Highlighted by Lao Zi's words in the front material of this book ("One must act on what has not happened yet"), the consequences of human decisions both in personal and in business life reach into the uncertain future. Economists refer to this situation as *decisions under uncertainty*. The notion of risk – as opposed to uncertainty – comes in when the decision maker possesses a probability distribution of future events – either objective, i.e. statistical, or at least subjective. This classification of probabilities into objective or subjective probabilities was foreshadowed by Bachelier [6, 7]:

> "One can consider two kinds of probabilities: 1. The probability which might be called 'mathematical', which can be determined a priori and which is studied in the games of chance. 2. The probability dependent on future events and, consequently, impossible to predict in a mathematical manner. This last is the probability which the speculator tries to predict."

Many business decisions also must rely on subjective probabilities. Systematic scenario analysis usually helps to go some way from subjective to objective probability, i.e. from 2. to 1. In contrast, uncertainty describes situations where the likelyhood of outcomes is unknown, and cannot even be estimated.

More precisely then, *uncertainty* refers to situations where it is only known that one of several outcomes will be realized. *Risk* describes situations where we know that a particular outcome will be realized with a certain – objective or subjective – probability. *Certainty*, of course, describes situations with a deterministic outcome.

Risk then may be looked at from three different perspectives:

1. Planning perspective: Failure to reach targets set for the future.
2. Decision perspective: Wrong decisions.
3. Financial perspective: Losses.

All three perspectives are important in banking. Given the focus of this book on the description of financial markets and asset prices, we usually implied the last meaning when speaking about risk.

In all three perspectives, risk refers to the deviation of the actual outcome of a decision from its planned consequences. When such a situation can be described in terms of a numerical variable, risk describes the deviations of the future realizations of this variable from a target or expected value. These values can be set either by a strategic management decision or a business plan ("targets") or by statistical techniques. Examples of the latter include statistical expectation values $\langle x(t) \rangle$, the explicit or implicit assumption of martingale properties, forecasts derived from autocorrelated stochastic processes or, perhaps an extreme case, the predictions of crashes in financial markets discussed in Chap. 9. Deviations can be positive or negative. In a more narrow sense, risk is understood as the negative deviations whereas the positive deviations often are referred to as chance or reward. In banking practice, this restricted focus on negative deviations is common. In the special cases when probability distributions are symmetric around zero, a case often encountered to a good approximation in this book, risk and chance cannot be separated, and both are measured by the same quantities.

In quantitative finance, we hence define risk as the negative deviations of the future value (return) of a portfolio (possibly a single asset) from its expectation or predicted value.

Risk management can be reduced to two main questions:

1. How can one ensure that the actual outcome of an action/investment is as close as possible to the expected outcome or, more pragmatically, that the consequences of the actual outcome are as close as possible to (those of) the expected outcome?
2. What provisions can one take for the case that risk strikes, i.e. that the outcome of an action (investement) significantly differs from the expected outcome?

This chapter focusses on the first question, and on instruments to measure risk. The second question is the subject of Chap. 11.

Etymologically, the term *risk* apparently is derived from *risco* in medieval Italian and Spanish, meaning cliff. It is established that *risk* was used in maritime insurance in fourteenth century Italy – quite naturally then, in view of the elevated rates of loss of vessels at those times.

10.3 Measures of Risk

Once risk has been defined, we must find quantitative measures of risk. Risk is defined and must be measured at various levels of hierarchy: the risk of an individual position, empirically derived, e.g., from a certain time series. Next comes portfolio risk. Again, risk can be measured based on the time series of portfolio values, similar to the risk of an individual position. However, the time series of portfolio values is an aggregation of the individual time series of the assets held in the portfolio. Consequently, we expect the risk measure of a portfolio to be generated from the risk measures of the constituent assets by some process of aggregation. This process of aggregation can be continued hierarchically, until on the last level, the total bank-wide risk, aggregated from all portfolios and risk types, is determined. The aggregation of individual time series and subsequent determination of portfolio risk from the aggregated time series, rarely is a practical process. Consequently, in practice, one is forced to aggregate risk measures taken on individual time series. Aggregation resp. the opposite process, disaggregation, present formidable challenges for the definition of risk measures, and for practical risk management.

In the following, as often before, we first will take a pragmatic approach and explain standard risk measures. We then will look more in depth and discuss properties that coherent risk measures should possess. We will show which risk measures fall short of them, and which measures pass the test.

10.3.1 Volatility

The standard measure of risk in finance apparently is volatility, i.e., the standard deviation σ_τ of a time series of price changes on a time scale τ. This is certainly true for the more basic aspects or quick information on a financial product. Volatility is often found in the characterization of the variability of stocks and funds in magazines and on internet sites for investors. The advanced risk management of professional financial institutions, however, often is based on the risk measures described later in this chapter.

For a historical time series containing $N+1$ data points S_i spaced in time by τ, the (historical) volatility, or the standard deviation, of the returns is estimated as

$$\sigma_\tau = \sqrt{\frac{1}{N-1} \sum_{i=1}^{N} \left(\left[\frac{S_i - S_{i-1}}{S_{i-1}} \right]^2 - \left\langle \frac{S_i - S_{i-1}}{S_{i-1}} \right\rangle^2 \right)} \tag{10.1}$$

with $\langle \ldots \rangle = \sum_{i=1}^{N} \ldots$. Various related definitions, e.g., for continuous-time processes, have been given elsewhere in this book. For a Gaussian process, the discrete-time volatility σ_τ is related to the continuous-time volatility rate by $\sigma_\tau = \sigma \sqrt{\tau}$. In this situation, (10.1) provides an estimator for σ from a historical realization of the process.

The importance of σ for risk measurement is certainly due to at least two factors. On the one hand, the central limit theorem seems to guarantee a Gaussian limit distribution for which σ is appropriate (we have seen, however, in Sect. 5.4 that the Gaussian obtains only when the random numbers are drawn from distributions of finite variance – but this seems to be the case in real-world markets). The other factor is the technical simplicity of variance calculations.

In a Brownian motion model, there is a practical interpretation of σ. Given a generalized Wiener stochastic process, (4.37), one can ask after what time the drift has become bigger than the standard deviation. The answer is

$$\mu t > \sigma \sqrt{t} , \quad t > (\sigma/\mu)^2 . \tag{10.2}$$

After that time, it is improbable that profits due to the drift μ in the stock price will be lost completely in one fluctuation. For geometric Brownian motion, (4.53), one can make the same argument for the drift and fluctuations of the return rate dS/S.

As an example, assume $\mu = 5\%y^{-1}$, $\sigma = 15\%y^{-1/2}$ ($y^{-1} \equiv$ p.a.). Then, $t > 9y$. Or consider the Commerzbank stock in Fig. 4.5. From the difference of end points, one has a drift $\mu = 58\%y^{-1}$, and a volatility $\sigma = 33.66\%y^{-1/2}$. Then $t > 4$ m only.

For strictly Gaussian markets, σ is the only relevant quantity. All other risk measures, in one way or another, can be reduced to σ. It may apply either to a position in stock, or bond, or derivative. With a probability of 68%, price changes $\Delta S_i / S_i$ are contained in the interval between $\pm \sigma$ around $\langle \Delta S_i / S_i \rangle$, while they fall outside this range with 32% probability. The confidence levels for multiples of σ for Gaussian processes are listed in (5.6). For more general processes, historic volatility is defined and estimated through (10.1).

Some of the (serious) problems related to the use of σ for risk measurement have been discussed earlier. Here are some more:

- The limit $N \to \infty$ underlying the central limit theorem, is unrealistic, even when one ignores or accepts the restriction that the random variables to be added must be of finite variance. With a correlation time of $\tau \approx 30$ minutes, a trading month will produce only about 320 statistically independent quotes.
- Extreme variations in stock prices are never distributed according to a Gaussian. There are simply not enough extreme events – by definition. The central limit theorem then no longer justifies the use of volatility for risk measurement. On the other hand, these extreme events are of particular importance for investors, be they private individuals or financial institutions.
- The volatility σ as a measurement for risk is tied to the Gaussian distribution. For stable Lévy distributions, it does not exist. In Chap. 5, we have seen, however, that the variance of actual financial time series presumably exists. On long time scales, they may actually converge towards a Gaussian.
- For fat-tailed variables, σ is extremely dependent on the data set. The convergence of the estimator (10.1) as the length N of the time series increases, is the worse the fatter the tails of the underlying probability distribution. Ultimately, when $\mu \leq 2$ in the equations following (5.41) or (5.59), volatility diverges when the length of the time series increases without bounds, and otherwise is extremely sample-dependent. Consider again the Commerzbank chart in Fig. 4.5: how much of the volatility of is due to the period July–December 1997?
- For non-Gaussian distributions, the relation of volatility to a specific confidence level of the statistics of returns is lost.

10.3.2 Generalizations of Volatility and Moments

Two other aspects should be kept in mind when σ is used for measuring risk. The first is that volatility, together with the likelihood of a negative fluctuation, also measures the positive ones. These we would not consider as a risk.

Of course, for symmetric distributions – and most of the return distributions of financial time series we have seen in this book are nearly symmetric – a risk measure operating on the negative fluctuations will inevitably also give an equivalent characterization of its positive fluctuations. For the skewed distributions characterizing credit and operational risk (cf. below), however, it is important to have (possibly additional) risk measures depending on the negative fluctuations only.

An immediate generalization of volatility is the *lower semivariance*. Variance, in (5.14), has been defined as

$$\sigma^2 = \int_{-\infty}^{\infty} dx \, (x - \langle x \rangle)^2 \, p(x) \,. \tag{10.3}$$

A consistent definition for the lower semivariance, measuring the negative deviations from the expectation value, is

$$\sigma_<^2 = \int_{-\infty}^{\langle x \rangle} dx \, (x - \langle x \rangle)^2 \, p(x) \,. \tag{10.4}$$

For a symmetric distribution, $\sigma_<^2 = \sigma^2/2$. This equation is also consistent with our definition of risk in the sense of Sect. 10.2, i.e. risk being defined as the probability of negative deviations from expectation, independent of sign(x). $\sqrt{\sigma_<^2}$ has the same dimension as volatility. As an alternative generalization of σ^2, the upper limit of the integral in (10.4) could, in principle, be set to zero so that only negative fluctuations are sampled, cf. below, (10.7). However, the version given in (10.4) is closer to banking practice where a terminology of *expected losses* and *unexpected losses,* to be explained further below, is common.

As shown in (5.14) and (5.15), the variance essentially is the second moment of the distribution. Lower semivariance is the below-expectation part of variance. A generalization of the lower semivariance to higher moments leads to the definition of the *lower partial moments* of the probability density function

$$m_{<,k}(r) = \int_{-\infty}^{r} dx \, (r - x)^k \, p(x) \,. \tag{10.5}$$

The lower semivariance is

$$\sigma_<^2 = m_{<,2}(\langle x \rangle) \,. \tag{10.6}$$

In the same way as the higher moments of a symmetric distribution, e.g. kurtosis, give indications of the fatness of the tails of the distribution, the lower partial moments are sensitive to extreme negative fluctuations. Their sensitivity to extreme tail risk increases with the order k of the moment.

For completeness, we list two further generalizations of these risk measures, Stone's risk measures

$$R_{S1}(k, r, q) = \int_{-\infty}^{q} dx \, |r - x|^k \, p(x) \,, \tag{10.7}$$

$$R_{S2}(k, r, q) = \sqrt[k]{R_{S1}(k, r, q)} \,. \tag{10.8}$$

R_{S1} allows for the fluctuation range to be included in the risk measure and the range of negative deviations from expectations to differ, while R_{S2} reduces the dimension of the risk measure to that of the risky variable. The direct generalization of the standard deviation, or volatility, to negative deviations from expectation only, thus is

$$\sigma_< \equiv \sqrt{\sigma_<^2} = R_{S2}\left(2, \langle x \rangle, \langle x \rangle\right) = \sqrt{\int_{-\infty}^{\langle x \rangle} dx \, |\langle x \rangle - x|^2 \, p(x)} \,. \tag{10.9}$$

Semivariance (10.4), singles out fluctuations below the expectation value for risk measurement. The lower partial moments weigh fluctuations below a threshold r, depending on their degree k. They, together with the more general Stone measures, allow to focus on the big fluctuations.

10.3.3 Statistics of Extremal Events

The emphasis of our thinking on risk on big adverse events best is illustrated by the example of car insurance. We contract an insurance because we want to eliminate the risk associated with major accidents, destroying the vehicle or causing damage to persons – not because there is a chance of small damage to the bumper. Risk is associated with large negative events.

We therefore first deal with the statistics of extremal events. Consider N realizations x_i of a random variable x. What is the maximal value contained in $\{x_i\}$?

This question only has a probabilistic answer. The probability for $x_{\max} < \Lambda$, some threshold, is

$$P(x_{\max} < \Lambda) = [P_<(\Lambda)]^N = [1 - P_>(\Lambda)]^N$$
$$\approx \exp\left[-N P_>(\Lambda)\right] \quad \text{for } P_>(\Lambda) \ll 1 \,. \tag{10.10}$$

$P_<(\Lambda)$ and $P_>(\Lambda)$ are defined as

$$P_<(\Lambda) = \int_{-\infty}^{\Lambda} dx \, p(x) \,, \quad P_>(\Lambda) = \int_{\Lambda}^{\infty} dx \, p(x) \,. \tag{10.11}$$

Λ is the lower $P_<$-quantile, or the upper $P_>$-quantile of the probability distribution of x, respectively. In the first equality in (10.10), we use the fact that, in order for $x_{\max}$ to be smaller than Λ, each of the N realizations of x, drawn from the same distribution, must be smaller than Λ. The last equality in (10.10) relies on the first-order expansions of both terms. For the median, i.e. at the 50% confidence level, $\Lambda_{1/2}$, with

$$P(x_{\max} < \Lambda_{1/2}) = \frac{1}{2}, \quad \text{i.e.,} \quad P_>(\Lambda_{1/2}) \approx \frac{\ln 2}{N} , \tag{10.12}$$

we have a 50% probability that the maximal value of N random numbers from the same distribution will indeed be below the threshold $\Lambda_{1/2}$. At the p-confidence level,

$$P(x_{\max} < \Lambda_p) = p, \quad \text{i.e.,} \quad P_>(\Lambda_{1/2}) \approx -\frac{\ln p}{N} \sim \frac{1}{N} , \tag{10.13}$$

this probability is $100 \times p\%$.

This was completely general. The practically important value of Λ_p, however, depends sensitively on the underlying distribution, i.e., the functional form of $p(x)$. Let us illustrate this with some examples.

- The exponential distribution is mathematically very simple. From

$$p(x) = \frac{\alpha}{2} e^{-\alpha|x|} , \tag{10.14}$$

we find

$$P_>(\Lambda_p) = \frac{e^{-\alpha\Lambda_p}}{2} \quad \text{and} \quad \Lambda_p = \frac{\ln N}{\alpha} - \frac{\ln(-2\ln p)}{\alpha} , \tag{10.15}$$

where the second term is completely negligible.
- The Gaussian distribution

$$p(x) = \frac{e^{-x^2/2\sigma^2}}{\sqrt{2\pi}\sigma} \tag{10.16}$$

gives

$$P_>(\Lambda) = \frac{1}{2}\operatorname{erfc}\left(\frac{\Lambda}{\sqrt{2}\sigma}\right) \approx e^{-\Lambda^2/2\sigma^2} \frac{\sigma}{\sqrt{2\pi}\Lambda}\left[1 - \frac{\sigma^2}{\Lambda^2} + \cdots\right] . \tag{10.17}$$

$\operatorname{erfc}(x)$ denotes the complementary error function. Equating this with (10.13), we find

$$\Lambda_p < \sqrt{\ln N} . \tag{10.18}$$

- For a stable Lévy distribution,

$$p(x) \sim \frac{\mu A^\mu}{|x|^{1+\mu}} , \tag{10.19}$$

one obtains

$$P_>(\Lambda) \sim \frac{A^\mu}{\Lambda^\mu} \quad \text{and} \quad \Lambda_p \sim A\left[\frac{N}{-\ln p}\right]^{1/\mu} . \tag{10.20}$$

- We can get a feeling for the difference in probability of extreme events between the different distributions by taking $N = 10000$. The thresholds Λ_p such that numbers smaller than $-\Lambda_p$ (or bigger than Λ_p) occur only with a probability p, are then determined by

$$
\begin{aligned}
\ln 10\,000 &\approx 9.21 \text{ (exponential)}, \\
\sqrt{\ln 10\,000} &\approx 3.03 \text{ (Gaussian)}, \\
(10\,000)^{2/3} &\approx 464 \text{ (Lévy with } \mu = 3/2) .
\end{aligned} \tag{10.21}
$$

More instructive even are the changes when N is decreased to 5000:

$$
\begin{aligned}
\ln 5000 &\approx 8.52 \text{ (exponential)}, \\
\sqrt{\ln 5000} &\approx 2.92 \text{ (Gaussian)}, \\
(5000)^{2/3} &\approx 292 \text{ (Lévy with } \mu = 3/2) ,
\end{aligned} \tag{10.22}
$$

i.e., approximately 7%, 3%, and 50% for exponential, Gaussian, and Lévy distributions. Changing the number of realizations does not cause big changes for the threshold Λ_p for the exponential and Gaussian distributions, but significantly changes it for Lévy distributions. This is a consequence of their fat tails. Notice also that for both exponential and Gaussian distributions, Λ_p becomes independent of p for any reasonably large number. The p-dependence remains, however, for the Lévy distribution. (Of course, the above comparison ignores all kinds of prefactors in $P_>(\Lambda)$. While this may change the numbers, the trends both with changing N and changing the distributions, are independent of these details.)

10.3.4 Value at Risk

A good manager must be prepared to face bad events. On the other hand, a good manager cannot afford to become paralyzed by a constant preoccupation of extreme catastrophies which could hit his firm. He therefore requires a clear definition of the realm of his management activities: what is to be managed and what is not? A practical and wide-spread approach can be built on the ideas of Sect. 10.3.3, and leads to the notion of *value at risk* [245, 246]. Value at risk, roughly speaking, measures the amount of money at risk over a given time horizon τ with a certain probability P_{var}.

Define $\Lambda_{\text{var}}(P_{\text{var}}, \tau)$ by

$$
P_{\text{var}} = \int_{-\infty}^{-\Lambda_{\text{var}}(P_{\text{var}}, \tau)} \mathrm{d}(\delta S_\tau)\, p(\delta S_\tau) . \tag{10.23}
$$

The value at risk $\Lambda_{\text{var}}(P_{\text{var}}, \tau)$ is the negative of the one-sided P_{var}-quantile of the return distribution function. $1 - P_{\text{var}}$ then is the confidence level of the underlying return distribution.

$\Lambda_{\mathrm{var}}(P_{\mathrm{var}}, \tau)$ as defined in (10.23) measures the percentage amount which a portfolio can loose over a time scale τ with probability P_{var}. An alternative and equivalent definition is

$$
\begin{aligned}
P_{\mathrm{var}} &= \int_{-\infty}^{-\overline{\Lambda}_{\mathrm{var}}(P_{\mathrm{var}}, \tau)} \mathrm{d}(\Delta S_\tau) \, p(\Delta S_\tau) \\
&= \int_{-\infty}^{S(t)-\overline{\Lambda}_{\mathrm{var}}(P_{\mathrm{var}}, \tau)} \mathrm{d}S(t+\tau) \, p[S(t+\tau)|S(t)] .
\end{aligned}
\tag{10.24}
$$

Here, $\overline{\Lambda}_{\mathrm{var}}(P_{\mathrm{var}}, \tau)$ measures the dollar amount which can be lost over a time scale τ with probability P_{var}. $\Delta S_\tau(t) = S(t) - S(t-\tau)$, and strictly speaking, in (10.23) and (10.24), $\delta S_\tau(t+\tau)$ and $\Delta S_\tau(t+\tau)$ should be understood. We neglect this subtlety assuming that the statistical properties of returns and price changes do not change over the time scale τ considered. In the following, we will work with (10.23) to keep consistency with the remainder of this book. On the other hand, a portfolio manager or banker likely is more interested in knowing $\overline{\Lambda}_{\mathrm{var}}$ of his portfolio, resp. his bank.

For P_{var} small enough, $\Lambda_{\mathrm{var}}(P_{\mathrm{var}}, \tau)$ usually is a large positive number. We have chosen this sign convention to keep consistency with the preceding section on the one hand, and with the frequent association of value at risk to losses on the other hand. When working with returns as we do consistently throughout this book, the explicit minus sign in (10.23) is necessary. Statistically, $-\Lambda_{\mathrm{var}}(P_{\mathrm{var}}, \tau)$ is the $100 \times P_{\mathrm{var}}\%$-quantile of the return distribution. Financially, it is the biggest return expected over a time scale τ during the $100 \times P_{\mathrm{var}}\%$ worst periods τ. For $P_{\mathrm{var}} = 0.01$, e.g., and $\tau = 1$ d, one will expect the biggest daily return from the one percent worst trading days to be $-\Lambda_{\mathrm{var}}(0.01, 1\mathrm{d})$. Conversely, with (10.23) rewritten as

$$
1 - P_{\mathrm{var}} = \int_{-\Lambda_{\mathrm{var}}(P_{\mathrm{var}}, \tau)}^{\infty} \mathrm{d}(\delta S_\tau) \, p(\delta S_\tau)
\tag{10.25}
$$

in 99% of the trading days, the daily returns would be expected to be bigger, i.e. the outcome to be better, than $-\Lambda_{\mathrm{var}}(0.01, 1\mathrm{d})$. Value at risk then is the lowest return expected with a probability $1 - P_{\mathrm{var}}$ over a period τ.

Equivalently, in a picture based on losses $\ell_\tau = -\delta S_\tau \Theta(-\delta S_\tau)$,

$$
P_{\mathrm{var}} = \int_{\Lambda_{\mathrm{var}}(P_{\mathrm{var}}, \tau)}^{\infty} \mathrm{d}(\delta \ell_\tau) \, \tilde{p}(\ell_\tau) .
\tag{10.26}
$$

the value at risk $\Lambda_{\mathrm{var}}(P_{\mathrm{var}}, \tau)$ is the smallest loss over a time scale τ expected to be incurred during the $100 \times P_{\mathrm{var}}\%$ worst periods τ. In (10.26), $\tilde{p}(\ell_\tau) = p(\delta S_\tau)\delta(\ell_\tau + \delta S_\tau)\Theta(-\delta S_\tau)$. With the numbers from above, $\Lambda_{\mathrm{var}}(0.01, 1\mathrm{d})$ is the smallest loss expected during the 1% worst trading days, resp. the biggest daily loss expected with a probability of 99%.

Transforming (10.23) into (10.25) is permissible only for a continuous distribution. For discrete distributions or distributions with discrete or piecewise continuous support, or discontinuous distributions, the definition of value at risk must be generalized to

$$\Lambda_{\mathrm{var}}(P_{\mathrm{var}}, \tau) = \inf\left(\Lambda \geq 0 \mid P(\delta S_\tau \geq -\Lambda) \geq P_{\mathrm{var}}\right), \tag{10.27}$$

$$\Lambda_{\mathrm{var}}(P_{\mathrm{var}}, \tau) = \inf\left(\Lambda \geq 0 \mid P(\ell_\tau \leq \Lambda) \geq P_{\mathrm{var}}\right), \tag{10.28}$$

for general returns and for losses, respectively. When the probability distribution or its underlying support are not continuous, the interpretations of value at risk as, e.g., "the smallest loss during the 1% worst trading days" and "the biggest loss during the 99% best trading days", resp. "the worst daily loss expected with a 99% probability" no longer are equivalent. Only the first interpretation, based on (10.23) and (10.26) is the correct one is such cases, and the one of general validity.

Of course, as is clear from the discussion in Sect. 10.3.3, for given P_{var}, the value of Λ_{var} sensitively depends on the probability density function. For a Gaussian distribution with width $\sigma\sqrt{\tau}$, e.g. generated by geometric Brownian motion

$$\Lambda_{\mathrm{var}}(P_{\mathrm{var}}, \tau) = \sqrt{2}\sigma\sqrt{\tau}\,\mathrm{erfc}^{-1}\left(2P_{\mathrm{var}}\right), \tag{10.29}$$

justifying the use of σ for risk measurement in this case. $\mathrm{erfc}^{-1}(x)$ is the inverse complementary error function. Value at risk in units of the standard deviation for different confidence levels $1 - P_{\mathrm{var}}$ is given in Table 10.1. On the other hand, for a stable Lévy distribution,

$$\Lambda_{\mathrm{var}}(P_{\mathrm{var}}, \tau) \sim A\left(P_{\mathrm{var}}\right)^{-1/\mu}. \tag{10.30}$$

Value at risk measures the probability of individual extreme realizations of the underlying random variable. It does not make statements about the risk of accumulating many unfavorable subsequent realizations. The consideration of N subsequent realizations, however, for IID random variables reduces to a sum of N independent realizations and, at the same time, amounts to changing the time scale $\tau \to N\tau$. The question then is about the scaling of

Table 10.1. Value at risk $\Lambda_{\mathrm{var}}(P_{\mathrm{var}}, 1)$ of a driftless Gaussian process with unit time scale as the one-sided P_{var}-quantile

P_{var}	$1 - P_{\mathrm{var}}$	$\Lambda_{\mathrm{var}}(P_{\mathrm{var}}, 1)$
0.16	0.84	$1.0\,\sigma$
0.1	0.9	$1.28\,\sigma$
0.05	0.95	$1.65\,\sigma$
0.02	0.98	$2\,\sigma$
0.01	0.99	$2.33\,\sigma$
0.001	0.999	$3.09\,\sigma$

$\Lambda_{\mathrm{var}}(P_{\mathrm{var}}, \tau)$ with time scale τ. Answers can be given for stable distributions, i.e. Gaussian or Lévy distributions, which obey definite aggregation laws.

As discussed in Sect. 5.4.2, a sum $x^{(N)} = \sum_{i=1}^{N} x_i$ of N IID random, normally distributed variables x_i is distributed again according to a normal distribution with rescaled parameters

$$\mu^{(N)} = N\mu, \quad \sigma^{(N)} = \sqrt{N}\sigma. \tag{10.31}$$

Packaging the N independent realizations into a rescaled time scale $\tau^{(N)} = N\tau$, it follows from (10.29) that

$$\Lambda_{\mathrm{var}}(P_{\mathrm{var}}, N\tau) = \sqrt{N}\Lambda_{\mathrm{var}}(P_{\mathrm{var}}, \tau) \tag{10.32}$$

for the same P_{var}-quantile.

For a sum $x^{(N)} = \sum_{i=1}^{N} x_i$ of N IID random variables x_i drawn from a stable Lévy distribution, we know from Sect. 5.4.3 that $x^{(N)}$ is described again by a stable Lévy distribution with the same tail exponent μ (not to be mixed up with the Gaussian drift parameter μ from the previous paragraph), and an N-fold tail amplitude, (5.50). With (10.30), we have

$$\Lambda_{\mathrm{var}}(P_{\mathrm{var}}, N\tau) = A\left(\frac{P_{\mathrm{var}}}{N}\right)^{-1/\mu} = N^{1/\mu}\Lambda_{\mathrm{var}}(P_{\mathrm{var}}, \tau) \tag{10.33}$$

for the same P_{var}-quantile.

In all realistic situations such as those described in Chaps. 5 and 6, the scaling of value at risk with time scale cannot be deduced easily. Moreover, it may depend both on the time scale in question and on the quantile examined. Nonstable distributions of IID random variables are governed by the central limit theorem and approach either a Gaussian or a stable Lévy distribution, depending on the finiteness of the variance. The scaling then depends on whether the time scale is large enough for statements to be based on the central limit theorem, and on whether the quantile examined is in the range of values for which the statements of the central limit theorem hold. Most likely, for specific propositions, numerical simulations are required.

The preceding discussion was concerned with value at risk as derived from the statistical properties of a single time series. In practice, one often is interested in the value at risk of portfolios involving many different assets. The number of assets of a portfolio may vary from a few up to several thousands. In those circumstances, the aggregation of the individual data into a single portfolio time series may not be practical. Moreover, a portfolio manager would often like to estimate the change in portfolio value at risk when assets are added to or liquidated from the portfolio. Correlation then is an important issue.

For uncorrelated identically distributed assets, the results used for time scaling can be used directly (relying only on the fact that the N assets considered are statistically independent)

$$\Lambda_{\text{var}}^{N,\,\text{Gauss}}(P_{\text{var}}, \tau) = \sqrt{N}\Lambda_{\text{var}}^{\text{Gauss}}(P_{\text{var}}, \tau) \,,$$
$$\Lambda_{\text{var}}^{N,\,\text{Lévy}}(P_{\text{var}}, \tau) = N^{1/\mu}\Lambda_{\text{var}}^{\text{Lévy}}(P_{\text{var}}, \tau) \,. \tag{10.34}$$

The scaling with portfolio size is very different in the opposite limit of perfect correlation (correlation coefficient unity)

$$\Lambda_{\text{var}}^{N,\,\text{Gauss}}(P_{\text{var}}, \tau) = N\Lambda_{\text{var}}^{\text{Gauss}}(P_{\text{var}}, \tau) \,,$$
$$\Lambda_{\text{var}}^{N,\,\text{Lévy}}(P_{\text{var}}, \tau) = N\Lambda_{\text{var}}^{\text{Lévy}}(P_{\text{var}}, \tau) \,. \tag{10.35}$$

Both for the Gaussian and for the Lévy stable processes, perfect correlation reduces the aggregation of N identically distributed random variables x to the time series of a single random variable Nx. For perfect correlation, value at risk scales linearly with portfolio size, a result valid not only for stochastic processes governed by stable distributions but more generally for all perfectly correlated time series. For intermediate correlations, numerical simulations are necessary for an accurate determination of the value at risk of complex portfolios in general. In addition, numerous approximations have been developed which may be useful in practice [245, 246].

For identically distributed asset returns not described by one of the stable distributions, or with a correlation coefficient smaller than unity, an equality for the aggregation of value at risk can no longer be derived. However, the inequality

$$\Lambda_{\text{var}}^{N}(P_{\text{var}}, \tau) < N\Lambda_{\text{var}}(P_{\text{var}}, \tau) \tag{10.36}$$

still holds. It only depends on the absence of perfect correlation.

Correlation strongly increases the portfolio risk, as shown by the different scaling of value at risk with portfolio size in (10.34) and (10.35). In other words, adding assets to a portfolio which are weakly correlated with those already held, leads only to a weak increase of the portfolio's risk. However, the portfolio return is the sum of the returns of its constituent assets, independent of correlations, i.e. the return of the asset added simply adds to the return of the portfolio held previously. This effect – linear increase of returns combined with sublinear increase of risk – is known as diversification and is an important tool for risk management. An even stronger effect is achieved by negative correlations between assets which will *reduce* the portfolio risk while *increasing* its returns. Negative correlations tend to hedge the risk of a portfolio. We will come back to these points in Sect. 10.5.5 when we deal with the techniques of risk management and portfolio selection.

The definitions of value at risk in (10.23) and (10.26) imply that value at risk is measured with respect to the present position. Another terminology common in risk management in banks and in banking regulation [238] is closer to our definition of risk in terms of the negative deviations between realizations and expectations. It uses the same general ideas but expresses the value at risk as defined above, in two separate terms, "expected losses" and "unexpected losses". The origin of these terms lies in the area of credit risk which will be briefly described in Sect. 10.4.2 below. In credit risk, one usually considers separately the losses from credit default (the risk that a obligor is unable to repay his credit either in part or entirely, i.e. counterparty risk) and the interest payments which, statistically, must compensate these losses. The losses from defaulted credits in a portfolio are represented by skewed, usually fat-tailed distributions with finite expectation values. Quite generally, expected losses simply represent the expectation value of the losses over a time horizon τ under their probabilitly distribution

$$\mathrm{EL}(\tau) = \int_0^\infty \mathrm{d}\ell_\tau \, p(\ell_\tau) \, \ell_\tau \, . \tag{10.37}$$

Unexpected losses at a certain confidence level, say $(1 - P_{\mathrm{var}}) \times 100\%$, are the difference of the P_{var} value at risk $\Lambda_{\mathrm{var}}(P_{\mathrm{var}}, \tau)$ and the expected losses,

$$\mathrm{UL}(P_{\mathrm{var}}, \tau) = \Lambda_{\mathrm{var}}(P_{\mathrm{var}}, \tau) - \mathrm{EL}(\tau) \, . \tag{10.38}$$

Of course, (10.24) can be used to find equivalent formulations giving the dollar amount of expected and unexpected losses.

The notion of expected losses is clear and consistent. Strictly speaking, the notion of unexpected losses is a misnomer. It should be thought of as a semantic rule to label values of a variable or quantiles of its which differ from its expectation value. Of course, any size of losses is expected under a given probability distribution so long as it is consistent with its support. Also, with a probability P_{var}, losses of the size of the "unexpected losses" are expected under the given probability distribution. Worse, even losses much bigger than the "unexpected losses" still are expected for a given probability distribution – they only would occur at probabilities still smaller than P_{var}. Truly unexpected losses would be inconsistent with the underlying probability distribution, i.e. would reject, at a certain confidence level, the null hypothesis of the portfolio losses being drawn from the prespecified loss distribution. This rejection of a null hypothesis is usually *not* implied by the notion of unexpected losses in banking jargon.

The legitimation of the decomposition of value at risk into expected and unexpected losses comes from banking practice. In a well-run bank, expected losses should be included in the cost calculation for the banking services provided (e.g. credits) by the department acquiring the customer, e.g. sales or corporate finance. As we explain below, unexpected losses can only be covered by provisions, i.e. they bind capital ("risk capital", "economic capital") which cannot be used for other profitable business.

The cost of this capital is its interest rate in the market. In a holistic approach to bank management, this cost should be billed by risk management to the department generating the business, as an insurance premium for the coverage of "unexpected losses".

Both, value at risk and unexpected losses, are consistent with the management requirement set out above. Losses smaller than the value at risk, resp. the unexpected losses, are covered by capital. Losses bigger than the value at risk are accepted, even expected with a certain small probability P_{var}. Such losses can threaten the ability of a bank to meet its contractual requirements with counterparties, or even its existence. Risk management then requires (i) to fix an acceptable confidence level $1 - P_{var}$ underlying the definition of value at risk and determining the expected frequency of such disastrous losses, and (ii) to select a portfolio with an acceptable value at risk. A risk strategy would set this value at risk to an amount consistent with the financial ressources of the bank, and its business objective, e.g. to attain a certain rating score.

The consistency with management requirements may be one reason for the popularity of value at risk as a risk measure. Value at risk, though, has a series of fundamental shortcomings. They do not manifest themselves at the level of the preceding discussion which was concerned with the risk of a single position. They turn up, however, when value at risk is calculated for complex portfolios involving derivatives where the probability density may not be unimodel or which may possess a discontinuous support.

10.3.5 Coherent Measures of Risk

One may wonder if a generalization of (10.36) to the case where the returns of the constituents of a portfolio are no longer identically distributed, is availabe. In other words, how does the risk of a portfolio vary when arbitrary assets are added? And how does value at risk change in such a situation?

Apart common sense ("Don't put all eggs into one basket"), an economic argument makes clear that quite generally

$$\rho\left(\sum_i \Pi_i\right) \leq \sum_i \rho(\Pi_i) . \tag{10.39}$$

In (10.39), $\rho(\ldots)$ is a risk measure, and Π_i is the *value* of the ith position in the portfolio Π. (In this section, we switch our presentation from returns to values/prices.) The property (10.39) is called *subadditivity*. The inequality (10.39) holds independent of the stochastic properties of the asset prices, correlations, the time horizon over which risk is assessed, etc. The argument is based on contradiction and goes as follows. It was mentioned above that a financial institution must hold an appropriate amount of economic capital to cover the unexpected losses, i.e. the risk, of a portfolio. Now suppose that contrary to (10.39), $\rho\left(\sum_i \Pi_i\right) > \sum_i \rho(\Pi_i)$. This implies that the capital to be held for the aggregate portfolio is bigger than the sum of the capital requirements of the individual positions. In such a situation, it would be advantageous to open separate accounts for each portfolio position in order to minimize the total capital requirement. Portfolio composition then would be useless, and equations like (10.39) need not be considered. Portfolios of assets precisely are composed in order to reduce their risk below the bound fixed by the right-hand side of the inequality (10.39).

Unfortunately, value at risk *does violate* (10.39) when portfolios more complex than in (10.34) are set up. One example is provided by two out-of-the-money short positions, one in a call and the other one in a put option [247]. We assume that $t = T - \tau$ where T is the maturity of the options and τ the time scale over which value at risk is calculated. The payoff profile of short option positions at maturity was sketched in Figure 2.2. The short put is at a loss when $S_T < X_{\text{put}} - P$ where X_{put} is the strike price of the put and P its price at $T - \tau$. Similarly, the short call is at a loss when $S_T > X_{\text{call}} + C$ in terms of the call's strike price and present value. For the positions described, the strike prices of the options are very far from the present price of the underlying, $X_{\text{put}} \ll S_T \ll X_{\text{call}}$, thus the probability of incurring a loss in one of the option positions is low.

To be definite, assume that a 95% confidence level is set for a value-at-risk calculation ($P_{\text{var}} = 5\%$). Assume further that the X_{put} and X_{call} are such that

$$\int_{-\infty}^{X_{\text{put}}-P} dS_T\, p(S_T) = 4\% , \quad \text{and} \quad \int_{X_{\text{call}}+C}^{\infty} dS_T\, p(S_T) = 4\% . \quad (10.40)$$

In this case, the risk of a loss in every option position alone is 4%, and goes undetected in a value at risk based on a 95% confidence level. On the other hand, the value at risk at the 95% confidence level certainly is finite, the probability of an unfavorable evolution of the market being close to 8% (when $P, C \ll X_{\text{put,call}}$). By adding two positions which are riskless at the 95% confidence level, we generate one which is risky at the same confidence level, i.e. violate (10.39). The violation of subadditivity is not a specific feature of this example – more examples can indeed be produced [248, 250]. It is due to the specific properties of value at risk as a risk measure.

What then makes up a good risk measure? Given the long history of risk management, it is surprising that an in-depth answer to this question was only given at the very end of the past millenium. A set of four mathematical axioms defining a *coherent measure of risk* was formulated [247, 248] which describe the minimum set of conditions a risk measure must satisfy in order to behave economically reasonable. Let $\rho(\Pi)$ be a risk measure and Π the random value of a portfolio (or position). A time scale τ is implied. $\rho(\Pi)$ is a coherent risk measure if and only if it satisfies the axioms

$$\rho(\Pi_1 + \Pi_2) \leq \rho(\Pi_1) + \rho(\Pi_2) \quad \text{(subadditivity)} , \quad (10.41)$$
$$\rho(\lambda\Pi) \qquad = \lambda\rho(\Pi) \quad \text{(homogeneity, scale invariance)} , \quad (10.42)$$
$$\rho(\Pi_1) \qquad \geq \rho(\Pi_2) \text{ if } \Pi_1 \leq \Pi_2 \quad \text{(monotonicity)} , \quad (10.43)$$
$$\rho(\Pi + ne^{r\tau}) = \rho(\Pi) - n \quad \text{(risk − free condition)} . \quad (10.44)$$

Axiom (10.41) requires the risk measure to be subadditive when two positions are added. It is the same as (10.39), and the preceding discussion shows that value at risk as well as other popular risk measures (e.g. standard deviation [247]) are not subadditive. Subadditivity guarantees that one can conservatively estimate the risk of a portfolio by adding the risks of its individual positions. An upper bound for the risk to which a financial institution is exposed, can be found by adding the risks of its various business lines, etc. In this way, a decentralized calculation of risk becomes safe and feasible. A complete centralized calculation in a major bank, on the other hand, would require prohibitive computational and data management resources. Finally, and perhaps most importantly, the subadditivity axiom (10.41) guarantees that diversification as a tool of risk management works: investing 1000\$ into two different assets is less risky – independent of the splitting ratio – than investing the 1000\$ into a single asset.

The homogeneity (or, as physicists would prefer, scale invariance) axiom (10.42) states that the risk of a given position scales with the size of the position. The monotonicity axiom (10.43) assigns the greater risk to the "smaller" position. Two random variables Π_1 and Π_2 are ordered in size through their cumulative probability distributions

$$\Pi_1 < \Pi_2 \quad \text{if} \quad P(\Pi_1 < a) > P(\Pi_2 < a) \,. \tag{10.45}$$

The position Π_1 then is more risky than Π_2 if it more often realizes small or negative values.

Finally, the risk-free condition (10.44) states that n units of capital invested into a risk-free asset with return r, reduce the risk of the position by n. It guarantees that capital invested into a risk-free asset lowers the risk of the aggregate position (naked position plus capital cover). Consequently, putting aside risk capital as a cushion to cover risk is reasonable. We shall come back to this point in Sect. 11.2 below. More specifically, (10.44) states that the effect of n units of capital invested into at the risk-free interest rate r on the portfolio risk is the same as that of a rigid shift of the random portfolio value Π by the capital invested including interests, $e^{rr}n$. It also follows that $\rho[\Pi + e^{rr}\rho(\Pi)] = 0$. Consequently, $n = \rho(\Pi)$ is the right amount of capital to cover the portfolio under consideration.

Equation (10.44) also embodies translational invariance. It allows to assume a position covered by capital today when measuring the risk of future variations, as is done in the seminal work of Artzner, Delbaen, Eber, and Heath [247, 248]. Acceptable positions then are those for which $\rho(\Pi) \leq 0$, i.e. there is enough capital to cover the risk of future variations of the position (equality), resp. capital can even be withdrawn (inequality). On the other hand, capital has to be added to an "inacceptable position", $\rho(\Pi) > 0$. If free capital is not available, risk management has to become active.

"Risk measures" which fail to satisfy all axioms (10.41)–(10.44) do not measure risk correctly and, in the first place, should not be called risk measures at all. Unfortunately, as shown at the beginning of this section, value at risk does not satisfy subadditivity in general, and thereby does not qualify as a risk measure – in spite of its popularity in financial institutions [245, 246] and even among bank regulators [238, 249].

On the positive side, coherent risk measures can be constructed from generalized scenarios. A generalized scenario is a probability measure on the states of nature. A simple example might be "The price of the asset falls by 1%", or "There is a 30% probability of the asset price moving up by 1%, a 40% probability of a fall by 1%, and a 30% probability of a fall by 3%". Of course, reality is more complex than these simple examples, and this should be taken into account in practical work. We can also specify the probabilities of future asset prices from a model or from a historical probability distribution. A coherent risk measure $\rho(\Pi)$ of the portfolio then is given by [247, 248]

$$\rho(\Pi) = \sup_{\text{all scenarios}} \left\langle -\Pi e^{-r\tau} \right\rangle_{\text{scenario}} . \tag{10.46}$$

The important point here is that, unlike value at risk, the coherent risk measure is defined through an *expectation value* over scenarios. The supremum operation guarantees that if several scearios are evaluated, risk is measured by the worst result obtained.

The downside risk of the two simple scenarios above is 1% in both cases. When a scenario is defined in terms of a model or a historical probability distribution, the risk measure is just the expectation value of this distribution. If all the scenarios mentioned are considered in the definition of the risk measure, it would obtain as the biggest of the scenario risks.

Of course, the preceding scenarios were discussed only to illustrate the principle of building a coherent risk measure, not for their intrinsic value. To make a step towards reality, though, let us look at the following scenario "Losses bigger than the historical 5% value at risk are realized with probabilities determined by their historical probability distribution". This scenario may be included into the set of scenarios on which (10.46) is evaluated and certainly yields a bigger risk estimate than those discussed before.

10.3.6 Expected Shortfall

In the previous scenario, only realizations of the random value of the portfolio below its 5% quantile were considered. Applying (10.46) to this scenario and assuming $p(\Pi)$ to be continuous, would give the expectation value below the quantile, the tail conditional expectation (tail value at risk) [248]

$$\rho(\Pi) = -\langle \Pi | \Pi < -\Lambda_{\text{var}}(0.05, \tau) \rangle = -\int_{-\infty}^{-\Lambda_{\text{var}}(0.05, \tau)} d\Pi\, p(\Pi) . \tag{10.47}$$

The tail conditional expectation depends on the probability distribution, and formulating a scenario with a different probability distribution will produce a different scenario risk. For continuous distributions, the tail conditional expectation is a coherent risk measure. For discontinuous distributions, there are some mathematical subtleties which can destroy the subadditivity property required for coherence [251].

To be specific, assume a continuous probability density function $\tilde{p}(x)$ for a random variable x plus at least one delta-function peak

$$p(x) = \tilde{p}(x) + p_0 \delta(x - x_0) . \tag{10.48}$$

The cumulative probability distribution

$$P(x) = \int_{-\infty}^{x} dx'\, p(x') = \int_{-\infty}^{x} dx'\, \tilde{p}(x') + p_0 \Theta(x - x_0) \tag{10.49}$$

possesses a discontinuity of strength p_0 at x_0. Define Λ_α as the lower α-quantile

$$\Lambda_\alpha = \inf \{x \mid P(x) \geq \alpha\} . \tag{10.50}$$

This definition is more general than the integral definition used in (10.11) which applies only to continuous distributions, and has been used in (10.27) and (10.28) above. It may now happen that, due to the discontinuity in $p(x)$,

$$P(\Lambda_\alpha) \equiv P(x \leq \Lambda_\alpha) > \alpha \tag{10.51}$$

when $\Lambda_\alpha = x_0$ (delta function in the integrand at the upper limit of the integral).

Returning to the notation used before, we now can formulate a definition of *expected shortfall* as

$$\begin{aligned} \mathrm{ES}(P_{\mathrm{var}}) = -\frac{1}{P_{\mathrm{var}}} \{ \langle \Pi \mid \Pi \leq -\Lambda(P_{\mathrm{var}}) \rangle \\ + \Lambda_{\mathrm{var}}(P_{\mathrm{var}}) \left[P(\Pi \leq -\Lambda_{\mathrm{var}}(P_{\mathrm{var}}) - P_{\mathrm{var}} \right] \} . \end{aligned} \tag{10.52}$$

$1 - P_{\mathrm{var}}$ is the confidence level underlying the value at risk $\Lambda_{\mathrm{var}}(P_{\mathrm{var}})$, and $\langle \Pi \mid \Pi \leq -\Lambda_{\mathrm{var}} \rangle$ denotes the expectation value of the portfolio value Π conditioned on being smaller than the value at risk $-\Lambda_{\mathrm{var}}$. The term in square brackets vanishes for continuous distributions, and is finite for discontinuous distributions whenever a quantile happens to coincide with the location of a discontinuity. The time scale τ used in the value-at-risk definition has been left implicit here. Also, an explicit minus-sign appears in front of Λ_{var} keeps consistency with the definition (10.23) resp. (10.24). Expected shortfall as defined in (10.52) is a coherent risk measure [250, 252]. Acerbi and Tasche [251] expand on mathematical properties of expected shortfall, and related coherent risk measures.

Expected shortfall is being implemented in practical risk management applications. It is not consistent, though, with the management perspective on risk discussed earlier. Unlike value at risk, it does not draw a clear boundary line between what is to be steered by a risk manager, and what is beyond the realm of his activity. A connection between the expected shortfall of a financial institution and and its rating is difficult to establish. On the other hand, expected shortfall provides important information on what is not managed when value at risk is used ("How bad is bad?"). Moreover, its definition as an expectation value makes it easily applicable in risk-based capital allocation, a topic to be discussed in the next chapter.

10.4 Types of Risk

In almost every department of a bank, the outcome of a decision may negatively deviate from its expected consequences. Drivers of risk lurk around
every corner. In the following sections, we briefly describe the most important
types of risk encountered in banking.

10.4.1 Market Risk

Market risk describes the negative deviations of the positions of traded assets
from their expected values, or of positions dependent on traded assets. Market
risk, of course, includes the risk from investments in stocks, bonds, currencies,
commodities, traded derivatives, etc. Market risk, however, also includes the
risk from OTC derivative positions, from investments into mutual funds,
funds of funds, hedge funds, etc. In terms of risk types, the preceding chapters
of this book only treated aspects of market risk!

The above items probably constitute the biggest contributions to the market risk of an investment bank. For a commercial bank, or a credit union,
there are more, and more important, contributions to the market risk, primarily interest rate risk related to credits. When a loan is extended to a
client with variable interest rates, say LIBOR + x% (LIBOR is the London
InterBank Offered Rate, one of the interest rate benchmarks available), the
interest payments received by the bank vary and constitute a source of risk.
When a loan is given to a client with a fixed interest rate, say 8% per year,
the instantaneous value of the credit depends on the market interest rates,
which are variable.

There are investments whose inclusion into or exclusion from market risk
is ambiguous. One example is private equity. Another one is real estate. In
both cases, the investment products are not traded regularly, and do not
depend directly on the values of a regularly traded asset. On the other hand,
both can be valued in principle, though perhaps not very precisely, and their
value sensitively depends on certain market conditions.

10.4.2 Credit Risk

There are two drivers of risk for a bank giving a loan to a customer:

1. Interest rate risk. Assuming that the borrower meets all of of his payment
   obligations in due course, as specified in the credit contract, the bank
   either faces a variable inflow of cash (credit with variable interest rate,
   e.g. LIBOR+x%) or receives a deterministic cash flow but faces a variable
   valuation of the credit (fixed interes rates) following the variability of
   market driven interest rates. As explained above, interest rate risk is a
   part of market risk.

2. Credit default risk, also termed counterparty risk. The assumption that a borrower meets all of his payment obligations exactly as specified in the credit contract, unfortunately is an unrealistic one. It often happens that debtors either pay their interests and repayments too late, or do not deliver their supposed payments at all. The obligors *default*, resp. the credit is foul. Credit risk usually is understood to be synonymous to credit default risk.

Buying a bond essentially is equivalent to writing a credit. The emitter of a bond is the debtor to bond holders. Bond holders therefore also face both interest rate risk (i.e. a market risk) and default risk (i.e. credit risk).

With the similarity between buying a bond and giving a loan we can show how fixed interest rates on the loan or bond lead to a variable value of the loan or bond. For a zero-coupon bond, all interest rate payments of the bond emitter are accumulated into a discount of the emission price with respect to the nominal. Assume that a zero-coupon bond with nominal X and a maturity T is emitted at $t = 0$. Let the interest rate on the bond be fixed at r_{ZC}. The emission price of the bond then is

$$S(0) = Xe^{-r_{ZC}T} . \tag{10.53}$$

At maturity, the nominal X is repaid to the bond holder. When the interest rates on the open market vary, the bond holder must revalue the bond in his portfolio. The new bond price is such that, when the market interest rates for zero-coupon bonds with maturity T are accrued, the nominal X is repaid at maturity. The instantanous value of the bond at time t with interest rate $r(t)$ is

$$S(t) = Xe^{-r(t)(T-t)} . \tag{10.54}$$

Because fixed interest rates have been agreed upon for the bond, its daily value varies as a function of market interests. Although details are different for a coupon-carrying bond, or for loans, the basic mechanism explained here also works for these products.

For commercial banks active in the credit business, and for credit unions, credit risk usually is considered to be the biggest risk in the bank, more important than market risk, and the risk types to be described below. Credit risk repeatedly has led to big write-offs in large banks, and led to the collapse at least of smaller banks. Readers in Germany may remember the case of Schmidt bank, a small, privately owned bank active in a limited regional market, in 2001. The default of the state of Argentina in meeting its obligations from a variety of bonds has gained universal prominence as many private and institutional bond holders have lost a fortune.

The moratorium of Russia on its debt repayments in 1998 also constitute a case of default (late payment). While to the best of my knowledge, all payment obligations have been honoured by Russia at later times, the consequences of the moratorium have spread far beyond the bond markets. They have affected stock markets worldwide, as shown for the DAX stock market index e.g. at the right end of Figure 1.1. This case points to an important issue: different types of risk often are not independent but correlated. In the case of the Russian debt crisis, market risk was driven by credit default risk. We will see below that credit default risk may also be driven by market risk, or be a consequence of operational risk.

Credit risk was not treated in this book, so a brief digression may be justified. There are two basic approaches to quantifying credit risk. One is based on rating. Big, publicly listed companies regularly are rated for their creditworthiness by rating agencies. The best known agencies are Moody's, Standard & Poor's, and Fitch. Rating systems, however, can be set up by any bank, or any company more generally, and can be applied to any type of customer (company, non-profit organization, private individual, etc.). Also private individuals are regularly rated, e.g., by their telephone companies.

Rating is a statistical procedure which attempts to estimate the probability of credit default of a customer from a combination of quantitative information (e.g. salary, balance sheet, cash flow, future pension payment obligations) and qualitative information (degree of innovation of product line, market perspectives, management experience, etc.). Its results most often are communicated as marks such as AAA, BB, etc. A bank then would adjust its credit spread, i.e. the (positive) difference between the interest rate charged to a customer and the risk-free rate, according to the rating information. We will come back to the issue of rating in Sect. 11.3.5, because it plays an important role in the new capital adequacy framework Basel II.

An alternative approach more in the spirit of this book is provided by the mapping of credit default onto option pricing theory [42, 239, 60]. Assume that there is a company A which takes a credit. Its ability to repay the credit will depend on its value at the time of maturity (in principle also on its value at all times where interests are due). However, the value of company A is difficult to quantify: it comprises the value of common stock it may have emitted, the value of machines and factories it possesses, its human capital, its brand names, etc.

In order to make progress, assume that company A has issued stock and introduce a company B whose sole purpose is to hold the stock of A. While the firm value of A is difficult to measure, the firm value of B is simply the number of shares of A it holds multiplied by the share price. Under the standard model of quantitative finance, the value of B therefore would follow geometric Brownian motion

$$dV_B = \mu V_B dt + \sigma V_B dz \ . \tag{10.55}$$

Notice that this is an assumption made for simplicity, and to show the argument. The body of this book emphasizes that this assumption *not* satisfied by actual share prices, i.e. firm values!

In order to keep things simple, we simplify to the extreme and assume that taking a credit is essentially the same as issuing a bond. Moreover, we assume that the bond/credit is a zero-coupon bond, i.e. there are no interest payments during the lifetime of the bond/credit. All interests are discounted into the price of the bond which is lower than its nominal, to be repaid at maturity. At time $t = 0$, company B thus issues a zero-coupon bond with nominal X, priced $X - P$ where P contains the interests, possibly including a spread with respect to the risk-free rate. The bond matures (the credit must be paid back) at $t = T$. If the firm value $V_B(T) > X$, the bond/credit is repaid in full. However, if the firm value $V_B(T) < X$, company B defaults: it cannot pay back the entire bond X but only the fraction corresponding to its value V_B. The obligor or bond emitter (holder of the stock of company A) therefore has acquired the right to sell company B to the bond holder at the price X even though it may be worth less. Of course, this right is exercised only when $V_B(T) < X$, i.e. default has occurred. This right carries a price tag of P.

Taking a credit, resp. a short position in the bond, therefore is equivalent to a long position in a (european style) put option on the company value. When (10.55) is satisfied, the put option may be priced by the Black-Scholes formula, (4.85). Stock prices, however, do not generally satisfy (10.55), and all the problems of (and solution paths for) option pricing in a non-Gaussian world outlined earlier, e.g. in Chap. 7, also apply to credit default risk valuation. Bonds/credits with regular interest payments correspond to nested series of options, i.e. an option on an option on ..., etc., and can likely be solved once the basic problem of valuing the option on the firm value has been solved.

Much less work has been done on non-Gaussian price processes, asset correlation and default correlation in the area of credit risk than for derivatives on underlyings exposed to market risk.

10.4.3 Operational Risk

Operational risk is defined as the "risk of losses resulting from inadequate or failed internal processes, people and systems or from external events" and has been highlighted in the new Basel II Capital Accord [238], finalized during 2004. Banks will be required to hold a capital cushion as a provision against operational losses in the future. Examples of operational risk in banking include rogue traders, limit violations, insufficient controlling, fraud, IT-failures and attacks, system inavailability, catastrophies such as fire, earthquakes, floods, etc. An important trigger for including operational risk into the regulatory framework for banking, and a prime example for this category of risk, certainly was the ruin of Barings bank by the activities of their Singapore-based trader Nick Leeson [240]. Initially, his losses on derivatives in Osaka had been classified as a case of market risk. The case was recognized as operational risk, however, when later it became clear later that Leeson could build up his positions only because of the absence of separation of duties between front and back office on Barings' Singapore desk, and because of the insufficient controlling at Barings in general. While the perception of operational risk is new in banking, it is rather well known in industry where often hazardous processes are involved in the production or transport of goods, e.g. in the chemical industry.

The principal challenges faced when attempting to describe operational risk are its latent character, the absence of data, and the rarity of high-impact events. While for market risk, plenty of data are publicly available, and for credit risk, sufficient data are available in banks internally, there are very few data available on operational risk. Moreover, data on very large losses which determine the tail of a loss distribution function, are even rarer. Worse even, however, for a given bank, stationary data time series may be an impossibility: Usually risk management is improved, in particular in response to losses suffered.

The modeling of operational risk comprises two important aspects: (i) the frequency with which operational losses occur, and (ii) the size (dollar amount) of the loss suffered in the case of an event. Of course, both quantities will be stochastic. One therefore is interested in determining their probability density functions. Many operational risks can be insured. Some inspiration can thus be gained from the standard model of actuarial science [242]. It postulates that the frequency of events (insurance claims) in a given time interval, e.g. one year, is random and drawn from a Poisson distribution. The distribution of the time interval between two claims then follows an exponential distribution with a well-defined life time. Also the size of insurance claims is random and drawn from a log-normal distribution!

Data collection therefore is an important focus of operational risk controlling. One typically would build up data bases of operational risk losses across a bank. When loss data are collected by a single bank, such a data base is of limited value, though, due to the infrequency of losses. E.g., a typical number for small banks, say with a balance sheet of 3×10^9 Euro as a proxy for size, is 25 loss events per year in excess of 1,000 Euro. The frequency of losses increases with the size of the bank, giving good statistics for the largest banks. These organizations, in practice, are so complex, though, that a statistical analysis at the highest level of hierarchy is too crude to give reliable information for risk management.

Data collection can be assisted by including data external to the bank. There are one or two commercial databases which systematically gather descriptions of those operational loss cases made public, e.g. in the press [241]. As an alternative, homogeneous groups of banks pool their loss data according to well-defined rules, to increase the data base upon which statistical analyses can be built, and the statistical significance of the results derived. Examples known to the author are the ORX (Operational Risk EXchange) consortium of European banks, the data pooling initiative of savings banks in Germany led by the German Association of Savings Banks, or a data pooling project led by the Italian Bankers' Association. These data bases contain standardized information on the date of an operational risk loss and on the size of the loss (gross, net, recoveries, etc.), a description of the scenario underlying the loss, its categorization in terms of causes and event types, and possibly additional information on various parameters characterizing the bank where the event occurred. Frequency and loss distribution functions then are generated and convoluted by Monte Carlo simulation and analyzed by standard statistical methods. The goal is to derive the established risk measures such as value at risk or expected shortfall, on a specified time horizon, e.g. one year.

An important unsolved problem in the inclusion of external loss data into a bank's risk model is the rescaling of the external information, to fit the bank in question. Both the relevant parameters and the functional scaling relations for the loss frequency and the loss amounts are largely unknown today. However, as the size of the data pools increases with time, research into these problems likely will lead to interesting results in the near future. Also, data seem to indicate that the tails of the loss distributions are much fatter than expected for a lognormal distribution. The frequency distribution, on the other hand, apparently is quite well described by a Possonian although evidence seems to be accumulating in favor of more complex two-parameter distributions.

A good operational risk controlling programme will, however, not rely on data alone. Apparently extreme views even would suggest not to rely primarily on loss data at all. One problem is that data necessarily describe the past whereas risk management would prefer to have a more dynamic picture including the consequences of management action on future risks. More serious, however, is the problem that there is risk even without data: a bank may face significant operational risks but may not have suffered large losses in the past – either because of sheer luck or due to the low event probability of some scenarios. A data-based operational risk measure would grossly underestimate the risk situation of the bank. Worse even, risk measures such as value at risk are strongly affected by extreme losses which, hopefully, occur seldom enough to prevent good data quality in that range. A qualitative self-assessment, i.e. expert workshops and interviews where the risk of certain scenarios is estimated by knowledgable members of staff, are a way out of this problem. When optimized in view of psychometric evaluation, such questionnaires may provide more realistic risk estimates than data-based approaches. Of course, methods such as fuzzy logic and Bayesian networks also allow to integrate loss data with expert-based risk estimates for consolidated risk measures.

Very recently, statistical models for operational risk management have appeared in the physics-oriented literature [243, 244].

10.4.4 Liquidity Risk

Liquidity risk is the risk that a bank is unable to satisfy all claims of payment against it, i.e. becomes illiquid. The bank thus would default on some payments. Liquidity risk in essence appears very similar to credit default risk.

Market conditions often are drivers of liquidity risk for investors. When a market participant wants to buy or sell an asset, situations may occur where no counterparty is willing to settle the trade proposed. A standard example are small cap stocks, either on their home markets or worse, on foreign markets. Another example are liquid markets turning illiquid in stress situations, e.g. the crashes discussed in the preceding chapter. Illiquid markets arise when the complete market hypothesis fails.

Other drivers of liquidity risk may be massive (correlated) credit defaults, the inability to liquidate collateral taken in to secure credits, etc.

10.5 Risk Management

Suppose that a speculator, or the trading desk of a financial institution, has taken a position, resp. a set of positions in a market. However, the market turns against the speculator, and the position looses in value. What should he do?

As another example, assume that, as a part of its business activities, a bank has extended a set of loans to its corporate customers, and/or written a set of options for them. From that moment on, the bank carries a huge risk: The customers may default on their loans. Or the options may increase in value, i.e. the obligations of the bank at expiration increase. What action must the bank take?

10.5.1 Risk Management Requires a Strategy

Ideally, every investment is the result of a strategy and involves opinions on the evolution of the markets. This strategy should contain statements as to why the asset was bought, the target value to be reached and time span needed. Most importantly, an investor must fix the amount of loss he is willing to accept on his investment when the asset does not follow his view of the market. This is the starting point of risk management. For a single position, the point of non-acceptance is a limit on the value of the asset. For a complex portfolio of traded assets, it may be a limit on the value of the portfolio, or on the value at risk of the portfolio, or on any other risk measure.

The situation is slightly different for positions in financial instruments which are taken for business objectives, and not for speculative purposes. The bank writes an option or extends a loan to satisfy the needs of its customers. Its business objective is to make money on the fees charged for those services. It does not intend to hold a risky position in those assets. Here, the strategy is obvious at first: Eliminate as much risk as possible by a compensating investment. However, a complete elimination of risk is rarely possible in real market, and the bank needs a strategy for dealing with the residual risk it is ready to accept.

10.5.2 Limit Systems

Limit systems provide a classical way to cope with these situations. Consider the speculator who holds a single asset, e.g. in late 1996 a number of stocks on Hocchst bought at 35 Euro. The chart of Hoechst corporation can be found in Fig. 8.4. The stock rises to above 40 Euro during 1997 but, in late 1997 falls below 30 Euro. If the investor cannot accept more than 15% loss on his position, it seems wise to place a stop-loss order at 30 Euro. The order is triggered when the price quoted falls below 30 Euro and then acts as an unlimited sell order.

There are two problems with this strategy of risk limitation. Firstly, it is not guaranteed that the price at which the order is executed, is 30 Euro, or even close to that value. This problem is not very serious, perhaps, in a Gaussian market but can cause large unexpected losses in stress situations in real markets where the tail of the return distribution is much closer to a stable Lévy distribution. This point was made by Mandelbrot, cf. Sect. 5.3.3.

The second problem is: What to do next, in particular if an investment in the Hoechst stock continues to appear promising on longer time scales? When enter the position again? The straightforward strategy of placing a stop-buy order at 30 Euro is dangerous, at least. The stop-buy order is triggered when the stock price exceeds 30 Euro and then behaves as an unlimited buy order. Again, it is uncertain if the order is executed at or close to 30 Euro. The difference between the actual buy and sell prices, augmented by the transaction fees, is a systematic loss due to the strategy.

The same problem arises with a naïve strategy to cover a short option position [10]. However, the losses usually are bigger due to the leverage of the options. Stop-loss and stop-buy limits are definitely not advised to cover short positions in options.

For a complex portfolio, one faces similar limitations. The naïve limit strategy outlined above would imply to liquidate several positions in the portfolio which are the main drivers of the limit violation. Both objections made above, apply here again.

Implementing limits on loan portfolios may be a difficult task because loans cannot be traded easily. A bank has very few options when, e.g., the value at risk of a credit portfolio exceeds a pre-set limit. The termination of loans may be feasible in some instances when the contracts permit. In general, tough, one can only resort to some of the methods outlined in the following sections. Notice that a quick remedy to the problem is unlikely because very often, litigation on contracts may be involved. On the other hand, credit risk limits often are violated due correlation: A group of borrowers, e.g. from one industrial sector, is perceived as more risky in their ability to honor their obligations. In such a case, a bank can stop extending new loans to any member of that group of clients. Instead, it could increase lending to those clients with zero or negative correlations with the risky cluster, and thereby lower its value at risk back to acceptable levels.

Limit systems for operational risk are considered to be of speculative nature, due to a variety of causes. The lack of reliable data makes any estimate of risk measures, to be held against a limit, extremely imprecise. Consequently, a limit violation most often is ambiguous. Secondly, operational risk is driven by the processes in a bank, and the big "portfolio of processes" typically operating in any bank, renders difficult the assignment of a putative limit violation to a single process which could be improved in the following. On the other hand, if a sufficiently clear picture of a limit violation due to operational risk can be obtained, remedy, even quick, may be available: As mentioned before, many operational risks can be insured. When an insurance is contracted, the bank transfers part of its operational risk to the insurance company. The risk of the bank is reduced promptly.

Traffic light systems are a more flexible form of limit systems. When the risk measure of a portfolio is far from its limit, the light is green, and no action is required. When the risk measure approaches the limit, the light switches to yellow. This is the time to closely monitor the portfolio, to analyse which components are responsible for the increased risk, and to evaluate various possible actions. Should the limit be violated, the light turns red, and immediate action is required.

In spite of the shortcomings mentioned before, as a last line of defense, every investor should fix a limit where he will liquidate his position or take any other action suitable to avoid further losses on his portfolio.

10.5.3 Hedging

The Black–Scholes analysis of Sect. 4.5.1 was based on offsetting the stochastic component in a short option position by a suitable long position in the underlying. The price of the option could then be calculated because the portfolio constructed was riskless, and its evolution deterministic.

For every option shorted, Δ shares of the underlying were required to form a riskless portfolio. This prescription ("Δ-hedging") precisely tells the bank which has written options for its clients, how to eliminate the risk associated with the option position. For such a "Δ-neutral" portfolio, we have

$$\frac{\partial \Pi}{\partial S} = -\frac{\partial f}{\partial S} + \Delta = 0 , \qquad \frac{\partial \Pi}{\partial t} = r\Pi . \qquad (10.56)$$

f is the value of the derivative. The portfolio is immune against small changes of the price of the underlying and therefore riskless for short times.

Δ, however, depends on the price of the underlying, and the hedge must be adjusted as soon as the price changes. The dependence of Δ on the price of the underlying has been discussed in Sect. 4.5.5. In the Black–Scholes analysis, a continuous adjustment of the position in the underlying is assumed, and the transaction costs associated with this adjustment are neglected. In practice, only a periodic adjustment of the hedge is possible. During the adjustment period, the portfolio no longer is riskless. Bigger price changes in the underlying may occur, and volatility and interest rates may change. The time to maturity certainly changes.

A Δ-neutral portfolio can be hedged further against these risk factors. Γ (Sect. 4.5.5) is the second derivative of the option value with respect to the underlying. If a Δ-neutral portfolio is hedged to be Γ-neutral in addition, it is made immune against bigger changes in the price of the underlying. For a Δ-neutral portfolio, we have [10]

$$\Theta + \frac{1}{2}\sigma^2 S^2 \Gamma = rf , \tag{10.57}$$

where Θ has been defined in (4.105). A portfolio with a certain Γ can be made Γ-neutral by adding $-\Gamma/\Gamma_T$ traded options, where Γ_T is the Γ of the traded options. After these options have been added, the portfolio is no longer Δ-neutral. An iterative adjustment in the number of shares of the underlying and in the traded options is necessary to achieve Δ- and Γ-neutrality at the same time. Even then, the portfolio is Δ- and Γ-neutral only instantaneously.

The last important risk driver of a Δ-neutral portfolio is volatility. The sensitivity of an option price to changes in volatility is measured by Vega, (4.109). A Δ-neutral portfolio with $\mathcal{V}$ can be hedged against changes in volatility by adding $-\mathcal{V}/\mathcal{V}_T$ traded options with $\mathcal{V}_T$. Again, the Δ- and Γ-neutrality of the portfolio must be restored iteratively. Although Γ and $\mathcal{V}$ are quite similar, a Γ-neutral portfolio, in general, in not $\mathcal{V}$-neutral at the same time. When a Δ-neutral portfolio is hedged against Γ and $\mathcal{V}$ at the same time, two traded options must be added to the portfolio.

Θ is special among the Greeks as it measures the time decay of an option value. Time is not a stochastic variable. Therefore, a hedge against Θ makes no sense.

10.5.4 Portfolio Insurance

A portfolio manager may be interested in protecting his portfolio against falling below a certain limit value X during a certain time span T. Holding a long position in put options with strike X and maturity T gives the desired protection.

When the portfolio is well-diversified and mirrors an index, put options on the index should be bought. For other portfolios, one can determine the correlation of the portfolio with an index or a benchmark asset (the β-parameter introduced in the next section) on which traded options have been written. Then a long position in β put options on the index provides the desired insurance.

When traded options suitable for the portfolio insurance desired are not available or the options markets cannot absorb the trades required, the portfolio manager can synthetically create the options required. The principle of synthetic replication of options has been explained in Sect. 4.5.6. In the specific case of insuring a portfolio worth Π against a drop below X, the portfolio manager must invest, at any time, a fraction $-\Delta(\Pi, X)$ of the portfolio in a riskless asset. As the value of the stock portfolio declines, the fraction invested in riskless assets increases. Conversely, when the value of the stocks increases, part of the cash must be used to repurchase stocks.

Of course, portfolio insurance comes with a cost which is the higher the smaller the amount of losses which the investor is ready to accept. E.g., when insuring a portfolio representing the DAX (quoted 4343.6 on March 24, 2005) against dropping below 4200 or 4000 points by year end 2005, the cost of the put options required was the equivalent of 154 resp. 103 DAX points. Notice that these options expire on December 8, 2005 already. When protection against losses effective to December 30, 2005 is required, the put option must be created synthetically. The cost of an option created synthetically is due to the fact that the portfolio manager sells low and buys high, in this scheme.

This kind of portfolio insurance has also been implemented in "absolute return" investment strategies and products which have become popular with investors after the strong decline of the world stock markets in the years 2000–2003. In a benchmark-related investment strategy, the portfolio manager, by active management, tries to generate an outperformance of his portfolio with respect to a benchmark. However, in bear markets, the portfolio still may decline in value. The strategy was successful when the porfolio decline is less than the decline in the benchmark. On the other hand, absolute return strategies attempt to achieve a minimal absolute performance, independent of the evolution of a benchmark. E.g., when the minimum return targeted is zero, we have an investment where the protection of the capital invested is attempted. The implementation of the absolute return strategy can be costly, though, and lowers the performance of the investment.

Notice that the portfolio insurance scheme discussed in Sect. 8.3.1 also is a rough way of creating an option synthetically.

10.5.5 Diversification

Correlation between assets is extremely important in risk management. The hedging of option positions discussed before, relies on the negative correlation between a short position in a call option and a long position in the underlying asset. More specifically, Δ measures the correlation between the option and the underlying, and the sign of Δ and of the option position (long/short) determine how a riskless hedge can be constructed.

We have seen another important example of the influence of correlation. For the special case of N time series of identically distributed uncorrelated assets (10.34) gives the evolution of the portfolio value at risk from the equivalent risk measure of a single time series. The corresponding evolution for identically distributed, perfectly correlated time series is given in (10.35). It turns out that the value at risk of the perfectly correlated portfolio exceeds that of the uncorrelated portfolio by a factor $\sqrt{N}$. Apparently then, a systematic optimization of the tradeoff between risk and return in a portfolio should be feasible.

Markowitz was the first to show that in portfolios containing several assets, one can optimize (within limits) a tradeoff between risk and return [253]. His quanitative theory derives the essential parameters for this optimization – not surprisingly correlation. Markowitz' theory essentially relies on Gaussian markets, and volatility as the measure of risk. The application to non-Gaussian markets is taken from Bouchaud and Potters [17].

In the following, we consider a portfolio with value Π, constituted by M risky assets with values S_i and one riskless asset with value S_0. p_i denotes the fraction of portfolio value contributed by the asset i, and $p_i < 0$, i.e., short selling, is allowed. Then,

$$\Pi = \sum_{i=0}^{M} p_i S_i , \quad \sum_{i=0}^{M} p_i = 1 . \tag{10.58}$$

Uncorrelated Gaussian Price Changes

Each of the assets has a return μ_i and a variance σ_i^2. Then, the return of the portfolio is

$$\bar{\mu} = \sum_{i=0}^{M} q_i \mu_i , \tag{10.59}$$

and its variance is

$$\sigma^2 = \sum_{i=1}^{M} q_i^2 \sigma_i^2 , \tag{10.60}$$

where $q_i = p_i S_i / \Pi$ accounts for the different values of the assets in the portfolio. One can now choose a return rate $\bar{\mu}$ of the portfolio and then minimize its variance σ^2 at fixed $\bar{\mu}$, using the method of Lagrange multipliers. Taking the derivative

$$\frac{\partial}{\partial q_i} (\sigma^2 - \lambda \bar{\mu}) \bigg|_{q_i = q_i^\star} = 0 , \quad (i \neq 0) , \tag{10.61}$$

leads to

$$q_i^\star = \lambda \frac{\mu_i - \mu_0}{2\sigma_i^2} , \quad \lambda = \frac{2(\bar{\mu} - \mu_0)}{\sum_{j=1}^{M} \left(\frac{\mu_j - \mu_0}{\sigma_j} \right)^2} . \tag{10.62}$$

The riskless asset has $q_0^\star = 1 - \sum_{i=1}^{M} q_i^\star$, and the optimal $p_i^\star$ are obtained by solving the linear system of equations relating them to the q_i through S_i. The minimal variance is then

$$\sigma^2 = \frac{(\bar{\mu} - \mu_0)^2}{\sum_{j=1}^{M} \left(\frac{\mu_j - \mu_0}{\sigma_j}\right)^2} . \tag{10.63}$$

The variance of the optimal portfolio therefore depends quadratically on the excess return over a riskless asset. This is shown as the solid line in Fig. 10.1. The optimization procedure may also be carried out with constraints (e.g., no short selling, $p_i > 0$, etc.). This leads to more Lagrange multipliers for equality constraints, or more complex problems for inequality constraints. Quite generally, the curve moves upward, say to the dashed line, when more constraints are added. The region below the solid line (the "efficient frontier") cannot be accessed: there are no portfolios with less risk than the optimal ones just calculated.

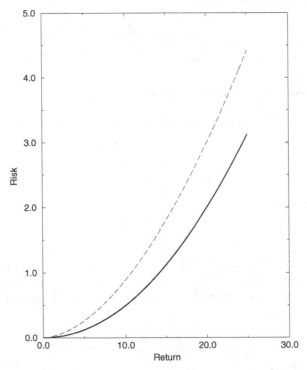

Fig. 10.1. Risk-return diagram of a mixed portfolio. In the absence of constraints, the optimal portfolios have a quadratic dependence of variance on return (*solid line*). In the presence of constraints, or for non-Gaussian statistics, the curve moves upward (*dashed line*). The region below the solid line is inaccessible. Reprinted from J.-P. Bouchaud and M. Potters: *Théorie des Risques Financiers*, by courtesy of J.-P. Bouchaud. ©1997 Diffusion Eyrolles (Aléa-Saclay)

Uncorrelated Lévy Distributed Price Changes

We now assume that the price variations of the assets in our portfolio are Lévy distributed (5.44), and follow Bouchaud and Potters [17]. In order to use the generalized central limit theorem, Sect. 5.4, we must further assume that all exponents are equal to μ, so that the main difference of the distributions is the amplitude A_i^μ of their tails,

$$p(\delta S_i) \to \frac{\mu A_i^\mu}{|\delta S_i|^{1+\mu}} \quad \text{as } \delta S_i \to -\infty . \tag{10.64}$$

Then, we can rescale the asset variables as $X_i = p_i S_i$, and these variables are drawn from distributions $p(\delta X_i) = p_i^\mu p(\delta S_i)$, and the convolution theorem can be applied to a sum of these random variables. The value of the portfolio is precisely such a sum (10.58). Then, its variations are distributed according to

$$p(\delta \Pi) \sim \frac{\mu A_\Pi^\mu}{|\delta \Pi|^{1+\mu}} \quad \text{with } A_\Pi^\mu = \sum_{i=1}^M p_i^\mu A_i^\mu . \tag{10.65}$$

Minimal value at risk Λ_{var} is equivalent to minimal amplitude A_Π^μ, at fixed return $\bar{\mu}$, (10.59). The optimization condition is

$$\frac{\partial}{\partial q_i} \left(\sum_{i=1}^M q_i^\mu A_i^\mu - \lambda \bar{\mu} \right) \Bigg|_{q_i = q_i^\star} = 0 . \tag{10.66}$$

It follows that

$$q_i^\star = \left[\lambda \frac{\mu_i - \mu_0}{\mu A_i^\mu} \right]^{1/(\mu-1)} , \quad \lambda = \frac{\mu(\bar{\mu} - \mu_0)^{\mu-1}}{\left(\sum_{j=1}^M \left[\frac{\mu_i - \mu_0}{A_i} \right]^{\mu/(\mu-1)} \right)^{\mu-1}} . \tag{10.67}$$

The effective amplitude $A_\Pi^\mu = \sum_{i=0}^M (p_i^\star)^\mu A_i^\mu$, where $p_i^\star$ is obtained from $q_i^\star$ by solving a linear system of equations, is then proportional to Λ_{var}. Λ_{var} vs. $\bar{\mu} - \mu_0$ behaves in a way similar to the dashed line in Fig. 10.1.

Correlated Gaussian Price Changes

Correlations between two or more time series, or between two or more stochastic processes, are measured by the covariance matrices introduced in Sect. 5.6.5. For two processes following geometric Brownian motion, (4.53), and representing the returns of two financial assets, the covariance matrix is

$$C_{ij} = \left\langle \frac{\delta S_i}{S_i} \frac{\delta S_j}{S_j} \right\rangle - \mu_i \mu_j . \tag{10.68}$$

The total variance of the processes is then

$$\sigma^2 = \sum_{i,j=1}^{M} q_i q_j C_{ij} \ . \tag{10.69}$$

In order to optimize the portfolio of M correlated assets, one can now follow the same strategy as in the absence of correlations, Sect. 10.5.5. The only difference is the replacement of the variance by the covariance matrix. As an example, the $q_i^{\star}$ determining the optimal fractions of the assets in the portfolio are given by

$$q_i^{\star} = \frac{\lambda}{2} \sum_{j=1}^{M} C_{ij}^{-1} (\mu_j - \mu_0) \ , \tag{10.70}$$

in analogy to (10.62).

This simplicity is due to the fact that the covariance matrix C_{ij} can be diagonalized [17]. One can therefore formulate, from the outset, a new set of stochastic processes obtained by linear combination of the original ones, so that they are uncorrelated. Their variances are the eigenvalues of the covariance matrix, and the transformation from the original to the new stochastic process is mediated by the matrix built from the eigenvectors, as in any standard eigenvalue problem. In this way, a portfolio of correlated assets is transformed into one of uncorrelated assets which, unfortunately, do not exist on the market but are constructed with the only purpose of simplifying the portfolio optimization problem. The procedure can also be generalized to correlated Lévy distributed assets [17].

In a Gaussian world, all optimal portfolios are proportional so long as they refer to the same market. Equation (10.62) shows why this is so. The optimal asset fractions in the portfolio $q_i^{\star} \propto \lambda$, and only λ depends (linearly) on the required excess return of the portfolio over the risk-free investment $\bar{\mu} - \mu_0$. Linear combinations of optimal portfolios are optimal, too. A market portfolio which contains all assets according to their market capitalization, is also an optimal portfolio. Of course, the returns and the risks of all these optimal portfolios may be different – but they all have the same risk-return relation, i.e. satisfy (10.63).

The practical definition of "the market" itself is not a trivial issue. In the US, the S&P500 index is generally taken as a benchmark for portfolio managers, indicating that it is taken as a proxy for "the market". With 500 stock included, it certainly is well diversified. Some argue, however, that the limitation to the 500 biggest stocks gives it a bias, and that small caps which often generate the biggest returns, are ignored. They would advocate that the Russell 1000 or Wilshire 5000 indices are much better representations of "the US market" [3]. The Dow Jones Industrial Average with 30 blue chips only certainly is not representative of the broader US market. In the same way, the Dow Jones Stoxx 50, DAX, CAC40, etc. indices are not representatives of the European, German, and French markets. For investors with world-wide portfolios, the MSCI World index is virtually the only benchmark available.

A market portfolio can therefore be taken to measure the performance of an individual financial asset, or of entire portfolios, by relating their returns μ_j to that of the market portfolio (value Π, return $\bar{\mu}$)

$$\mu_j - \mu_0 = \beta_j(\bar{\mu} - \mu_0) , \quad \beta_j = \frac{\left\langle \left(\frac{\delta S_j}{S_j} - \mu_j \right) \left(\frac{\delta \Pi}{\Pi} - \bar{\mu} \right) \right\rangle}{\sqrt{\left\langle \left(\frac{\delta S_j}{S_j} - \mu_j \right)^2 \right\rangle \left\langle \left(\frac{\delta \Pi}{\Pi} - \bar{\mu} \right)^2 \right\rangle}} , \quad (10.71)$$

where β_j is the covariance of the asset or portfolio j with the market portfolio. This is the basis of the Capital Asset Pricing Model (CAPM) which relates the returns of assets to their covariance with a market portfolio. It cannot be generalized to non-Gaussian markets.

10.5.6 Strategic Risk Management

Risk management, in the first place, starts with a selection of those asset classes whose risk is deemed acceptable given the risk appetite, resp. risk tolerance, of an investor. There may be major differences, e.g., in the risk tolerance of the trading operations of a universal bank, a focussed investment bank, an insurance company or an industrial corporation. There is thus no clear distinction between strategic risk management and asset management, in general.

The different classes of assets: bonds, stocks, commodities, currencies, real estate, private equity, etc. carry different risk and return expectations. Strategic risk management will select those classes of assets which can be used for investment, and those which are excluded. This selection usually is followed by more detailed rules which may set limits on the fraction of assets to be held as stocks (in the insurance business, this fraction may even be set by a regulator), the use of derivatives for speculation, or their use for hedging purposes. Strategic risk management may differ between the assets held for trading purposes, and the positions entered for business purposes [254], i.e. in the case of banks, between the trading book and the banking book.

In many industrial sectors, though not in banking, natural hedging of foreign exchange risk is an important strategic consideration. Foreign exchange risk comes from buying goods in one currency and selling products in another currency. The exposure to currency fluctuations for a corporation is less when products are manufactured and sold in areas using the same currency in which most of the raw materials bought are billed. For many corporations deciding on the opening of new plants in foreign countries, the opportunities for natural hedging are an important consideration.

Strategic risk management is all the more important the less tradable products are available for use in risk management. Although credit derivatives have been created and now are traded regularly, an important part of credit risk management simply consists in defining how many loans can be extended to specific classes of clients, in order to optimize the risk-return profile. Also the participation in credit pooling initiatives by which several financial institutions swap parts of their credit risks, requires extensive preparation and thus strategic decisions. As mentioned earlier, once the loans have been given, there are only limited options for acting on the portfolio. In the area of operational risk, the operational risk associated with all new products should be assessed systematically, before the final decision on the introduction of the product is taken.

11. Economic and Regulatory Capital for Financial Institutions

Suppose that at time t, a speculator invests a capital amount of $S(t)$ dollars in the stock market. He expects a return $\langle \delta S_\tau(t) \rangle$ on a time scale τ. In the preceding chapter, we discussed the risk associated with this position, i.e. to what extent the actual return $\delta S_\tau(t)$ may deviate from the expected outcome $\langle \delta S_\tau(t) \rangle$.

The present chapter is concerned with the inverse problem. Given a certain risk of a bank, how can the bank ensure that it can safely take this risk, i.e. that the risk poses no threat to the prosperity or even the survival of the bank. This is all the more important as the risk may not necessarily arise from speculative proprietary trading but simply from the bank's day-to-day business with its customers.

11.1 Important Questions

In this chapter, we discuss the following important questions:

- What is the relation between risk and capital requirements for a financial institution?
- How much capital does a bank need?
- Which factors determine the capital requirement of a bank?
- How much capital does a business line in a bank need?
- What is the relation between the capital requirement of a bank and those of its various business lines?
- Can capital be used as a tool for risk management?
- Are banks free in the determination of their capital requirements?
- What is the difference between economic and regulatory capital?
- What is the current framework for regulatory capital calculations?
- To what extent is regulatory capital risk-sensitive?
- What is "Basel II", and how will it affect the determination of regulatory capital in the future?
- What is to come after "Basel II"?

11.2 Economic Capital

When Nick Leeson's positions on the Japanese derviatives market blew up
Barings, the bank did not have enough money to cover the losses, and went
bankrupt. When Long Term Capital Management got out of control, more
than three billion US$ were provided by a consortium of banks to cover the
losses and unwind LTCM in an orderly fashion. These examples show that a
capital cushion is needed to protect a bank (or any other business) against
unexpected events – risk. Capital, or better *economic capital,* turns out to be
the central concept determining how much risk a bank can take. *Capital allocation,* i.e. the attribution of certain fractions of the total capital available to
business units, is an important tool in bank management. Capital allocation
sets limits on how much risk individual business lines, departments, or trading desks can take. Here, we do not go into the subtleties of defining capital.
We take its existence for granted, and discuss its use in bank management.

11.2.1 What Determines Economic Capital?

Every time risk strikes, bank capital is used to cover up the losses. Both
the times of the losses and amounts lost, are stochastic variables. How much
capital should a bank put aside to cover its losses? A look at the Barings
case helps to give an answer. A bank needs enough capital to guarantee its
survival when the worst case within the management horizon hits. We opened
Sect. 10.3.4 with the observation that a good manager needs a clear definition
of the realm of his management activity, i.e. what is to be managed, and what
should not be managed. This boundary determines the capital requirement
of a bank, resp. an individual business line within a bank.

  More formally, the economic capital requirements of a bank therefore are
determined by the survival probability which its management targets. In the
language of credit risk, the complement of the survival probability is the
default probability which, itself, is indicated by a bank's creditworthiness
rating. If, e.g., a bank is rated by Moody's as A1, its implied annual default
probability is estimated to be about 0.05% (cf. Table 11.2 below). Its survival
probability for the next year is 99.95%. It sets the confidence level on the
evaluation of the risks which must be covered by capital, and thus on the
amount of capital required. When senior management wants to conserve the
A1 rating, economic capital must equal at least the 0.05% value at risk of
the bank. If senior management wishes to improve further the bank's rating,
an even higher confidence level should be set.

  While individual realizations of losses are unpredictable, statistics on the
loss histories provides the expected losses defined in (10.37). In a statistical
sense, losses of this order of magnitude are predictable over the time horizon
used, and a prudent banker will build up loss provisions for these events.
These loss provisions are approximately constant in time and are better balanced by the income generated by the bank's operations, rather than taken

out of the bank's capital base regularly. Ultimately, the expected losses should be included in the pricing of the bank's products and services.

The actual losses differ almost always from their expectation values. When they exceed the expected losses, capital indeed must be used to cover them. However, if loss provisions and the pricing of products and services are made correctly, economic capital is only used to cover the unexpected losses, defined in (10.38) at the confidence level set by the bank. More capital than this value may be held in practice, e.g., to take into account possible stress scenarios (e.g. reduced liquidity) associated with catastrophic events.

11.2.2 How Calculate Economic Capital?

The principle of an economic capital calculation is simple: calculate the value at risk at the chosen confidence level, and subtract the expected losses. The practice of economic capital calculation, however, presents almost unsurmountable challenges. All risk types (market, credit, operational, etc.) for all portfolios in all businesses of the bank must be aggregated to a single number. Aside many other challenges, one important issue is the estimation of the relevant correlation matrices between the various assets held. An impression of the consequences of correlations can be gained by comparing (10.34) and (10.35).

In practice, the problem is solved only partially, and at a very low level of aggregation. Economic capital may be determined systematically for individual portfolios, and individual risk types. A variety of approximate techniques is available to estimate the (mostly market) value at risk (or related risk measures) for reasonably complex portfolios [245, 246, 255].

Current bank research is focussed on the integration of market risk and credit risk into an overarching risk model, and thus capital framework. The integration of operational risk has not been attempted to date. The importance of correlations is seen easily when thinking about an economic downturn: stock market prices fall, and at the same time, due to the bad economic conditions, the number of credit defaults rises. Also, there may be correlations between the variation of interest rates, and the number of defaulting loans. These correlations are an important driver of economic capital needs.

Another fundamental challenge becomes apparent when attempting to integrate market and credit risk: the widely disparate time scales of the data sets used. Market data are available at high frequency. Credit risk data are certainly available on an annual basis, for large-volume credits perhaps quarterly. The standard time scale for an economic capital determination is one year. On the other hand, the time scale of risk management non-defaulted credits is somewhere between the quarter and a year, depending on the exposure. The time scale of market risk management finally varies from the intraday range to ten trading days, perhaps. Often, approximations such as the $\sqrt{T}$-law, exact only for uncorrelated Gaussian assets, are used to relate the different time scales involved. Similarly, in some instances capital figures may simply be added, implying perfect correlation between the assets in the various classes, to produce the total economic capital. Much research still needs to be done in order to develop accurate economic capital numbers for realistic situations.

11.2.3 How Allocate Economic Capital?

The inverse problem to risk aggregation, capital allocation, is as important from a practical point of view and as much unsolved from a fundamental standpoint. Moreover, capital allocation is a problem in its own right, and often a practical necessity even when the risk aggregation problem has not been solved satisfactorily. Capital allocation can be understood as an investment or budgeting process. It is done in any industry, enterprise and even private households more or less consciously. Business administration provides concepts for capital allocation resp. budgeting from an investment perspective.

Risk-based capital allocation attempts to allocate the capital of the bank to its businesses, portfolios and risk drivers. Let us first assume a stationary environment. Then the capital for the next period can be allocated on the basis of the present risk profile. The challenge is that capital is an additive quantity whereas risk is a subadditive quantity.

Assume that risk has been aggregated over all portfolios, businesses, and risk drivers. Unless all assets are perfectly correlated, subadditivity resp. diversification will guarantee that the total risk is less than the sum of all partial risks. This is independent of the risk measure used provided that it is coherent. The total risk of the bank therefore is known, and we assume that it is balanced by the bank's capital. How much capital should be allocated to each of the bank's businesses?

To be more specific, we assume that there are three businesses only: A, B, and C, and that the bank has fixed target rating of A1 with a 99.95% confidence level. At that confidence level, the capital requirement (unexpected losses) of business A is assumed to be 2×10^8\$, that of business B is 10^8\$, and business C claims 5×10^8\$. With the additional assumptions of vanishing correlation and normal distribution, the total capital required by the bank is 5.48×10^8\$. The effect of diversification is clearly visible. With this amount of capital available, the bank as a whole can safely balance the risk of its businesses. However, the capital is not sufficient to give every business the amount required so that it could, on its own, balance its risk at the desired confidence level. This would require 8×10^8\$.

Several ways out of this dilemma are conceivable.

- Every business receives 68.5% (=5.48/8) of its initial capital request. In this case, there are two options for proceeding:
  - Every business reduces its operations by the amount necessary to make the capital allocated appropriate for its risk at the 99.95% confidence

level. While this makes the risk management of every individual business safe, bank capital is wasted, as the aggregate reduced risk only requires 3.75×10^8\$ of capital. With an assumed return on capital of 10%, the bank wastes 1.73×10^7\$ of income – a disadvantageous strategy indeed.

– Business operations are not reduced following the reduced capital allocation. The full amount of risk continues to be managed at the 99.95% confidence level in every business. Each individual business is undercapitalized but the bank as a whole is capitalized correctly. A system of risk sharing agreements must be elaborated between the businesses because in some years, one business may need more capital than it has to cover its losses. However then, the other businesses are expected to have excess capital with respect to their realized risks which can be transferred to the suffering unit.

- Capital allocation is used as a book-keeping device only but capital is not allocated physically. Each business unit must behave as if it had been given the amount of capital requested. Business A, e.g., must follow its management strategy based on a capital of 2×10^8\$, include the cost of this capital amount in its profit and loss statement, etc. However, the sum rule on capital is no longer operational, and the risk is balanced against physical capital only at the bank level, not on the business unit level.

- By extending this idea, one can set up a central "insurance" function which takes over the unexpected losses of the various businesses against an insurance premium. Operating within a bank, a fair price plus a profit margin can be charged for such a service. Business A, e.g., can "sell" its unexpected losses up to a cap set by the 99.95% confidence level to this insurance function. In exchange it pays a premium equal to the cost of this capital, say 5% + margin, to the insurance department. Here, all risks are aggregated effectively, and balanced by capital.

- The rules of the risk management game can be changed so that risk measures become additive. Then a strict proportionality between risk and capital can be implemented. We discuss this path in the following.

The preceding discussion, and that in Chap. 10 started from the fixing of a common confidence level for all businesses, portfolios, and risk drivers. Then risk was aggregated bottom-up, referring at every aggregation layer to the common confidence level. The rating of the bank, say A1, attached to this confidence level, implicitly was transferred to all business units.

One can take a different approach, though, and not require the same confidence level for each of the bank's units. Instead, one can allocate capital based only on the contribution made by each business unit to the aggregate risk of the bank at the chosen confidence level. Here, the reference is made to the numerical value of the the risk measure, e.g. value at risk or expected shortfall, at the appropriate confidence level on the bank level. In the above example, with an A1 target rating and a 99.95% confidence level, the unexpected losses of the bank are 5.48×10^8\$. The capital allocation scheme then is based on the contribution of the individual businesses to bank-wide losses of precisely this order or magnitude.

In the following, we take expected shortfall (Sect. 10.3.6) as the risk measure of choice because the scheme can be implemented straightforwardly only with a risk measure which can be represented as a mathematical expectation value. Value at risk is not suitable for this purpose. Moreover, to keep the discussion simple, we neglect expected losses. For a continuous distribution, the expected shortfall for a portfolio Π, (10.52), simplifies to

$$\mathrm{ES}(P_{\mathrm{var}}; \Pi) = -\frac{1}{P_{\mathrm{var}}} \langle \Pi \mid \Pi \leq -\Lambda_{\mathrm{var}}(P_{\mathrm{var}}) \rangle . \tag{11.1}$$

The specific expression appropriate to our example is

$$\mathrm{ES}(0.05\%; \Pi) = -2 \times 10^3 \langle \Pi \mid \Pi \leq -\Lambda_{\mathrm{var}}(5 \times 10^{-4}) \rangle . \tag{11.2}$$

In the present context, Π is taken to be the entire bank. In terms of its three businesses A, B, and C, and their I_A, J_B, and K_C respective subportfolios, the bank portfolio is

$$\Pi(t) = \Pi_A(t) + \Pi_B(t) + \Pi_C(t) \tag{11.3}$$

$$= \sum_{i=1}^{I_A} \Pi_{Ai}(t) + \sum_{j=1}^{J_B} \Pi_{Bj}(t) + \sum_{k=1}^{K_C} \Pi_{Ck}(t) . \tag{11.4}$$

Now simulate a very large number of scenarios at least at the level of the business units and determine the 0.05% value at risk $\Lambda_{\mathrm{var}}(5 \times 10^{-4})$ of the bank. Next calculate the expected shortfall of the bank, $\mathrm{ES}(5 \times 10^{-4}; \Pi)$ by summing over those scenarios whose losses exceed $\Lambda_{\mathrm{var}}(5 \times 10^{-4})$,

$$\mathrm{ES}(5 \times 10^{-4}; \Pi) = 5.48 \times 10^8 \$ = \mathrm{ES}\left[\Lambda_{\mathrm{var}}(5 \times 10^{-4}); \Pi\right] . \tag{11.5}$$

The first equality in (11.5) emphasises the definition of expected shortfall in terms of a preselected default probability (resp. confidence level), whereas the second equality relates it to the dollar value of the bank-wide 0.05% value at risk. This relation to the bank-wide value at risk is important in the following. $\Pi(t)$ is additive over businesses, and the expectation value $\langle ... \rangle$ is additive over scenarios. We therefore can write

$$\begin{aligned}
\mathrm{ES}\left[\Lambda_{\mathrm{var}}(P_{\mathrm{var}}); \Pi\right] &= -\frac{1}{P_{\mathrm{var}}} \langle \Pi \mid \Pi \leq -\Lambda_{\mathrm{var}}(P_{\mathrm{var}}) \rangle \\
&= -\frac{1}{P_{\mathrm{var}}} \langle \Pi_A + \Pi_B + \Pi_C \mid \Pi \leq -\Lambda_{\mathrm{var}}(P_{\mathrm{var}}) \rangle \\
&= -\frac{1}{P_{\mathrm{var}}} \{ \langle \Pi_A \mid \Pi \leq -\Lambda_{\mathrm{var}}(P_{\mathrm{var}}) \rangle \\
&\quad + \langle \Pi_B \mid \Pi \leq -\Lambda_{\mathrm{var}}(P_{\mathrm{var}}) \rangle \\
&\quad + \langle \Pi_C \mid \Pi \leq -\Lambda_{\mathrm{var}}(P_{\mathrm{var}}) \rangle \} .
\end{aligned} \tag{11.6}$$

The three terms in (11.6) are sometimes called *risk contributions* and have the desired property of being an additive decomposition of the bank's risk, as measured by expected shortfall. Based on these risk contributions, an easy capital allocation is possible.

Notice, however, that

$$-2 \times 10^3 \langle \Pi_A \mid \Pi \leq -\Lambda_{\text{var}}(5 \times 10^{-4}) \rangle \neq \text{ES}(5 \times 10^{-4}; \Pi_A) = 2 \times 10^8 \$ \,, \quad (11.7)$$

because $\Lambda_{\text{var}}(5 \times 10^{-4})$ is the bank's value at risk and not the 0.05% value at risk of business A. The risk contribution sums the contributions of business A to the most catastrophic scenarios for the bank as a whole, no matter what their relevance for business A. For this same reason,

$$-2 \times 10^3 \langle \Pi_A \mid \Pi \leq -\Lambda_{\text{var}}(5 \times 10^{-4}) \rangle \neq \text{ES} \left[\Lambda_{\text{var}}(5 \times 10^{-4}); \Pi_A \right] \,, \quad (11.8)$$

although, most likely, scenarios contributing to the right-hand side of the inequality will also contribute on the left-hand side.

Quite generally, the risk contribution of business A to the bank-wide expected shortfall is different from the stand-alone expected shortfall of business A, both when computed at the bank-wide confidence level of 99.95%, and when computed at the bank-wide value at risk $\Lambda_{\text{var}}(5 \times 10^{-4})$. Once the risk contribution of business A has been determined, the process can be iterated to reallocate business A's capital to its I_A subportfolios Π_{Ai}.

11.2.4 Economic Capital as a Management Tool

Economic capital is an important management tool. In the preceding section, a stationary environment was assumed for capital allocation, and capital allocation was discussed with the focus of balancing actual risk by economic capital. The argument can easily be turned around: When there is an imbalance between allocated capital and current risk, a powerful incentive for change is created. When capital allocated to a business is reduced, the business is forced to either reduce its operations, or to engage in less risky or better diversified operations. When economic capital is increased, a business may take more risk, either from expansion, or from trading riskier products, etc.

How can one come to a decision about increasing or decreasing the capital cushion of individual businesses? The central question is: Which of the three businesses A, B, or C of the bank generates the highest return from the capital allocated? Models for bank performance can support such decisions. One such model (among many others) is the RORAC system [256]. **RORAC** is the abbreviation of **R**eturn **O**ver **R**isk-**A**djusted **C**apital, and is defined as

$$\text{RORAC} = \frac{\text{return}}{\text{allocated risk capital}} \,. \quad (11.9)$$

Table 11.1. Performance numbers of two regional divisions of a bank

	Eastern	Western
Assets	1,000	1,000
Income	10	11
Return on Assets	1.0 %	1.1 %
Economic Capital	75	51
RORAC	13.3%	21.6%

In our simplified framework where we deliberately neglect investment budgets, risk-adjusted capital equals risk capital. There is a considerable flexibility in the definition of the terms in (11.9). Capital allocated for investments, e.g. in infrastructure modernization, may be included in the denominator. In the numerator, return may be corrected by expected losses from the risky business, may be understood before or after taxes, etc. Basically, these subtleties are quite irrelevant so long as a system is consistently rolled out in the entire bank.

RORAC is a standard measure of bank performance. The kind of insight it provides for senior management is best illustrated by an example [257]. Suppose that a bank has an Eastern and a Western Division, and that they report the figures summarized in Table 11.1 to their board of directors.

Standard performance measures such as income, or the return on assets, are pretty comparable for both divisions. Economic capital analysis changes this simple picture. Economic capital reflects the different level of risk associated with both divisions, and leads to strikingly different numbers for the return over risk-adjusted capital. The Western division earns an excellent 21.6% RORAC while the Eastern division sticks at 13.3% not too far above common values for the hurdle rate where senior management starts wondering about the future of the business.

Given the difference in RORAC between the two divisions, one can (i) inquire about its origins in terms of business, (ii) perform a similar analysis within the Eastern division, perhaps in terms of districts, to understand if there is a similar heterogeneity of performance, (iii) change the capital allocation between the two divisions (and/or within the divisions), and (iv) give guidelines for the managers of the badly performing division/districts on how to improve their results.

From Table 11.1, it is clear that the Western division earns about 50% more from each dollar of capital invested than the Eastern division. If the primary aim of the bank is to maximize its return over risk-adjusted capital, a transfer of capital from the Eastern to the Western division should be considered. If additional capital is available for investment, it should only be invested in the Western division.

Assume that both divisions are only active in the credit sector. An analysis of their portfolios might show that the Eastern division has a higher fraction of commercial lending and a lower fraction of retail lending, a higher average probability of default, and a higher average maturity than the Western division. These observations shed light on the differing RORAC numbers. Commercial lending is significantly more risky than retail lending, in general. A commercial portfolio of the same size as a retail portfolio contains a smaller number of loans with higher notional amounts, and thus higher exposures at default. In addition, the risk is increased by the observed higher default frequency, and the longer maturity. The longer the maturity of a loan, the more likely is a default of the borrower, all other things being equal. This analysis shows how the business of the Eastern division could be changed in order to raise its performance numbers: more retail lending, shorter maturities, only well-rated debtors acceptable.

The RORAC analysis can be taken one level deeper, in addition. If performance differences similar to those between the divisions are uncovered at the level of districts, too, similar measures (capital allocation changed, business focus changed) can be set up by the managers of the Eastern division. Ultimately, the system can be extended down the entire hierarchy of the bank to the level of the individual transaction. Every transaction then can be analyzed to check if is adds value to the bank.

Other performance measures are constructed using different formulae. They may differ in details of emphasis on specific factors. However, they all follow the basic principle of comparing risk and return in a single number, illustrated by RORAC. Finally, they all serve the same purpose of quantitatively supporting management decisions.

11.3 The Regulatory Framework

11.3.1 Why Banking Regulation?

Banking is one of the most heavily regulated activies in the economy rivalled, perhaps, only by air traffic. From birth to death, a bank is subject to a plethora of regulation acts. The founding of a bank is subject to regulation. Its operations are subject to regulation (One purpose of regulation is to prevent the elimination, by competition, of badly performing banks from the marketplace). When the "unthinkable" happens despite regulation, regulation also governs the closing down of a banking operation.

It is not our task to discuss if regulation to the extent practised today is reasonable. Banking regulation follows two main purposes. An immediate purpose is to protect the deposits of customers, and thereby the stability of the economy. Unlike other industries, an important part of the financial resources of a bank is contributed by the deposits of a very large number of people who mostly are inexperienced in financial matters. The depositors are unable both conceptually and economically, to supervise the bank in its role as a borrower of money, and to protect themselves against business practices of banks not in their interest. A regulatory institution thus steps in to ensure that banks operate in the interest of their depositors.

A second purpose of banking regulation is to ensure the safety and stability of the financial system by limiting the risks a bank can take. In fact, new developments in banking regulation often have followed the breakdown of financial institutions is the wake of excessive risk taking. While the regulatory acts are decreed and enforced by national regulators, the globalization of the financial industry has also led to an increasing international harmonization of regulatory frameworks.

Banking regulation is imposed along two avenues. One is direct rule writing, describing what is permissible and what is not. The other is the setting of certain capital requirements which, explicitly or implicitly, depend on the riskiness of the banks' businesses. The first avenue is the field of lawyers and internal and external auditors. The second avenue is tightly related to risk management, and we will discuss it in the following.

11.3.2 Risk-Based Capital Requirements

Capital plays a significant role in the risk-return tradeoff at banks. Increasing capital reduces the risk of default of the bank by increasing its cushion against losses or, more generally, earnings volatility. Firms with greater capital can take more risk.

Capital also influences growth opportunities, profits, and the returns to shareholders. Banks with more capital can borrow at lower interest rates and can make larger loans. Both normally yield higher income. With more capital, a bank can more easily invest in growth and acquisitions, creating the seeds for increase future profits. On the other hand, the holding of greater capital decreases the returns to shareholders. Finding the optimal capital level is an important task of bank management. We will not pursue these topics here. We also discard a number of further important questions on the use of capital to cover risk, such as "What constitues capital?", "How do capital requirements impact a bank's policies and business practices?", or "What are the advantages or disadvantages of various sources of external and internal capital?". These important topics are treated in the standard literature on bank management [256].

Instead, we focus on the important quantitative problem of risk management: "How much capital is adequate given the exposure of the bank?", rephrased here as "How much capital do regulators require to hold given the exposure of the bank?". With reference to both the main body of this book in general, and to the preceding chapter in particular, one quickly might come up with the suggestion to tie regulatory capital requirements to the unexpected losses of a bank at a certain confidence level. While economically reasonable (cf. the discussion in the preceding section on economic capital), bank regulators apparently do not trust the ability of banks to accurately and reliably determine the capital numbers in question. Consequently, the internal determination of capital requirement ("internal models") is allowed only in the often less important domains of market and (in the future) operational risk. The procedures which regulators impose on banks for the most important area of credit risk work according to a very different logic: divide your assets into certain classes, and attach to them risk weights and capital numbers fixed in advance by the regulators. These regulations have been in practice for about two decades and are being loosened somewhat in the near future with the arrival of the new Basel II Capital Accord. Abandoning them completely in favor of an bank-wide internal model covering all risky assets will certainly take another one or two decades.

Banking regulation by risk-based capital requirements is responsible for many job openings in ,the financial industry. For this reason, a discussion of these practices, though not comprehensively based on rigorous scientific methods, is mandatory.

Historically, in many countries in the 1970s, national regulators imposed capital requirements on certain assets held by a bank, resp. fixed limits on the volume of assets a bank was allowed to hold, depending on its capital. International harmonization of supervisory rules was one of the tasks of the Basel Committee on Banking Supervision, located at the Bank of International Settlements (BIS) in Basel, Switzerland. In 1988, a first international capital accord (now dubbed "Basel I") [258] was reached by the Committee which represents members of the Group of Ten Countries' (G 10) central banks (i.e. the most important countries of western Europe plus the United States and Canada) and and their regulatory authorities. Despite the limited representativity of the Basel Committee, in the years following its publication, the accord has been implemented in the national legislation and rule making of more than 100 countries worldwide, in particular in all countries members of the Organization of Economic Cooperation and Development (OECD).

A second round of many years of negotiation by the Basel Committee has led to the publication of a final version of a new international capital accord, Basel II, in July 2004. Its purpose is to set rules for more a risk-sensitive determination of regulatory capital and to create incentives for the implementation of better risk management procedures in banks. The rules agreed upon in Basel II are scheduled to become effective on January 1, 2007, and January 1, 2008, depending on the sophistication of the procedures adopted by a bank. In the meanwhile, the accord must be transferred into national legislation in the countries represented in the Basel Committee. Based on the experience with Basel I, it is expected that Basel II will set the risk management and capital standards for the financial industry worldwide, for the one or two decades to come. By early 2005, more than 100 countries had committed themselves to the implementation of Basel II in the years to come.

11.3.3 Basel I: Regulation of Credit Risk

The first Basel Accord in 1988 marked the birth of risk-based capital standards in banking regulation [258]. The Basel I agreement only covers credit risk. For many banks except investment banks, the biggest of their risks is credit risk. The basic procedure to determine the bank's capital involves four steps.

1. Classify all your loans in one of five risk categories appropriate to the obligor, the collateral, or the guarantor of the asset. These five asset categories are described below and are distinguished according to the order of magnitude of the default probability of the assets. A bank carries a big risk from the default of a debtor, i.e. her failure to correctly deliver all payments due, cf. Sect. 10.4.2 on credit risk. However, the notion of a default probability has not been used in Basel I. It is introduced in Basel II, and reverse engineering can be done to estimate the numerical default probabilities of the five classes.

2. Convert off-balance sheet commitments to their on-balance sheet equivalents, and proceed as in 1.
   We will not dive into the practices of moving assets off the balance sheet of a bank, nor into the conversion procedure required here. It is sufficient to mention, at this point, that when a bank can generate income from assets which are "expensive" to hold (e.g. in terms of risk capital), it may be tempted to keep part of this income while avoiding the cost of the risk. Securitization is one way of doing this. Assets, e.g. loans, may be packaged into a new kind of security (e.g., collaterized debt obligations, CDOs) and sold to the capital markets. The counterposition opened by this security makes the loans disappear from the bank's balance sheet. A reader of the financial statement will not be able to correctly assess the riskiness of the bank's business practices based on that information alone. Many derivative positions do not appear in a bank's financial statement. Long term loan commitments are another example of off-balance sheet activities. When such a commitment is pending, the bank has not yet given out a loan to be covered by capital. Nevertheless the bank carries a risk because the obligor may take the loan, in particular when its creditworthiness declines. Dramatic examples of these practices have been given by Enron and Kmart in late 2001/early 2002, just before filing for bankruptcy.

3. Multiply the amount of assets (in home currency) in each risk category by an appropriate risk weight factor. The sum over all five risk categories gives the risk-weighted assets.

4. Mulitply the risk-weighted assets by a minimum capital percentage to obtain the capital required to hold against the assets. The capital ratio is 8%. (Here, this rule is simplified somewhat to avoid a discussion related to subtleties of the definition of capital.)

Asset category 1 contains assets of the best quality available: direct obligations from the US government or other OECD governments, currencies and coins, gold, government securities, and unconditional government guaranteed claims. These assets do not carry a default risk – the US government is not considered to default, and neither are the other OECD governements. The risk weight of this category is zero. No capital must be held against these assets.

Category 2 contains claims on public sector entities excluding central government, and loans guaranteed by such entities. At national discretion, a risk weight of 0, 10, 20, or 50% is attached to these assets.

Category 3 consists of obligations of multilateral development banks or guaranteed by these banks, obligations of banks incorporated in the OECD and loans guaranteed by these banks, obgligations of banks incorporated outside the OECD with a residual maturity of less than one year and loans with residual maturity up to one year guaranteed by these institutions, and obligations by non-domestic OECD public sector entities and loans guaranteed by such entities. Assets in this category carry a risk weight of 20%. The capital to be held against these assets is 1.6% of the asset valuc.

Category 4 contains loans fully secured by mortgage on residential property. Its risk weight is 50%, implying a capital charge of 4% effectively.

Category 5 contains, among others, obligations of the private sector, of banks outside the OECD with residual maturities of more than a year, real estate loans other than first mortgages, premises and other fixed assets, capital instruments issued by other banks, etc. The risk weight of this category is 100%, i.e. all assets carry a capital requirement of 8%. Off-balance sheet activities are converted into on-balance sheet assets with conversion factors similar to the risk weights, and then entered into category 5. Some national supervisors chose more conservative risk weights for the five categories.

The main difference between the the five categories is the likelihood of default of the assets. Category-1 assets are approximated as risk-free. Their interest rates do not contain an adjustment to compensate for the possibility of a default. When, e.g., in the Black–Scholes equation, (4.85), the risk-free interest rate is sought, the interests paid by these Category-1 assets should be used. Assets in the other categories are risky and can default. The capital ratio of 8% on risk-weighted assets has not been derived from a model or a theoretical framework. Most likely, it is a result of both good guessing and political bargaining.

There is no direct relation of the capital numbers determined to the risk of a bank's credit portfolio. This, in fact, has been the main criticism of the Basel I framework: The capital charge levied on a portfolio is independent of its risk.

It is not permissible to estimate a default probability of the assets in the five categories from their risk weights and the overall capital ratio. Strictly speaking, capital is used only to cover unexpected losses in the sense of

(10.37). Expected losses should be contained in the credit spread, the difference in interest earned by the asset and the risk-free rate. Notice that, in the Basel I framework, capital scales linearly with asset volume. This is not a usual property of risk measures which, except in special circumstances, scale sublinearly with asset volume. In an uncorrelated Gaussian world, risk scales as the square root of asset volume, cf. Sect. 10.3. The failure of the regulatory capital requirement to scale sublinearly with asset volume points to its two major shortcomings: (i) the lack of a scientific basis for its determination, and (ii) the love of regulators to assume worst-case scenarios. When perfect correlation is assumed, i.e. all assets in a portfolio default at the same time, a linear dependence of risk on asset volume is expected. Apparently, such a scenario is at the origin of the regulatory credit risk capital determination.

The capital determination process in the Basel I Accord appears rudimentary and gross. Apparently, it ignores all the fine statistics and physics-inspired analysis presented in the main body of this book. It has been presented here to give an impression of the current state of banking regulation, and of the kind of details risk management and accounting experts in banks have to go into.

Basel I should not be blamed, though, for its rudimentary character in terms scientific credit risk modeling. Market risk modeling using advanced statistics was well developed at the time Basel I was negotiated. Advanced credit risk modeling, on the other hand, only developed during the 1990s. Today, sophisticated financial institutions are able to manage their credit risk according to an internal (statistical) model. However, even in developed countries, many banks limit their formal treatment of credit risk to a framework such as that set out by Basel I.

11.3.4 Internal Models

Market risk has not been regulated by Basel I. A 1996 paper by the Basel Committee defines market risk and both sets up a standardized framework for regulatory capital requirements for market risk and allows the recognition of an internal model for the determination of market risk capital [259]. Market risk is subdivided into interest rate risk, equity position risk, foreign exchange risk, and commodities risk and includes the risk from derivative positions with these assets as underlyings. The standardized measurement method for market risk follows a philosophy similar to the Basel I treatment of credit risk discussed in the preceding section, and is not covered here. A capital charge is imposed only on the market risk in the trading book, i.e. for those assets which the bank holds for short-term trading purposes. There is no capital charge for market risk in the banking book.

Instead of the standardized risk-weighting procedure, a bank can elect to use an internal model to determine its regulatory capital requirement [249, 255, 259]. In some countries, depending on the size of its trading book, it may be obliged to do so. An internal model is an internally built risk measurement model which has received supervisory approval. Banks with important trading activities will develop such a model for their risk management and economic capital allocation, anyway. The point here is that, when implementing regulatory restrictions and parameter settings, this model may be used to determine the regulatory capital. It is expected that the capital numbers based on such an internal model will come out lower than those from the standardized risk-weighting procedure. As some of the regulatory settings for the internal model may be overly conservative, banks often run two structurally similar models, one with the regulatory settings to determine regulatory capital, and one for economic capital and risk management with the settings which internally are deemed most appropriate.

A bank's capital charge for market risk essentially is the value at risk of its trading assets as well as foreign exchange and commodity positions, whether or not they are in the trading book. The regulators do not prescribe a particular type of model nor a specific computational methodology. The internal model must, however, satisfy a number of general requirements:

1. Value at risk should be computed on each business day and should be based on a one-sided 99% confidence level.
2. The holding period underlying the value-at-risk calculation is fixed to ten days.
3. The model must measure all material risks of the institution.
4. The model may utilize historical correlations within broad categories of risk factors (equity and commodity prices, foreign exchange, interest rates), but not among these categories. The consolidated value at risk is the sum of the value-at-risk numbers of the categories, i.e. a perfect correlation is assumed between the categories.
5. The nonlinear price characteristics of options must be adequately addressed.
6. The historical observation period used to estimate future price and rate changes must have a minimum length of one year.
7. The data history must be updated at least once every three months, and more frequently if market conditions require.
8. Each yield curve in a major currency must be modeled using at least six risk factors appropriate to the interest-rate sensitivity of the traded assets. The model must also include spread risk.

The modeling is further complicated by the distinction made between general and specific market risk, and event risk. General market risk refers to all changes in the market value of assets resulting from broad market movements. It is approximated, e.g., by the variation of a representative market index. Specific market risk is the residual risk associated with individual securities, not reflected by broader market moves. It is related to the β-factors (10.71) of the Capital Asset Pricing Model discussed in Sect. 10.5.5 and measured the return dynamics of an asset *relative* to a broad market index. Event risk denotes rare events affecting an individual security. An example often cited is the rating downgrade of a bond issuer. The distinction of general and specific market risk and event risk certainly is somewhat arbitrary (just think about the reflection of a rating downgrade of a listed company in its share price and the ramifications this may have on the stock market as a whole). It may become important, though, when approximations are used in the model-building process.

In summary, the value at risk $\Lambda_{\mathrm{var}}(0.01, 10\mathrm{d})$ as defined in (10.25) must be calculated every business day based on the preceding prescriptions. Regulatory capital for market risk is related to this value at risk by a number of add-ons [259]. Firstly, the capital to be held on day t is the higher of the value at risk on the preceding business day $t-1$, and the moving average of the value at risk over the last 60 business days. Secondly, a multiplication factor smaller than 5 is applied to this number based on the regulator's "assessement of the bank's risk management system" (i.e. a somewhat subjective quantity), and the model's performance in backtesting. To be specific, in the implementation of internal models in Germany, the multiplication factor is decomposed into a fixed basic value of 3, and two add-ons for backtesting and the subjective evaluation by the supervisors which both vary between 0 and 1 [249]. Moreover, additional add-on charges are implemented for banks which do not include explicitly specific risks and event risks into their internal model [249, 259].

Notice that regulatory capital is determined with reference to value at risk, and not with reference to the unexpected losses (10.38), as economic capital would be. Likely, for a financial institution with a good pricing framework, where expected losses (10.37) are included in the prices of products and services, there is some double counting of the expected losses in capital and in the prices.

Stress testing and backtesting are important steps in the introduction of an internal model. Stress testing makes a model using parameters estimated from historical time series, more forward-looking. Stress testing is the study of model behavior under extreme scenarios which have not been realized in that past time used for estimating the model's parameters. To formulate these scenarios, one may recur to past events such as the crashes described in Chap. 9. Such scenarios are either given by the supervisors or developed by the bank itself. In the end, the sufficiency of the bank capital with respect to the losses incurred is evaluated.

Backtesting is the process of running a completed model on long historical time series, before going live. In this way, one can check that the model performs according to expectation before it is actually used in day-to-day risk management. E.g., when the value at risk of the entire bank is determined at the 99% confidence level, the actual frequency of losses bigger than the 99% value at risk is expected be 0.01. The time series must be long enough that this frequency, as well as its uncertainty can be estimated with acceptable precision.

One may only speculate on the reasons why internal models have not been recognized for the credit risk capital determination in Basel I (and continue not to be recognized under Basel II). Credit risk by far needs most capital in almost all banks. Moreover, the data situation in credit risk is not as good as for market risk: no bank would reevaluate their credit portfolio on a daily basis (there is simply not enough new information to warrant such an action). In addition, the 8% capital ratio has been determined quite arbitrarily. Most likely, in view of these uncertainties, regulators did not, and still do not have enough confidence in these models to allow banks to determine the biggest portion of their capital requirements by an internal model, independently of the quality and intensity of the regulatory examination.

11.3.5 Basel II: The New International Capital Adequacy Framework

The financial world has changed enormously during the 15 years since the implementation of the Basel I Accord. Financial instruments have become more complex, perhaps more so in the important area of credit risk than in market risk. Financial operations and technology have increased in complexity. In parallel, methods in risk management have become more sophisticated. Consequently, a new, more risk-senssitive framework for regulatory capital is called for. Moreover, it has been realized that important risks or aspects of risk had been left out of Basel I. Thus at the same time, such a new framework could be formulated to include broader risk categories into the regulatory capital calculation.

After more than five years of negotiation, the second Basel Capital Accord ("Basel II") was finalized in summer 2004. It is scheduled to be implemented in the G10 countries by 2007 (some of the more sophisiticated approaches by 2008 only) and will be adopted by many other countries subsequently. Basel II essentially refines the treatement of credit risk and introduces operational risk as a new risk type to be covered by a capital charge. Moreover, it formalizes criteria for the supervisory review process of banks as well as criteria for the disclosure of risk information towards the capital markets.

There are a few common principles underlying the Basel II accord.

- Basel II has been conceived as a compensation approach, i.e., *on the average,* banks should hold the same regulatory capital after the implementation of Basel II as before, when they use capital determination methods of comparable sophistication.
- Good risk management processes are most important in a bank, perhaps more important than the actual amount of risk taken by a bank.
- There should be incentives for banks to improve their risk management systems despite the investments necessary. Good risk management therefore should be rewarded by a significant capital reduction.

- The board of directors and the senior management are directly responsible for the risk management processes, and for the risk taken by a bank.
- Basel II rests on three pillars necessary to implement these objectives. The first pillar establishes quantitative minimum capital charges for the market, credit, and operational risks of a bank. The second pillar contains the criteria and guidelines for the supervisory evaluation of a bank's risk management systems. The third pillar is the requirement of formalized disclosure of information on a bank's risk management system and risk position towards the capital markets. Disclosure is meant to lead to "market discipline", i.e. it is expected that markets react unfavorably to information on substandard risk management procedures, thus providing a strong incentive for the banks.
- A "level playing field" should be established both between different nations and between different financial institutions. The regulation of financial institutions should be equitable and should not distort competition.

Pillar 1: Market Risk

In the area of market risk, no fundamental changes have been made with respect to the market risk amendment to Basel I [259]. The topic of interest rate risk in the banking book is raised though no formal capital charge is imposed. The banking book contains all positions in credits and deposits which are not held for trading purposes. The topic was transferred to Pillar 2, i.e. the supervisors should check that the bank has in place a sound system to measure these risks. The national supervisors may also impose a capital charge.

Pillar 1: Credit Risk

Credit risk by far is the most significant part of Basel II. It is in this area where the progress in financial risk management methods has led to the biggest changes in the regulatory framework. For the regulatory treatment of credit risk, a bank can choose between two fundamentally different approaches.

The Standardized Approach is directly derived from the Basel I framework. The philosophy of the approach is exactly the same as in Basel I: Classify your assets in terms of the originator of bonds resp. debtor in loans, in terms of collateral or guarantor. Then multiply the dollar values of the assets with risk weights preset by the regulator, and multiply the sum of all risk-weighted assets by 8% to obtain the capital charge for credit risk. What has changed considerably with respect to Basel I is the number of the special cases, the level of detail of the rules and the implementation issues. Also credit risk mitigation, i.e. the transfer of credit risk to the capital markets, has received much attention.

As an alternative, a bank can opt for an Internal Rating Based (IRB) Approach [238], provided it possesses an internal rating system approved by the regulators. In the IRB Approach, a bank classifies its assets resp. customers according to their internal rating, estimates statistical parameters characterizing different rating classes, and uses these internal parameter estimates in a set of formulae given by Basel II in order to calculate the regulatory capital for credit risk. In a true internal model, both the model and the parameters are set by the bank. In the IRB Approach, the "model" still is set by the supervisors but banks are allowed to use internally generated parameter values.

Rating is a statistical procedure to estimate, perhaps in terms of classes or marks, the creditworthiness of a borrower. An external rating is performed by a rating agency. The best known rating agencies are Standard & Poor's, Moody's and Fitch. The rating describes the likelyhood of payment, i.e. the capacity and willingness of the obligor to meet its financial commitments as they com due. The rating agencies express the results of their rating as a score, such as AAA, BB, or C for Standard & Poor's, or Aaa, A1, or Ba for Moody's. The agencies interpret the meaning of their rating scores in words. E.g., the Standard & Poors descriptions of A and B issuer credit ratings are "An obligor rated 'A' has strong capacity and willingness to meet its financial commitments but is somewhat more susceptible to the adverse effects of circumstances and changes than obligors in higher-rated categories", resp. "An obligor rated 'B' is more vulnerable than the obligors rated 'BB' but currently has the capacity to meet its financial commitments. Adverse business, financial or economic conditions will likely impair the obligor's capacity or willingness to meet its financial commitments" [260]. To a large extent, rating thus is relative information. Bonds rated BBB (Baa) or higher are called investment grade. Those rated BB (Ba) or lower are called junk bonds. The rating of a company often is not stable in time: It may improve or deteriorate. The migration from one rating grade to another is formalized by rating matrices. Their entries give the probability of migration of, e.g., AAA-rated borrowers to AA+, or to A−, etc.

External ratings can be made comparable through the quantitative information they imply. Rating scores, in fact, are indicative of an expected default probability. If sufficiently large numbers of default events are analyzed statistically, the average default probabilities implied by rating scores can be estimated. E.g., the S&P AAA rating seems to imply a default probability of 0.01% per year, or less, AA apparently implies a default probability of 0.03% per year. Table 11.2 compares the rating scores of Standard & Poor's and Moody's, and provides estimates of implied default probabilities PD_{imp}. Notice that these estimates are based on independent research and have not been supplied by the rating agencies. Moreover, there are examples where two different agencies gave scores implying different default probabilities for the same institution. The process of rating by one of the big rating agencies is both formal and costly. Only big companies active on the capital markets usually undergo an external rating.

Table 11.2. List of the rating scores of Standard & Poor's and Moody's. Their implied one-year default probabilities PD_{imp} were derived from independent statistical analysis of default events

S&P	Moody's	Implied PD
AAA	Aaa	$\leq 0.01\%$
AA+	Aa1	0.02%
AA	Aa2	0.03%
AA-	Aa3	0.04%
A+	A1	0.05%
A	A2	0.07%
A-	A3	0.08%
BBB+	Baa1	0.12%
	Baa2	0.17%
BBB		0.30%
BBB-	Baa3	0.40%
BB+		0.60%
	Ba1	0.90%
BB	Ba2	1.3%
BB-		2.0%
B+	Ba3	3.0%
	B1	4.4%
B	B2	6.7%
B-	B3	10.0%
CCC		20.0%
D		defaulted

Internal rating refers to a rating system built internally by a bank with the purpose of rating its customers and assets. For most companies and for all private individuals, a standardized and transferable rating process is neither practical, nor economical, nor possible. There are several reasons why a bank may want to possess information on the default probability of a client. One, of course, is for the decision of acceptance or rejection of a credit demand. Another one is for the correct pricing of a loan. Losses from a higher default probability should be compensated by income from a higher interest rate charged. A third reason is that more (economic and) regulatory capital must be held against risker loans. We shall come to that point below.

The description of an actual rating system is beyond the scope of this book. In addition, much information is classified. The principle of an internal rating can be illustrated based on public information on the system developed by the German Savings Banks' Association (DSGV) which, at present, is used by most Savings Banks (Sparkassen) in Germany [261]. The core of the rating is the analysis of the financial statement of the client company. It produces a small number of key figures characterizing the profitability, the financial situation and the equity value of the company which are aggregated to a financial rating score. Secondly, a variety of qualitative factors ranging from an evaluation of client accounts with the bank, the history of the banking relationship, formal decisions on management succession in the company, to a more subjective assessment of management quality and business prospects are condensed into a qualitative client score. This qualitative score is aggregated with the financial rating to a bare customer rating. Should there be any major irregularity in the business relation with the customer such as a violation of important agreements, returned checks or debit entries, or account seizure, the final stand-alone client rating is obtained by a downgrade by one notch (out of 15-20) with respect to the bare rating. Finally, should the client be part of a major conglomerate or holding structure, guarantees of a parent may change the rating mark once more, giving the final integrated client rating mark. Quite generally when building a rating system, the main challenge is the valid identification and aggregation of a sufficiently small number of discriminating factors. (As a side remark, notice that the development of this system has profited enormously from the participation of several physicists in the project.)

Subject to certain minimum conditions and after supervisory approval, banks may use the IRB Approach and rely on their own internal estimates of risk components for capital calculation. The risk components to be estimated internally include the probability of default (PD), the loss given default (LGD), the exposure at default (EAD), and an effective maturity (M) of the assets [238]. Exposure at default is what can be lost at default, i.e. the entire amount outstanding. Loss given default includes the utilization of collateral and other receivables, i.e. what actually has been lost when in the default of the counterparty. In practice, LGD is given as a fraction of EAD. Exposures are categorized into five asset classes: (a) corporate, (b) sovereign, (c) bank, (d) retail, and (e) equity, all of which are defined in quite some detail.

In the IRB Approach, only unexpected losses are to be covered by capital. Expected losses are treated in a different manner, depending on the volume of general loan loss provisions set aside by the bank. There are two variants of the IRB Approach: a foundation and an advanced approach. In the advanced approach, a bank can use internal estimates for the entire list of parameters given above. In the foundation approach, it can only use internal estimates for the default probability PD, and must recur to supervisory values for the remaining parameters. In both cases, the parameters must be injected into asset-class specific risk-weight functions to determine the risk-weighted assets which, in the end, are multiplied by 8% to determine the capital requirement for credit risk. The unexpected losses to be covered by capital under Basel

II therefore are *not* the specific unexpected losses of a bank credit portfolio but those of a standard supervisory portfolio used to determine the risk-weight functions. We do not discuss further the foundation IRB Approach, as the general principles are better illustrated by the advanced IRB Approach. Practical reasons for preferring the IRB foundation approach to an advanced approach include the cost of implementation and the amount of data available for a reliable estimations of LGD, EAD, etc. Compared with PD-estimation where all loans extended contribute to the statistics, EAD and LGD are estimated on the defaulted loans only. Samples for these quantities typically are one to two orders of magnitude smaller than for PD.

To give an impression of the world of Basel II formulae, we give the basic expression for the capital $K_{\text{B II}}^{\text{non-def}}$ to be held against a non-defaulted exposure in the classes of corporates, sovereigns, and banks,

$$
\begin{aligned}
K_{\text{B II}}^{\text{non-def}} &= \text{EAD} \times \text{LGD} \\
&\times \left\{ N \left[\sqrt{\frac{1}{1-R}} G(\text{PD}) + \sqrt{\frac{R}{1-R}} G(0.999) \right] - \text{PD} \right\} \\
&\times \frac{1 - \frac{3}{2}b + b(M-1)}{1 - \frac{3}{2}b} \, .
\end{aligned}
\tag{11.10}
$$

In (11.10), $N(x)$ is the cumulative normal distribution with zero mean and unit variance

$$
N(x) = \int_{-\infty}^{x} dx' \, p(x') = \frac{1}{2} \left[1 + \text{erfc}\left(\frac{x}{2}\right) \right] ,
\tag{11.11}
$$

where $p(x)$ was defined in (4.24), and the second equality gives the relation to the complementary error function, $\text{erfc}(x)$. $G(x)$ is the inverse cumulated normal distribution,

$$
G(x) = N^{-1}(x) , \quad \text{i.e.} \quad G\left[N(x)\right] = x ,
\tag{11.12}
$$

and may be understood as the quantile function. $N(x)$ measures the probability weight below x. When N is assigned a value $N(x) = P$, $G(P)$ returns the P-quantile $x = \Lambda(P)$. E.g., the second G-term in the argument of the cumulative normal distribution in (11.10) is the 99.9%-quantile of the normal distribution.

The capital formula depends on the correlation parameter R, defined as

$$
R = 0.12 \frac{1 - e^{-50\text{PD}}}{1 - e^{-50}} + 0.24 \left(1 - \frac{1 - e^{-50\text{PD}}}{1 - e^{-50}} \right) .
\tag{11.13}
$$

The weight of the maturity adjustment is determined as

$$
b = \left[0.11852 - 0.05478 \ln(\text{PD}) \right]^2
\tag{11.14}
$$

M is a cash-flow averaged effective loan or portfolio maturity,

$$M = \frac{\sum_{t=0}^{\infty} t\,\mathrm{CF}(t)}{\sum_{t=0}^{\infty} \mathrm{CF}(t)} ,$$ (11.15)

where $\mathrm{CF}(t)$ denotes the cash flow (interest payments, principal repayments, fees) at time t.

The capital requirement for a defaulted exposure is

$$K_{\mathrm{B\,II}}^{\mathrm{def}} = \mathrm{EAD} \times \max\left(0,\ \mathrm{LGD}_{\mathrm{def}} - \mathrm{EL}_{\mathrm{est}}\right) .$$ (11.16)

$\mathrm{LGD}_{\mathrm{def}}$ is the loss given default estimated for the specific defaulted exposure, and $\mathrm{EL}_{\mathrm{est}}$ is the bank's best estimate for the expected loss of the portfolio to which the exposure belonged before default.

Details as to how the expressions (11.10)–(11.16) and their numerous counterparts were derived by the Basel Committee, are not available. Crazy as they appear (but notice that in earlier consultative documents of the Basel II Accord, equations were decorated by funny exponents such as 0.44), the following derivation procedure for the formulae can be guessed, though. (i) Compose one or several model portfolios of loans corresponding to the asset class in question. (b) Simulate the evolution of losses from these portfolios using some assumptions about factors which are known to affect credit port- folios. (c) Determine both expected losses and unexpected losses for each portfolio and each set of parameter values. (d) Try to fit the unexpected losses against the various parameters, and change the fitting function until a good-looking fit is achieved. (e) Try to combine the individual fits into a mul- tidimensional fit by suitably changing parameters. (f) Bring the final result into the political arena and declare it open for negotiation. (g) Write down the result of the negotiations and publish it.

Despite the cynical tone in the description, it approximately corresponds to the generation and evolution of the Basel II formula world. While one may have a critical opinion about the numbers used and the specific dependences implemented in Basel II, there is an important background to each driving factor.

First set $R = 0$, i.e. assume uncorrelated counterparty defaults. Then, (11.10) becomes $K_{\mathrm{B\,II}}^{\mathrm{non-def}} = 0$. In the absence of counterparty default corre- lation, there is no capital to be held against a portfolio of corporate, sovereign, or bank obligations. The formal reason for the vanishing of regulatory capital when $R = 0$ is that capital is used only to cover unexpected losses. Of course, it is an idealization to assume that the loss amount of a loan portfolio is a sharp variable. On the other hand, this assumption may become a valid approximation for a highly diversified portfolio of many credits with small denominations. Then, the law of large numbers works, as it does, e.g., for the credit card business in the retail sector. Basel II precisely assumes well- diversified portfolios in its models. For less granular portfolios, unexpected losses certainly are bigger. In the intermediate stages of Basel II consultation, this effect was caught by a "granularity adjustment factor". This factor, however, was dropped later on during the political negotiations.

The next important message which emerges from the limit $R \to 0$ is that counterparty default correlation is the main driving factor of unexpected losses in a sufficiently granular loan portfolio. Intuitively, this is easy to understand. With large default correlation, the number of independent loans is reduced considerably, the portfolio effectively behaves as one with a few very large loans, and fluctuations become appreciable.

In principle, the correlation coefficient R should be measured in a portfolio, or for the entire banking book. Instead, in Basel II, it is fixed to the value implied by (11.13) by the regulators. It decreases from 0.24 to 0.12 as PD increases from zero to one. The value $R = 0$ used in our argument, is not permissible in Basel II! While the interpolation proposed certainly is largly guesswork, the important message is that the default correlation of very good loans is higher than that of badly rated loans. A simple-minded picture where risky loans are likely to default due to obligor-specific factors, e.g. bad management, but rather riskless loans would default mainly as a collective phenomenon, e.g. due to economic downturn, is consistent with the trend contained in (11.13).

Next, set the maturity, (11.15), $M = 1$. For a moment, ignore the exact definition of M as a cash-flow averaged maturity, and think about it simply as the lifetime of a loan. For $M = 1$, the maturity adjustment factor in (11.10) reduces to unity, i.e. the Basel II capital charge has been calibrated on a one-year lifetime of a loan (portfolio). It turns out that for a given one-year default probability, the unexpected losses of a portfolio depend on its effective maturity. The higher the maturity, i.e. the longer the lifetime of the loans, the bigger the unexpected losses, i.e. the default risk. More capital thus is required. However, the squared logarithmic dependence on default probability and the five-digit figures in the maturity adjustment factor (11.15) certainly are not to be taken too serious from a scientific point of view.

The regulatory capital requirement (11.10) is linear in the remaining open parameters, EAD and LGD. The exposure at default, EAD, is the total amount of loan outstanding at the time of default. Notice that even the definition of "default" is not unique in banking. The standard is a "90 days past due"-rule, i.e. the debtor is past due more than 90 days on a major credit obligation. The sum of all payments outstanding and expected until the maturity of the loan then is the exposure at default. EAD is measured in real currency, e.g. dollars.

LGD is the loss given default. It is less than EAD because usually, the bank is able to utilize collateral or other receivables, leading to a recovery. LGD is measured as a fraction.

In the advanced IRB Approach, banks may estimate internally all three open parameters of (11.10), PD, LGD, and EAD. M can be calculated from the cash flows. In the foundation approach, only PD may be estimated. The true challenge in Basel II is the estimation of these data. A rating system provides information on PD. LGD and EAD can only be estimated by analyzing a sufficiently large number of default events, and by extracting the parameters from the credit files. The length of the time series used to estimate the parameters must be five years, at minimum.

This paragraph was intended to summarize the basic logic of thought underlying Basel II. The main body of the documents, however, is filled with detailed instructions on the treatment of many particular cases and products. These details are beyond the scope of this book.

Pillar 1: Operational Risk

Basel II defines operational risk as the risk of loss resulting from inadequate or failed internal processes, people and systems, or from external events [238]. It includes legal risk but excludes reputational, business and strategic risk.

Operational risk is widespread, a fact which is obvious from the definition. Almost every industry is subject to operational risk, and private individuals are, too. Insurance companies make their living from operational risk. In fact, many operational risks can be insured. The financial services industry has been woken up on operational risk by Basel II only.

The attitude towards operational risk depends on the industry concerned. In a hospital, e.g., operational risk often is a matter of life and death. Consequently, every control possible is implemented to avoid *any* operational risk, if possible. Air traffic, or the chemical and nuclear industries, are other examples of extreme operational risk aversion. Many other industries can afford to have a more differentiated attitude as the consequences of operational risks striking are less dramatic. Controls may become a matter of cost considerations, and there may be trade-offs between implementing controls and subscribing to an insurance policy. In banks, controls help to avoid operational risk striking, and insurance may help to cover losses once a risk event happened. Moreover, in the future world of Basel II, banks will be required to hold regulatory capital against their operational risks.

Basel II provides three approaches to determine the regulatory capital charge for operational risk, a Basic Indicator Approach (BIA), a Standardized Approach (SA), and the Advanced Measurement Approaches (AMA). Both the Basic Indicator and Standardized Approaches are not risk sensitive, in analogy to the Standardized Approaches of Basel I and Basel II in credit risk. The Advanced Measurement Approaches, on the other hand, are risk sensitive and amount to building an internal model for operational risk.

In the Basic Indicator Approach, the regulatory capital for operational risk is given by [238]

$$K_{\mathrm{BIA}} = \alpha \, \mathrm{GI} \, , \quad \alpha = 0.15 \, . \tag{11.17}$$

GI denotes gross income, and includes net interest income plus net non-interest income. These quantities are determined by accounting standards.

The Basic Indicator Approach is very easy to use. All quantities required to calculate gross income are available from the annual financial statement of the bank. The prefactor $\alpha = 0.15$ was *not* calibrated on loss histories of banks but from two general requirements. Firstly, the average regulatory capital of an ensemble of banks should be left unchanged when banks use the Standardized Approach for credit risk and the Basic Indicator Approach for operational risk. Secondly, it was decided that about 12% of the total regulatory capital should be set aside for operational risk. This figure reflects some loss history by banks (collected in so-called "quantitative impact studies") but also much political bargaining (initially, the fraction of operational risk capital had been set to 20% of total capital).

Gross income, a priori, is not a risk-sensitive quantity. Its use also leads to perverse consequences: Banks with high income will have to hold much capital, those with low income need much less capital. Standard reasoning, however, suggests that high income only can be achieved by few risks striking while low income may be the consequence of a big exposure to all kinds of risks, operational risk among them. Life has shown, though, that *abnormally* high income often may be the consequence of too much operational risk taking. This was the case with Nick Leeson who ruined Barings bank. This rule is also confirmed on the failures or near-failures of smaller banks around the world.

The Standardized Approach follows the same philosophy. However, it attempts to introduce a minimum of risk-sensitivity by dividing the bank into eight business lines and by modulating the multipliers of gross income according to the riskiness of the business lines, as perceived by the regulators. The capital requirement under the standardized approach then is given by [238]

$$K_{\mathrm{SA}} = \sum_{j=1}^{8} \beta_j \, \mathrm{GI}_j \, . \tag{11.18}$$

The eight business lines, and their multipliers β_j, are summarized in Table 11.3. A bank which wants to determine its capital according to the standardized approach, must fulfill a list of qualitative requirements, and get a supervisory approval. When its main sources of income belong to the business lines with $\beta_j = 0.12$, it may expect a lower capital charge. On the contrary, the capital requirement increases when important income is generated in the $\beta_j = 0.18$-business lines. When conducting the third quantitative impact study among the German Savings Banks it turned out, however, that there is no systematic advantage or disadvantage in capital charge, of the Standardized Approach with respect to the Basic Indicator Approach. However, mapping the organizational structure of a bank onto the standard Basel-II business lines introduces a significant degree of complexity into the Standardized Approach.

Table 11.3. Business lines of the Standardized Approach and the Advanced Measurement Approaches to operational risk, and the gross-income multipliers β used in the Standardized Approach

Business Line	β_j
Corporate finance	0.18
Trading and sales	0.18
Retail banking	0.12
Commercial banking	0.15
Payment and settlement	0.18
Agency services	0.15
Asset management	0.12
Retail brokerage	0.12

Basel II also discusses an Alternative Standardized Approach (ASA) whose applicability, however, depends on the national supervisors. Under this approach, banks may calculate their capital charge for retail banking not based on the gross income but rather based on the total volume of outstanding retail loans and advances, $\mathrm{LA_{RB}}$. The capital for retail banking then is

$$K_{\mathrm{ASA}}^{\mathrm{RB}} = \beta_{\mathrm{RB}}\, m\, \mathrm{LA_{RB}}\ . \tag{11.19}$$

$m = 0.035$ is a multiplier calibrated to make the capital charge grossly comparable to a gross-income based calculation. $\beta_{\mathrm{RB}} = 0.12$ is the standard multiplier for retail banking. Banks may also aggregate their retail banking and commercial banking credit portfolio if they use a common multiplier of 0.15 for both. If the gross income in the remaining internal business lines cannot be separated clearly and mapped on the Basel business lines, they may also be taken as one cluster, at the expense of a prefactor of 0.18. Conceptually, the Alternative Standardized Approach is as questionable as is the Standardized Approach. However, it may be implemented more easily and more cost-effectively, if allowed.

Finally, the Advanced Measurement Approaches (AMA) do not set up a formula framework for capital calculation, but rather give the bank the freedom to construct an internal model for operational risk. Of course, there is a long list of qualifying criteria which a bank must satisfy, and it must get the approval of its regulators after a trial period which, at the start of Basel II, has been set to two years.

There are some regulatory constraints to the construction of an AMA which we now discuss. The main challenge of operational risk measurement lies in the scarceness of data. A measurement system for operational risk in line with an internal model in market risk would record actual loss events which happened in the bank. These are the equivalent of the (negative) price changes of securities recorded for market risk measurements. Prices for securities are available with frequencies of at least once daily down to one tick every couple of seconds for the high-frequency data used, e.g., in Chap. 5 for stock index quotes, and Chap. 6 for foreign exchange. On the other hand, loss events from operational risk happen quite seldom. A major public sector bank in Germany with size measured by a balance sheet of 3×10^{11} Euro, e.g., possesses a loss data collection with a few thousand entries, collected in more than five years. Typical numbers for German savings banks with a balance sheet of $3 - 5 \times 10^9$ Euro, are about 25–50 loss events per year with losses exceeding 1,000 Euro. However, capital for operational risk is not held to cover 1,000 Euro losses but large events, potentially threatening the survival of the bank. A broad distinction between such events is provided by the notions of "high-frequency low-impact events" (e.g., cash differences, typing errors on the trading desks, retail customer complaints, credit card fraud, etc.) and "low-frequency high-impact events" (e.g., kidnapping of the chairman, fire caused by lightning, rogue traders, unlawful business practices). In 2004/2005, some banks considered "Spitzer risk", the risk of New York federal attorney Elliot Spitzer investigating against them, to be their most severe operational risk exposure. Given the low probability of large losses, many more data (or complementary methods of risk estimation) are needed to capture this range of risk reliably. Regulators indeed require that the approach of a bank must cover these potentially severe "tail" events, and that the risk measure is based on a 99.9% confidence level.

The operational risk measurement system of a bank must be granular enough to determine the risk separately for the eight business lines listed in Table 11.3, and for seven event categories. These risk categories are listed in Table 11.4. Basel II also defines a second and third level of both the business lines and the risk categories, to make them more granular and more specific. They can be found in the Basel document [238]. Banks are free to use their internal categories for their risk measurement system but must be able to map their losses onto the Basel categories. Also, a bank may use an internal

Table 11.4. Event-based risk categories of Basel II

Risk Category
Internal Fraud
External Fraud
Employment Practices and Workplace Safety
Clients, Products & Business Practices
Damage to Physical Assets
Business Disruption and Systems Failure
Execution, Delivery & Process Management

definition of operational risk but, at the same time, must guarantee that it covers the same scope as the definition set forward by the Basel Committee.

At variance with credit risk and best practice in risk management in general, Basel II requires to hold regulatory capital both against the expected and unexpected losses from operational risk. Only when it is demonstrated explicitly that expected losses are included in product pricing, a reduction of capital to cover solely unexpected losses can be allowed. Unless a bank has reliable estimates for correlations, based on methodologies approved by the supervisors, it must add the exposure estimates across business lines and risk categories. This implies that a perfect correlation is assumed between events in different business lines and risk categories. Several quantitative models indicate that the capital requirement is essentially determined by the "low-frequency high-impact" scenarios. For those, the perfect-correlation assumption certainly leads to a significant overestimate of the actual risk incurred.

The modeling of operational risk must use internal loss data, relevant external loss data, scenario analysis and factors reflecting the business environment and internal control systems. Let us discuss the various data types in some more detail.

An internal loss database certainly is the anchor of every operational risk management system. It records in detail and in a standardized format every loss event due to operational risk. From such a loss database, a time series of losses can be constructed. In principle, this time series can be used for a risk estimate, in analogy to market risk. One problem with this approach has been discussed above: usually, there are not enough data available. Secondly, only for extremely long time series, i.e. when the "low-frequency high-impact" events have realized sufficiently frequently, can such a risk estimate be trusted. Otherwise, one must be concerned about the modeling of these tail events, i.e. the difference between loss history and actual risk. Thirdly, even when such long times series are available, the hypothesis of stationary environment underlying their use in a risk model, can rarely be justified in view of the dynamics of change in the financial industry. Fourthly, there is no forward-looking element in this extrapolation of the past into the future. On the other hand, after severe loss events, management will usually take the appropriate measures to prevent a repetition of the event. For these reasons, the Basel Committee requires the inclusion of additional data types into the operational risk model.

External loss data can complement internal data. They can help with the second problem noted before, the capture of "low-frequency high-impact" events. To the extent that time and ensemble sampling are equivalent, loss events materialized in another bank are indicative of risk incurred in the own institute, even though nothing has happened yet. However, the important challenge with external loss data is to determine the extent to which they are relevant for the own institute, resp. they can be made relevant by suitable rescaling. At the time of writing, no standard scaling model for operational risk losses was available. External loss data can be bought from commercial operational risk databases, or be collected in data consortia. In a commercial database, public information, mostly from the financial press, is collected and analyzed. In a data consortium, a group of banks agrees to contribute anonymized information on all operational loss events to a central collection facility. This information is grouped and then reflected back to the participating banks for use in their internal risk models. The importance attached to such external data in the risk models can be gauged from the fact that even banks directly competing with each other jointly have set up such data consortia. Without going into details, we add that there is no unique procedure for blending the internal and external loss data. Hence, a certain element of subjectivity is introduced in the model.

When performing scenario analysis, experts subjectively evaluate the frequency of a certain scenario, and the losses associated, based on their business experience and the knowledge of changes which have been introduced as a reaction to past loss events. The scenarios may either be formulated by the experts themselves, or be taken from a central scenario pool. Scenario analysis is a suitable tool to address the all-important "low-frequency high-impact" events which may have catastrophic consequences for a bank. In scenario analysis, one deliberately relies on the subjective information provided by the experts. The aim, though, is to derive almost objective information to be fed into a risk model. There are several approaches to limit the subjectivity of the estimates. One is to ask a group of experts, and to require consensus in the answer. The other one is the Delphi method (named after the famous greek oracle): Ask the same question to a number of people, then drop the highest and the lowest answer, and take the average of the rest. Finally, in social sciences, there is a branch called psychometrics which specifically deals with designing and evaluating questionnaires. Scenario analysis is valuable because it also possesses that forward-looking view which loss data collection misses. Changes in processes can be incorporated in the estimates a long time before they show up in changed parameters of a loss history.

The data type of factors reflecting the business environment and internal control systems is rather ill-defined, and is subject of controversy and confusion in the financial industry at the time of writing. There are several ways to evaluate the internal control system of a bank. One way, again, is to ask experts for an evaluation, e.g., in terms of school marks. While subjective, it quickly gives valuable information on the state of the controls. Another option is to systematically record the failure of processes, or process elements. It is only applicable with highly standardized processes, and economical at best when both the processes and the failure recording are automated. It is obvious that such information should be included in a management information system. What is less obvious is if and how it could be included in a quantiative risk model.

The same can be said about the business environment factors. Several interpretations have been discussed. One is to search for correlations between operational risk and certain high-frequency business variables such as the daily number of customer orders to be transmitted to the stock exchange, the work load of the IT systems, the fluctuation rate of staff, or the number of excess working hours. Such factors are correlated to operational risk by a hypothesis about their influence on the bank's processes. E.g., the number of typing errors in the transmission of customer orders could be proportional to the number of orders. The cost/loss associated with one typing error is N dollars, on the average. Risk thus could be calculated from these risk indicators, and capital could vary accordingly. The problem with this approach is that no significant correlation between these risk indicators and actual loss histories could be uncovered to date. Another, perhaps more promising interpretation is in terms of discriminating factors when considering a larger pool of banks. Such discriminating factors could be the real estate holdings of a bank (high/low), geographic spread (international/national/regional/city), the business lines supported, production depth (outsourcing significant or not), etc. While it is not clear how such factors determine the risk model of an individual bank, they can be used to form peer groups within a pool of banks, where external data are taken only from institutes of the same peer group.

It will be interesting to see how these data types are combined in actual AMA during the next years. At the time of writing, many banks worldwide were in the process of setting up the quantitative models for their AMA. None of them has a definite model yet, and none of them had obtained approval from its supervisors. Experience with the introduction of internal models in the area of market risk suggests that initially, the regulators could indeed give considerable freedom in the model construction and focus primarily on issues of data quality and completeness. If true, only when a broader experience on the performance of the various models has become available, stricter guidance on the structure of the models is expected.

Finally, many credit defaults may be due to operational risk. Examples are credits obtained in a fraudulent manner, breach of controls in the internal credit approval process, inappropriate use of the internal rating system with inappropriate credit pricing as a consequence. Basel II requires these events to be recorded as operational risk events, but to exclude them from the operational risk capital calculation. Instead, they should be flagged, and be included in the credit risk capital charge. This is mainly done to ensure continuity of the established credit default records.

Pillar 2: Supervisory Review

The first pillar of the Basel II regulatory framework requires banks to hold enough capital to cover that part of their risks which can be quantified, perhaps only approximately. The second pillar of banking regulation focusses on the risk management processes and their assessment by supervisory authorities [238]. Some regulators have made the point that it is the risk management processes that matter, more than the risks themselves.

The supervisory review is based on four key principles. Principle 1 states that banks should have a process for assessing their overall capital adequacy in relation to their risk profile, and a strategy for maintaining their capital levels. The paper also specifies the five main elements, according to the Basel Committee, of a rigorous Internal Capital Adequacy Assessment Process (ICAAP):

- Board and senior management oversight. Basel II emphasizes that the bank management is responsible for developing the internal capital adequacy assessment process, and for the bank taking only so much risk as the capital available can support. Conversely, bank management must ensure that the capital is adequate for the risk taken. Bank management must formulate a strategy with objectives for capital and risk, including capital needs, anticipated capital expenditures, desirable capital levels, and external capital sources. Moreover, the board of directors must set the bank's tolerance for risk.
- Sound capital assessment. Here, policies and processes must be designed to ensure that the bank identifies, measures, and reports all materials risks. Capital requirements must then be derived from the risk to which the bank is exposed, and a formal statement of capital (in)adequacy must be made. Notice that no reference is made to regulatory capital or any of the calculation schemes introduced under pillar 1. What is required is the bank's own assessment of its capital needs. ICAAP targets economic capital, although this is not spelled out explicitly. The next element requires banks to quantify or estimate all important risks they are exposed to. To determine the economic capital, these risks must be aggregated either using a quantitative (internal) model or by rough estimation. It must be guaranteed that the bank operates at sufficient levels of capital to support these aggregated risks. Finally, internal controls, reviews, and audits must ensure the integrity of the entire management process.
- Comprehensive assessment of risks. The bank must ensure that all significant risks are known to its management. The notion of risk here is not limited to those types of risk for which pillar 1 imposes capital charges, and may include reputational risk, strategic risk, liquidity risk, and finer details of market, credit, and operational risk which are not covered by pillar 1. Moreover, this element also requires risk identification when a bank uses one of the standardized, non-risk-sensitive approaches for the determination of its regulatory capital. When risk cannot be quantified, risk should be estimated.
- Monitoring and reporting. The bank should establish a regular reporting process and ensure that its management is informed in a timely manner about changes in the bank's risk profile. The reports should enable the

senior management to determine the capital adequacy against all major risks taken, and assess the bank's future capital requirements based on the changed risk profile.

- Internal control review. The bank should conduct periodic reviews of its control structure to ensure its integrity, accuracy, and reasonableness. Apart the review of the general ICAAP, this process should identify large risk concentrations and exposures, verify the accuracy and completeness of the data fed into the risk measurement system, ensure that the scenarios used in the assessment process are reasonable, and include stress tests.

The second principle asks supervisors to review and evaluate the bank's internal capital adequacy assessments and strategies, as well as their ability to monitor and ensure their compliance with regulatory capital ratios. Supervisors should take appropriate action if they are not satisfied with the results of this process. Again, four elements give more specific instructions to supervisors as to how implement this principle.

- Review of adequacy of risk assessment. Supervisors should assess the degree to which internal targets and processes incorporate all material risks faced by the bank. The adequacy of risk measures used and the extent to which they are used operationally to set limits, evaluate performance, and to control risks, should be evaluated.
- Assessment of the control environment. Supervisors are instructed to evaluate the quality of the bank's management information and reporting systems, the quality of aggregation of risks in these systems, and the managements record in responding to changing risks.
- Supervisory review of compliance with minimum standards. In order to apply certain advanced methodologies such as the IRB approach or the AMA, banks must satisfy a list of qualifying criteria. Here, supervisors are instructed to review the continuous compliance with these minimum standards for the approaches chosen.
- Supervisory response. Supervisors should take appropriate action if they are not satisfied with the bank's capital assessment and risk management processes.

According to the third principle, supervisors should expect banks to operate above the minimum regulatory capital ratios and should have the ability to require banks to hold capital in excess of the minimum. Here, it is recognized that the pillar 1 capital charges, conservative as they may appear, were calibrated on the average of an ensemble of banks. The individual capital requirements of a specific bank may be different and are treated under pillar 2. In particular, regulators may set capital levels higher than the pillar-1 capital when they deem appropriate for the situation of a bank.

In the fourth principle, supervisors are requested to intervene at an early stage to prevent capital from falling below the minimum levels required to support the risk characteristics of a particular bank, and should require rapid remedial action if capital is not maintained or restored. Supervisors have some options at their disposal to enforce appropriate capital levels. These may include intensifying the monitoring of the bank, restricting the payment of dividends, requiring the bank to prepare and implement a satisfactory capital restoration plan, and requiring the bank to raise capital immediately. The ultimate threat, of course, is the closure of the bank by the supervisory authority.

Pillar 3: Disclosure

Banks are required to disclose certain information on their risk management processes, the risks they face, and the capital they hold to cover it [238]. This requirement is established to complement pillars 1 and 2.

By pillar-3 disclosure, investors should be enabled to monitor the risk management of a bank and thus provide incentives for continuous improvement. Investors are assumed to prefer the shares of a bank with good risk management over one with poor risk management. Rating agencies will more highly value a bank with good risk management – according to Table 11.2, the rating score is directly related to the bank's default probability, and its creditworthiness. It determines its credit spread on the markets. Pillar 3 thus is designed to leverage the self-interest of the bank in good risk management.

The Basel II paper has detailed tables with the disclosure requirements for banks.

11.3.6 Outlook: Basel III and Basel IV

We have not touched upon the definition of bank capital and the different types of capital existing because this book is focused on the statistical aspects of banking and risk management. Capital has been defined and classified in the Basel I Accord [256, 258]. The capital definition was left unchanged by Basel II. It is expected that the next round of Basel negotiations leading to a Basel III, will provide new definitions of what constitutes bank capital.

At present, it is not expected that Basel III will fundamentally change the modeling of banking risks. Only a Basel IV agreement may bring the long-expected recognition of internal models for credit risk capital determination.

Both the volume of the Basel documents and the length of the negotiation rounds have increased strongly from Basel I to Basel II. If this trend continues, the time until the next fundamental innovations in international banking regulation will likely be measured in decades rather than in years. For the time being, the preceding sections give a brief though valid introduction.

Appendix: Information Sources

This appendix gives tables of some important information sources relevant for the topic of this book. Naturally, this list is extremely incomplete. They were up to date at the time of writing but may become outdated at any time thereafter. Moreover, they are somewhat biased towards European and more specifically German sources. This both reflects my own background and interests but also the fact that much of the research in financial markets with methods from physics actually takes place in the old world. I apologize for any inconvenience which this bias may cause.

Publications

These basically follow from statistics on the Reference section of this book.

Physics Publications

- Physica A
  http://www.elsevier.nl/inca/publications/store/5/0/5/7/0/2/
- European Physical Journal B
  http://www.edpsciences.com/docinfos/EPJB/OnlineEPJB.html
- Physical Review E
  http://pre.aps.org/
- Europhsics Letters
  http://www.edpsciences.com/docinfos/EURO/OnlineEURO.html
- International Journal of Theoretical Physics C
  http://www.wspc.com.sg/journals/ijmpc/ijmpc.html
- Nature
  www.nature.com
- Physical Review Letters
  http://prl.aps.org/

Physics–Finance Interface

- International Journal of Theoretical and Applied Finance
  http://www.wspc.com.sg/journals/ijtaf/ijtaf.html

- Quantitative Finance
  http://www.iop.org/Journals/qf

Finance

- Journal of Finance
  www.afajof.org/jofihome.shtml
- Journal of Banking and Finance
  http://www.elsevier.nl/inca/publications/store/5/0/5/5/5/8/
- Journal of Empirical Finance
  http://www.elsevier.nl/homepage/sae/econbase/empfin/menu.sht
- Finance and Stochastics
  http://link.springer.de/link/service/journals/00780/index.htm
- RISK Magazine
  http://www.riskpublications.com/risk/index.htm
- Applied Mathematical Finance
  www.tandf.co.uk/journals/routledge/1350486X.html
- Econometrica
  http://www.jstor.org/journals/00129682.html

Preprint Servers

- http://xxx.lanl.gov/archive/cond-mat located at Los Alamos National Laboratory is the central preprint server for condensed matter and statistical physics. Many of the papers published in the physics journals listed above have appeared on this server before publication, and can be retrieved there. Some other papers were listed on related servers, such as chao-dyn, adap-org, or physics. To access these, just replace cond-mat in the URL above by the appropriate server label.
- http://netec.wustl.edu/, located at Washington University, is a set of servers with economics related information. BibEc contains information on printed working papers, WoPEc data about electronic working papers, WebEc lists World Wide Web resources in economics, and JokEc is a list of jokes about economists and economics.

Computational Resources

- http://finance.bi.no/~bernt/gcc_prog/algoritms/algoritms/algoritms.html features *Financial Numerical Recipes,* by Bernt Arne Ødegaard. The intentions of this site are clear from its title: To provide an exhaustive discussion of important algorithms and computer code for advanced financial calculations, in a format that is similar to its big brother:

Numerical Recipes: The Art of Scientific Computing [174]. It contains algorithms, both basic and advanced, for option pricing, and some algorithms dealing with term structure modeling and pricing of fixed income securities. All computer code is in the C++ language, and implemented as self-contained subroutines that can be compiled on any standard C++ compiler.

- More links to computational resources can be found on the web sites listed in the following section.

Internet Sites

The central internet sites at the crossroads of physics and finance are:

- `http://www.ge.infm.it/econophysics/`, located at the University of Genova, provides extensive lists of research papers, conferences and schools, courses, job advertisements, and links to research institutes and companies.
- `http://www.unifr.ch/econophysics/` contains news, meeting announcement, book reviews, lists of recent preprints, a "paper of the month", opinions, and discussions. There is also a page with data sources and access to financial data and links to financial institutions. This site is host to the minority game web site, where plenty of useful information on this game can be found. There is also an interactive minority game where a visitor can play against the computer.
- `http://www.quantnotes.com` is a high-quality (though not always immediately responsive) web site providing selected publications. It features introductory articles where you will learn about various financial instruments and how mathematics you may be familiar with, is applied daily by banks to fairly price these instruments. In addition, there are book reviews, links to software and data sites, job and event listings, etc.
- `http://www.mailbase.ac.uk/lists/finance-and-physics/` contains a mailbase for discussion and information exchange.
- Finance-and-Physics-Services at `http://l3www.cern.ch/homepages/su-sinnog/finance/` is another site providing many links, papers, and data to the public. They have a list of preprints, many of them from the finance community, structured along topics. This distinguishes this site from the three sites above which are more physics oriented. I found particularly useful the link to `http://www.probability.net/` placed on this site in summer 2000.

I list a few more institutions where further links, working papers on subjects of interest, etc., can be found:

- `www.gloriamundi.org` is a site containing a wealth of material on value at risk and related topics. Many important papers on value at risk are available for download, and there is a good list of books covering this topic. The site also includes papers containing criticism of value at risk as well as work on coherent risk measures, expected shortfall, etc. In terms of types of risk, most material naturally covers market risk. Credit risk is less prominent, perhaps due to regulators' reluctance to recognize internal models, and a few papers address operational risk.

- Institut für Entscheidungstheorie und Unternehmensforschung at Karlsruhe university
  http://finance.wiwi.uni-karlsruhe.de/Hotlist/index.html
- Freiburger Institut für Datenanalyse und Modellbildung
  http://paracelsus.fdm.uni-freiburg.de/
- RiskLab, Zurich
  http://www.risklab.ch/
- The Santa Fe Institute
  http://www.santafe.edu/

Companies

- The Prediction Company, Santa Fe
  www.predict.com
- Science & Finance, Paris
  www.science-finance.fr
- Olsen & Associates, Zurich
  www.olsen.ch
- J. P. Morgan's RiskMetrics
  http://www.riskmetrics.com/
- Deutsche Bank Research
  http://www.dbresearch.de/
- Algorithmics, Inc.
  http://www.algorithmics.com

References on Banking Topics

For the readers who want to learn more on bank management and current topics in banking, I recommend

- T. W. Koch and S. S. MacDonald: *Bank Management* (Thomson South-Western, Mason 2004), and
- G. H. Hempel and D. G. Simonson: *Bank Management: Text and Cases* (Wiley 1998).

For those readers who have to dive into the Basel Capital Accord after reading this book, I recommend to start their reading with the 1996 *Amendment to the Capital Accord to Incorporate Market Risks* [259]. This makes easiest for the scientific mind the transition from a scientific text to regulatory prose. Then read the brief Basel I Accord [258] before struggling with the 250-page Basel II monster [238].

Nonscientific Books

These are a few nonscientific books which I liked reading:

- B. G. Malkiel: *A Random Walk Down Wall Street* (W. W. Norton, New York 1999) basically is an investment guide but contains a wealth of information of financial markets, and a good list of references to important papers in finance. The basic thesis of this book is that very few (professional!) investors succeed in consistently beating a reference index over long periods of time. Consequently, the author's best advice would be to invest in broadly structured low-load index funds.
- Nick Leeson: *Rogue Trader* (Little, Brown, London 1996) has the story of Nick Leeson, the Singapore based derivatives trader who ruined Barings Bank.
- Frank Partnoy: *FIASCO* (Penguin Books, New York 1999) is the inside story of a Wall Street Trader.
- Nicholas Dunbar: *Inventing Money* (Wiley, Chichester 2000) gives a nonscientific story of derivatives and derivatives trading, and the academic researchers involved in the modeling of derivatives, culminating in the breakdown of Long Term Capital Management, a hedge fund whose partners were, among others, Robert Merton and Myron Scholes.
- Ron S. Dembo and Andrew Freeman: *Seeing Tomorrow* (Wiley, New York 1998) promote forward-looking risk management including, in addition to concepts discussed in this book, scenario analysis, risk–return assessment, and the notion of "regret". Regret is a measure of the subjective pain or objective consequences of worst-case scenarios. Ron Dembo is president and CEO of Algorithmics, Inc., a Toronto-based firm for high-end risk management software.
- Peter L. Bernstein: *Against the Gods: the Remarkable Story of Risk* (Wiley, New York 1998) retraces the history of risk management from the times of the ancient Greeks to the present days of derivative trading. This book contains a lot of biographical information on the principal drivers of this development.

Notes and References

1. DAX, Deutscher Aktienindex, is a stock index composed of the 30 biggest German blue chip companies
2. Stop-loss and stop-buy orders are limit orders to protect an investor against sudden price movements. In a stop-loss order, an *unlimited sell order* is issued to the stock exchange when the price of the protected stock falls below the limit. In a stop-buy order, an *unlimited buy order* is issued when the stock price rises above the limit, cf. Sect. 2.6.1
3. B. G. Malkiel: *A Random Walk Down Wall Street* (W. W. Norton, New York 1999)
4. A. Einstein: Ann. Phys. (Leipzig) **17**, 549 (1905)
5. G. J. Stigler: J. Business **37**, 117 (1964)
6. L. Bachelier: *Théorie de la Spéculation* (Ed. Jacques Gabay, Paris 1995). This is a reprint of the original thesis which appeared in Ann. Sci. Ecole Norm. Super., Sér. 3, **17**, 21 (1900). An English translation is available in [7]
7. P. H. Cootner (ed.): *The Random Character of Stock Market Prices* (MIT Press, Cambridge, MA 1964)
8. M. F. M. Osborne: Operations Research **7**, 145 (1959), reprinted in [7]
9. Most papers of this kind have appeared on the condensed matter preprint server at Los Alamos, http://xxx.lanl.gov/archive/cond-mat, and are referred to as cond-mat/XXYYZZZ where XX labels the year, YY the month, and ZZZ the number of the preprint. Some of them can be found on related servers, such as chao-dyn, adap-org, or physics. To access these papers, just replace cond-mat in the above URL by the appropriate server name
10. J. C. Hull: *Options, Futures, and Other Derivatives* (Prentice Hall, Upper Saddle River 1997)
11. M. Groos, K. Träger, H. Hamann: *Capital-Handbuch Geld* (Mosaik-Verlag, München 1993) (in German). This book gives a very elementary, nonscientific introduction and is mainly written for investors. It often provides simple explanations for the most important notions. Similar but more advanced is E. Müller-Mohl: *Optionen und Futures* (Verlag Schäffer-Poeschel, Stuttgart 1995) (in German)
12. More material on derivatives, as well as the techniques for their valuation established in the financial community is contained in [10] as well as in N. A. Chriss: *Black–Scholes and Beyond* (Irwin Professional Publishing, Chicago 1997), and in Campbell, *et al.,* [13]
13. J. Y. Campbell, A. W. Lo, and A. C. MacKinlay: *The Econometrics of Financial Markets* (Princeton University Press 1997)
14. S. N. Neftci: *An Introduction to the Mathematics of Financial Derivatives* (Academic Press, San Diego 1996)
15. P. Wilmott: *Derivatives* (Wiley, Chichester 1998)
16. C. Alexander: *Market Models* (Wiley, New York 2001)

17. J.-P. Bouchaud and M. Potters: *Théorie des Risques Financiers* (Aléa-Saclay, Paris 1997, in French); *Theory of Financial Risk* (Cambridge University Press 2000)
18. R. N. Mantegna and H. E. Stanley: *An Introduction to Econophysics* (Cambridge University Press 2000)
19. B. Roehner: *Patterns of Speculation* (Cambridge University Press, Cambridge 2002)
20. M. Levy, H. Levy and S. Solomon: *Microscopic Simulation of Financial Markets* (Academic Press, San Diego 2000)
21. D. Sornette: *Why Stock Markets Crash (Critical Events in Complex Financial Systems)* (Princeton University Press, Princeton 2003)
22. B. B. Mandelbrot: *Fractals and Scaling in Finance* (Springer-Verlag, New York 1997)
23. M. M. Dacorogna, R. Gençay, U. A. Müller, R. B. Olsen, and O. V. Pictet: *An Introduction to High-Frequency Finance* (Academic Press, San Diego 2002)
24. W. Paul and J. Baschnagel: *Stochastic Processes: From Physics to Finance* (Springer Verlag, Berlin 2000)
25. H. Kleinert: *Path Integrals in Quantum Mechanics, Statistics, Polymer Physics, and Financial Markets*, 3rd ed. (World Scientific, Singapore 2002)
26. Int. J. Theor. Appl. Fin. **3**, 309–608 (2000); Eur. Phys. J. **20** 471–625 (2001); Physica A **287**, 339–691 (2001); Adv. Compl. Syst. **4**, 1–163 (2001); Hideki Takayasu (ed.): *Empirical Science of Financial Fluctuations - The Advent of Econophysics* (Springer Verlag, Tokyo 2002); Physica A **299**, 1–351 (2001)
27. *Xetra* Marktmodell Release 2, Aktien-Wholesale-Release, Version 1 (Deutsche Börse AG, Frankfurt 1997)
28. See Hull [10], Chriss [12] or Campbell, Lo, and MacKinley [13]
29. E.g. F. Reif: *Fundamentals of Statistical and Thermal Physics* (Mc Graw-Hill, Tokyo 1965)
30. E.g. W. Feller: *An Introduction to Probability Theory and its Applications* (Wiley, New York 1968).
31. N. Jagdeesh: J. Finance, July 1990, p. 881; J. A. Murphy: J. Futures Markets, Summer 1986, p. 175
32. D. R. Cox and H. D. Miller: *The Theory of Stochastic Processes* (Chapman & Hall, London 1972); P. Lévy: *Processsus Stochastiques et Mouvement Brownien* (Gauthier-Villars, Paris 1965); D. Revuz and M. Yor: *Continuous Martingales and Brownian Motion* (Springer-Verlag, Berlin 1994)
33. B. B. Mandelbrot: *The Fractal Geometry of Nature* (Freeman, New York 1983)
34. K. V. Roberts: in [7]
35. J. Perrin: *Les Atomes* (Presses Universitaires de France, Paris 1948)
36. E. Kappler, Ann. Phys. (Leipzig), 5th series **11**, 233 (1931). I am indebted to an anonymous referee for pointing out Kappler's work which was unkown to me
37. H. Risken: *The Fokker–Planck Equation* (Springer- Verlag, Berlin 1984)
38. P. Gaspard, M. E. Briggs, M. K. Francis, J. V. Sengers, R. W. Gammon, J. R. Dorfman, and R. V. Calabrese: Nature **394**, 865 (1998)
39. W. A. Little: Phys. Rev. **134**, A1416 (1964)
40. D. Jérome and L. G. Caron (eds.): *Low-Dimensional Conductors and Super-conductors* (Plenum Press, New York 1987)
41. G. Soda, D. Jérome, M. Weger, J. Alizon, J. Gallice, H. Robert, J. M. Fabre, and L. Giral: J. Phys. (Paris) **38**, 931 (1977)
42. F. Black and M. Scholes: J. Polit. Econ. **81**, 637 (1973)
43. R. C. Merton: Bell J. Econ. Manag. Sci. **4**, 141 (1973)

44. J. Honerkamp: *Stochastic Dynamical Systems* (VCH-Wiley, New York 1994); *Statistical Physics* (Springer-Verlag, Berlin 1998)
45. B. Mandelbrot and J. R. Wallis: Water Resources Res. **5**, 909 (1969)
46. J. A. Skjeltorp: Physica A **283**, 486 (2000)
47. B. B. Mandelbrot and J. W. van Ness: SIAM Review **10**, 422 (1968)
48. R. F. Engle: Econometrica **50**, 987 (1982)
49. T. Bollerslev: J. Econometrics **31**, 307 (1986)
50. R. P. Feynman and A. R. Hibbs: *Quantum Mechanics and Path Integrals* (McGraw-Hill, New York 1965)
51. B. E. Baaquie: J. Phys. I (Paris) **7**, 1733 (1997)
52. R. Cont: cond-mat/9808262
53. R. Hafner and M. Wallmeier: Int. Quart. J. Finance **1**, 27 (2001)
54. R. Cont and J. de Fonseca: Quant. Finance **2**, 45 (2002)
55. *Leitfaden zu den Volatilitätsindizes der Deutschen Börse, Version 1.8*, technical document (Deutsche Börse AG, Frankfurt 2004)
56. F. Black: J. Fin. Econ. **3**, 167 (1976)
57. *VIX CBOE Volatility Index*, technical document (CBOE, Chicago 2003)
58. K. Demeterfi, E. Derman, M. Kamal, and J. Zou: J. Derivatives **6**, 9 (1999)
59. S. Dresel: *Die Modellierung von Aktienmärkten durch stochastische Prozesse*, Diplomarbeit, Universität Bayreuth, 2001 (unpublished)
60. J. Voit: Physica A **321**, 286 (2003)
61. This database is operated by *Institut für Entscheidungstheorie und Unternehmensforschung*, Universität Karlsruhe, http://www-etu.wiwi.uni-karlsruhe.de/
62. P. Gopikrishnan, V. Plerou, L. A. N. Amaral, M. Meyer, and H. E. Stanley: Phys. Rev. E **60**, 5305 (1999)
63. L.-H. Tang and Z.-F. Huang: Physica A **288**, 444 (2000)
64. E. F. Fama: J. Business **38**, 34 (1965)
65. S. S. Alexander: Ind. Manag. Rev. MIT **4**, 25 (1964), reprinted in [7]
66. B. B. Mandelbrot: J. Business **36**, 394 (1963)
67. R. Mantegna: Physica A **179**, 232 (1991)
68. See, e.g., J. Teichmöller: J. Am. Statist. Assoc. **66**, 282 (1971); M. A. Simkowitz and W. L. Beedles: J. Am. Statist. Assoc. **75**, 306 (1980); J. C. So: J. Finance **42**, 181 (1987) and Rev. Econ. Statist. **69**, 100 (1987); R. W. Cornew, D. E. Town, and L. D. Crowson: J. Futures Markets **4**, 531 (1984); J. W. McFarland, R. R. Pettit, and S. K. Sung: J. Finance **37**, 693 (1980)
69. R. Mantegna and H. E. Stanley: Nature **376**, 46 (1995)
70. E. Eberlein and U. Keller: Bernoulli **1**, 281 (1995); K. Prause: working paper no. 48, Freiburger Zentrum für Datenanalyse und Modellbildung (1997)
71. R. Mantegna and H. E. Stanley: Phys. Rev. Lett. **73**, 2946 (1994)
72. V. Pareto: *Cours d'Économie Politique*. In: *Oeuvres Complètes* (Droz, Geneva 1982)
73. V. V. Gnedenko and A. N. Kolmogorov: *Limit Distributions of Sums of Independent Random Variables* (Addison-Wesley, Reading 1968)
74. I. Koponen: Phys. Rev. E **52**, 1197 (1995)
75. M. F. Shlesinger, G. M. Zaslavsky, and U. Frisch (eds.): *Lévy Flights and Related Topics* (Springer Lect. Notes Phys. **450**) (Springer-Verlag, Berlin 1995)
76. J.-P. Bouchaud and A. Georges: Phys. Rep. **195**, 127 (1990)
77. C. Tsallis: Phys. World, July 1997, p. 42
78. M. Ma: *Modern Theory of Critical Phenomena* (Benjamin/Cummings, Reading 1976)
79. P. Bak, C. Tang, and K. Wiesenfeld: Phys. Rev. A **38**, 364 (1988)

80. A. Ott, J.-P. Bouchaud, D. Langevin, and W. Urbach: Phys. Rev. Lett. **65**, 2201 (1990)
81. T. H. Solomon, E. R. Weeks, and H. L. Swinney: Physica D **76**, 70 (1994)
82. T. H. Solomon, E. R. Weeks, and H. L. Swinney: Phys. Rev. Lett. **71**, 3975 (1993); E. R. Weeks, J. S. Urbach, and H. L. Swinney: Physica D **97**, 291 (1996)
83. C.-K. Peng, J. M. Hausdorff, J. E. Mietus, S. Havlin, H. E. Stanley, and A. L. Goldberger: in Shlesinger, Zaslavsky, and Frisch [75]
84. D. Adam, F. Closs, T. Frey, D. Funhoff, D. Haarer, H. Ringsdorf, P. Schuhmacher, and K. Siemensmeyer: Phys. Rev. Lett. **70**, 457 (1993); see also D. Adam: Diskotische Flüssigkristalle – eine neue Klasse schneller Photoleiter. PhD thesis, Universität Bayreuth (1995)
85. E. Barkai, R. Silbey, and G. Zumofen: Phys. Rev. Lett. **84**, 5339 (2000)
86. L. Kador: Phys. Rev. E **60**, 1441 (1999)
87. L. Kador: J. Luminesc. **86**, 219 (2000)
88. K. Umeno: Phys. Rev. E **58**, 2644 (1998)
89. C. Tsallis, S. V. F. Levy, A. M. C. Sousa, and R. Maynard: Phys. Rev. Lett. **75**, 3589 (1995)
90. C. Tsallis: J. Statist. Phys. **52**, 479 (1988)
91. L. Borland: unpublished preprint (1998)
92. L. Borland: Phys. Rev. E **57**, 6634 (1998)
93. C. Beck: Phys. Rev. Lett. **87**, 180601 (2001)
94. M. Baranger: Physica A **305**, 27 (2002)
95. G. Kaniadakis, M. Lissia, and A. Rapisarda (eds.): *Non Extensive Thermodynamics and Physical Applications,* Physica A **305** (2002)
96. D.-A. Hsu, R. B. Miller, and D. W. Wichern: J. Am. Statist. Assoc. **69**, 1008 (1974); D. E. Upton and D. S. Shannon: J. Finance **34**, 131 (1979); D. Friedman and S. Vandersteel: J. Int. Econ. **13**, 171 (1982); J. A. Hall, B. W. Brorsen, and S. H. Irwin: J. Finance Quant. Anal. **24**, 105 (1989)
97. T. Lux: Appl. Finance Econ. **6**, 463 (1996)
98. B. M. Hill: Ann. Statist. **3**, 1163 (1975)
99. M. R. Leadbetter, G. Lindgren, and H. Rootzén: *Extremes and Related Properties of Random Sequences and Processes* (Springer-Verlag, Berlin 1983)
100. R. Cont: 'Modeling Economic Randomness: Statistical Mechanics of Market Phenomena'. In: *Statistical Physics on the Eve of the 21st Century: in Honor of J. B. McGuire on the Occasion of His 65th Birthday* (World Scientific, Singapore 1998)
101. B. LeBaron: Quant. Finance **1**, 621 (2001)
102. U. A. Müller, M. M. Dacorogna, and O. V. Pictet: in *A Practical Guide to Heavy Tails: Statistical Techniques for Analyzing Heavy Tailed Distributions,* ed. by R. J. Adler, R. E. Feldman, and M. S. Taqqu (Birkhäuser, Boston 1998)
103. P. Gopikrishnan, M. Meyer, L. A. Nunes Amaral, and H. E. Stanley: Eur. Phys. J. B **3**, 139 (1998)
104. V. Plerou, P. Gopikrishnan, L. A. N. Amaral, M. Meyer, and H. E.Stanley: Phys. Rev. E **60**, 6519 (1999)
105. F. Lillo and R. N. Mantegna: Phys. Rev. **62**, 6126 (2000)
106. F. Lillo and R. N. Mantegna: Eur. Phys. J. B **15**, 603 (2000)
107. Y. Liu, P. Gopikrishnan, P. Cizeau, M. Meyer, C.-K. Peng, and H. E. Stanley: Phys. Rev. E **60**, 1390 (1999)
108. G. O. Zumbach, M. M. Dacorogna, J. L. Olsen, and R. B. Olsen: preprint GOZ 1998-10-01 (Olsen, Zürich 1998); Int. J. Theor. Appl. Finance **3**, 347 (2000)

109. R. Cont: cond-mat/9705075
110. T. Lux: Appl. Econ. Lett. **3**, 701 (1996)
111. N. Crato and P. J. F. de Lima: Econ. Lett. **45**, 281 (1994); Z. Ding, C. W. J. Granger, and R. F. Engle: J. Emp. Finance **1**, 83 (1993)
112. N. Vandewalle and M. Ausloos: Physica A **268**, 240 (1999)
113. T. Ohira, N. Sazuka, K. Marumo, T. Shimizu, M. Takayasu, and H. Takayasu: Physica A **308**, 368 (2002)
114. V. Plerou, P. Gopikrishnan, L. A. N. Amaral, X. Gabaix, and H. E. Stanley: Phys. Rev. E **62**, 3023 (1999)
115. M. Potters, R. Cont, and J.-P. Bouchaud: Europhys. Lett. **41**, 239 (1998)
116. F. Black: in *Proceedings of the 1976 American Statistical Association, Business and Economical Statistics Section* (American Statistical Association, Alexandria, VA 1976) p. 177
117. J.-P. Bouchaud, A. Matacz, and M. Potters: Phys. Rev. Lett. **87**, 228701 (2001)
118. J. Perelló and J. Masoliver: cond-mat/0202203
119. A. A. Drăgulescu and V. M. Yakovenko: cond-mat/0203046
120. T. Guhr and B. Kälber: J. Phys. A: Math. Gen. **36**, 3009 (2003)
121. L. Laloux, P. Cizeau, J.-P. Bouchaud, and M. Potters: Phys. Rev. Lett. **83**, 1467 (1999)
122. V. Plerou, P. Gopikrishnan, B. Rosenow, L. A. N. Amaral, and H. E. Stanley: Phys. Rev. Lett. **83**, 1471 (1999)
123. M. L. Mehta: *Random Matrices* (Academic, Boston 1991); T. Guhr, A. Müller-Gröhling, and H. A. Weidenmüller: Phys. Rep. **299**, 190 (1998)
124. J. Kwapień, S. Drożdż, F. Grümmer, F. Ruf, and J. Speth: cond-mat/0108068
125. J. D. Noh: Phys. Rev. E **61**, 5981 (2000)
126. W.-J. Ma, C.-K. Hu, and R. E. Amritkar: Phys. Rev. E **70**, 026101 (2004)
127. S. Drożdż, J. Kwapień, F. Grümmer, F. Ruf, and J. Speth: Physica A **299**, 144 (2001)
128. R. N. Mantegna: Eur. Phys. J. **11**, 193 (1999)
129. G. Bonanno, N. Vandewalle, and R. N. Mantegna: Phys. Rev. E **62**, 7615 (2000)
130. G. Bonanno, F. Lillo, and R. N. Mantegna: cond-mat/0009350
131. H.-J. Kim, Y. Lee, I.-M. Kim, and B. Kahng: cond-mat/0107449
132. G. Cuniberti and L. Matassini: Eur. Phys. J. B **20**, 561 (2001)
133. G. Cuniberti, M. Porto, and H. E. Roman: Physica A **299**, 262 (2001)
134. U. Frisch: *Turbulence* (Cambridge University Press, Cambridge 1995)
135. B. Chabaud, A. Naert, J. Peinke, F. Chillà, B. Castaing, and B. Hébral: Phys. Rev. Lett. **73**, 3227 (1994)
136. R. Friedrich and J. Peinke: Phys. Rev. Lett. **78**, 863 (1997)
137. M. Ragwitz and H. Kantz: Phys. Rev. Lett. **87**, 254501 (2001)
138. J. Timmer: Chaos, Solitons, Fractals **11**, 2571 (2000)
139. A. LaPorta, G. A. Voth, A. M. Crawford, J. Alexander, and E. Bodenschatz: Nature **409**, 1017 (2001)
140. W. Breymann and S. Ghashghaie: in Proceedings of the Workshop on Econophysics, Budapest, July 21–27, 1997
141. S. Ghashghaie, W. Breymann, J. Peinke, P. Talkner, and Y. Dodge: Nature **381**, 767 (1996)
142. F. Schmitt, D. Schertzer, and S. Lovejoy: Appl. Stoch. Models Data Anal. **15**, 29 (1999)
143. U. Müller, M. M. Dacorogna, R. D. Davé, R. B. Olsen, O. V. Pictet, and J. E. von Weizsäcker: J. Emp. Finance **4**, 211 (1997)
144. A. Arnéodo, J.-F. Muzy, and D. Sornette: Eur. Phys. J. B **2**, 277 (1998)

145. R. Friedrich, J. Peinke, and C. Renner: Phys. Rev. Lett. **84**, 5224 (2000)
146. C. Renner, J. Peinke, and R. Friedrich: Physica A **298**, 499 (2001)
147. J. Timmer and A. S. Weigend: Int. J. Neural Syst. **8**, 385 (1997)
148. D. Sornette: Physica A **290**, 211 (2001)
149. R. N. Mantegna and H. E. Stanley: Nature **383**, 588 (1996) and Physica A **239**, 255 (1997)
150. W. Breymann, S. Ghashghaie, and P. Talkner: Int. J. Theor. Appl. Finance **3**, 357 (2000)
151. T. Tél: Z. Naturforsch. **43a**, 1154 (1988)
152. B. Mandelbrot, A. Fisher, and L. Calvet: *A Multifractal Model of Asset Returns,* Cowles Foundation for Research in Economics working paper (1997)
153. L. Calvet, A. Fisher, and B. Mandelbrot: *Large Deviations and the Distribution of Price Changes,* Cowles Foundation for Research in Economics working paper (1997)
154. A. Fisher, L. Calvet, and B. Mandelbrot: *Multifractality of Deutschemark/US Dollar Exchange Rates,* Cowles Foundation for Research in Economics working paper (1997)
155. B. Mandelbrot: Quant. Finance **1**, 113, 124, 427, and 641 (2001)
156. M. M. Dacorogna, U. A. Müller, R. J. Nagler, R. B. Olsen, and O. V. Pictet: J. Int. Money Finance **12**, 413 (1993)
157. E. Derman: Quant. Finance **2**, 282 (2002)
158. T. Lux: Quant. Finance **1**, 632 (2001)
159. B. B. Mandelbrot: J. Fluid Mech. **62**, 331 (1974)
160. S. Lovejoy, D. Schertzer, and J. D. Stanway: Phys. Rev. Lett. **86**, 5200 (2001)
161. F. Schmitt, D. Schertzer, and S. Lovejoy: in *Chaos, Fractals, Models,* ed. by F. M. Guindani and G. Salvadori (Italian University Press, Pavia 1998)
162. N. Vandewalle and M. Ausloos: Int. J. Mod. Phys. C **9**, 711 (1998); Eur. Phys. J. B **4**, 257 (1998)
163. J.-P. Bouchaud, M. Potters, and M. Meyer: cond-mat/9906347
164. J.-P. Bouchaud and D. Sornette: J. Phys. I (Paris), **4**, 863 (1994); J.-P. Bouchaud, G. Iori, and D. Sornette: Risk **9**, 61 (1996)
165. K. Pinn: Physica A **276**, 581 (2000)
166. F. A. Longstaff and E. S. Schwartz: Rev. Financ. Stud. **14**, 113 (2001)
167. M. Potters, J.-P. Bouchaud, and D. Sestovic: Physica A **289**, 517 (2001); Risk **13**, 133 (2001)
168. R. Osorio, L. Borland, and C. Tsallis: in *Nonextensive Entropy: Interdisciplinary Applications,* ed. by C. Tsallis and M. Gell-Mann (Santa Fe Studies in the Science of Complexity, Oxford, to be published); F. Michael and M. D. Johnson: cond-mat/0108017
169. L. Borland: Phys. Rev. Lett. **89**, 098701 (2002)
170. H. Kleinert: Physica A **312**, 217 (2002)
171. A. Matacz: University of Sydney and Science & Finance working paper (2000)
172. L. Ingber: Physica A **283**, 529 (2000)
173. G. Montagna, O. Nicrosini, and N. Moreni: Physica A **310**, 450 (2002)
174. W. H. Press, B. P. Flannery, S. A. Teukolsky, W. T. Vetterling: *Numerical Recipes in C++: The Art of Scientific Computing* (Cambridge University Press, Cambridge 2002). Similar volumes are available for the programming languages C, Fortran 77, and Fortran 90 **6**, 721 (1984)
175. G. Bormetti, G. Montagna, N. Moreni, and O. Nicrosini, cond-mat/0407321
176. G. Kim and H. Markowitz: J. Portfolio Management **16**, 45 (1989)
177. J. Coche: J. Evol. Econ. **8**, 357 (1998)
178. G. Caldarelli, M. Marsili, and Y. C. Zhang: Europhys. Lett. **40**, 479 (1997)

179. G. K. Zipf: *Human Behavior and the Principle of Least Action* (Addison-Wesley 1949)
180. M. Levy, H. Levy, and S. Solomon: J. Phys. I France **5**, 1087 (1995) and Econ. Lett. **45**, 103 (1994)
181. G. Iori: Int. J. Mod. Phys. C **10**, 1149 (1999)
182. D. J. Watts and S. H. Strogatz: Nature **393**, 440 (1998)
183. J. Sethna, K. Dahmen, S. Kartha, J. A. Krumhansl, B. W. Roberts, and J. D. Shore: Phys. Rev. Lett. **70**, 3347 (1993)
184. D. Stauffer and A. Aharony: *Introduction to Percolation Theory* (Taylor & Francis, London 1994)
185. M. Mézard, G. Parisi, and M. A. Virasoro: *Spin Glass Theory and Beyond* (World Scientific, Singapore 1987)
186. J. M. Karpoff: J. Fin. Quant. Anal. **22**, 109 (1987)
187. A.-H. Sato and H. Takayasu: Physica A **250**, 231 (1998); cf. also H. Takayasu, M. Miura, T. Hirabayashi, and K. Hamada: Physica A **184**, 127 (1992) for an earlier variant of this model
188. P. Gopikrishnan, V. Plerou, Y. Liu, L. A. N. Amaral, X. Gabaix, and H. E. Stanley: Physica A **287**, 362 (2000)
189. D. S. Scharfstein and J. C. Stein: Am. Econ. Rev. **80**, 465 (1990); B. Trueman: Rev. Fin. Stud. **7**, 97 (1994); M. Grinblatt, S. Titman, and R. Wermers: Am. Econ. Rev. **85**, 1088 (1995)
190. R. Cont and J.-P. Bouchaud: cond-mat/9712318, and p. 71 in [17]
191. D. Stauffer and T. J. P. Penna: Physica A **256**, 284 (1998)
192. D. Chowdhury and D. Stauffer: Eur. Phys. J. B **8**, 447 (1999)
193. T. Lux and M. Marchesi: Nature **397**, 498 (1999)
194. M. Marsili and Y.-C. Zhang: Physica A **245**, 181 (1997)
195. W. B. Arthur: Am. Econ. Assoc. Pap. Proc. **84**, 406 (1994)
196. D. Challet and Y.-C. Zhang: Physica A **246**, 407 (1997)
197. M. Hart, P. Jefferies, N. F. Johnson, and P. M. Hui: Physica A **298**, 537 (2001); M. Hart, P. Jefferies, P. M. Hui, and N. F. Johnson: Eur. Phys. J. B **20**, 547 (2001)
198. D. Challet, M. Marsili, and Y.-C. Zhang: Physica A **299**, 228 (2001)
199. M. Marsili: Physica A **299**, 93 (2001)
200. D. Challet, M. Marsili, and Y.-C. Zhang: Physica A **276**, 284 (2000)
201. D. Challet, M. Marsili, and R. Zecchina: Phys. Rev. Lett. **84**, 1824 (2000)
202. P. Jefferies, M. L. Hart, P. M. Hui, and N. F. Johnson: Eur. Phys. J. B **20**, 493 (2001)
203. N. F. Johnson, M. Hart, P. M. Hui, and D. Zheng: Int. J. Theor. Appl. Finance **3**, 443 (2000)
204. N. F. Johnson, D. Lamper, P. Jefferies, M. L. Hart, and S. Howison: Physica A **299**, 222 (2001); D. Lamper, S. Howison, and N. F. Johnson: cond-mat/0105258
205. G. P. Harmer and D. Abbott: Nature **402**, 864 (1999)
206. P. M. Garber: J. Portfolio Management **16**, 53 (1989)
207. Chap. 2 in *A Random Walk Down Wall Street* [3]
208. H. Dupuis: Tendences, 18 September 1997, p. 26 discusses the prediction by N. Vandewalle, M. Ausloos, Ph. Boveroux, and A. Minguet, of the 1997 crash. Their work is documented in [218]
209. N. Vandewalle, Ph. Boveroux, A. Minguet, and M. Ausloos: Physica A **255**, 201 (1998)
210. A. Johansen, D. Sornette, H. Wakita, U. Tsunogai, W. I. Newman, and H. Saleur: J. Phys. I (France) **6**, 1391 (1996)
211. C. Allègre, J. L. LeMouel, and A. Provost: Nature **297**, 47 (1982)

212. D. Sornette: Phys. Rep. **297**, 239 (1998)
213. D. Sornette and C. G. Sammis: J. Phys. I (France) **5**, 607 (1995)
214. K. Shimazaki and T. Nakata: Geophys. Res. Lett. **7**, 279 (1980)
215. J. Murray and P. Segall: Nature **419**, 287 (2002); R. S. Stein: Nature **419**, 257 (2002)
216. D. Sornette and A. Johansen: Physica A **245**, 411 (1997)
217. A. Johansen and D. Sornette: Eur. Phys. J. B **9**, 167 (1999)
218. N. Vandewalle, M. Ausloos, Ph. Boveroux, and A. Minguet: Eur. Phys. J. B **4**, 139 (1998)
219. D. Stauffer and D. Sornette: Physica A **252**, 271 (1998)
220. J. A. Feigenbaum and P. G. O. Freund: Int. J. Mod. Phys. B **12**, 57 (1998); see also J. A. Feigenbaum and P. G. O. Freund: Int. J. Mod. Phys. B **10**, 3737 (1996)
221. L. Laloux, M. Potters, R. Cont, J.-P. Aguilar, and J.-P. Bouchaud: Europhys. Lett. **45**, 1 (1999)
222. http://phytech.ddynamics.be/
223. A. Johansen and D. Sornette: Eur. Phys. J. B **17**, 319 (2000)
224. R. J. Barro, E. F. Fama, D. R. Fischel, A. H. Meltzer, R. Roll, and L. G. Telser: in R. W. Kamphuis, Jr., R. C. Komendi, and J. W. H. Watson (eds.) *Black Monday and The Future of Financial Markets* (Mid American Institute for Public Policy Research and Dow Jones-Irwin 1989)
225. A. Johansen and D. Sornette: in *Contemporary Issues in International Finance* (Nova Science Publishers 2003)
226. J.-F. Muzy, J. Delour, and E. Bacry: Eur. Phys. J. B **17**, 537 (2000)
227. E. Bacry, J. Delour, and J.-F. Muzy: Phys. Rev. E **64**, 026103 (2001)
228. D. Sornette, Y. Malevergne, and J.-F. Muzy, Risk **16**, 67 (February 2003)
229. A. Johansen, O. Ledoit, and D. Sornette: Int. J. Theo. Appl. Fin. **3**, 219 (2000)
230. B. M. Roehner: Int. J. Mod. Phys. C **11**, 91 (2000)
231. A. Johansen and D. Sornette: Int. J. Mod. Phys. C **10**, 563 (1999)
232. A. Johansen and D. Sornette: Int. J. Mod. Phys. C **11**, 359 (2000)
233. D. Sornette and W.-X. Zhou: Quant. Fin. **2**, 468 (2002)
234. W.-X. Zhou and D. Sornette: Physica A **330**, 543 (2003)
235. D. Sornette and W.-X. Zhou: Quant. Fin. **3**, C39 (2003)
236. N. Patel: Risk **16**, 10 (December 2003)
237. B. Gutenberg and C. F. Richter: Annali di Geofisica **9**, 1 (1956); S. K. Runcorn, Sir E. Bullard, K. E. Bullen, W. A. Heiskanen, Sir H. Jeffreys, H. Mosby, T. Nagata, M. Nicolet, K. R. Ramanathan, H. C. Urey, and F. A. Vening Meinesz, (eds.): *International Dictionary of Geophysics* (Pergamon Press, Oxford 1967)
238. *International Convergence of Capital Measurement and Capital Standards, A Revised Framework* The Basel Committee for Banking Supervision, Bank of International Settlements, Basel (2004), http://www.bis.org
239. R. C. Merton: J. Finance **29**, 449 (1974)
240. N. Leeson: *Rogue Trader* (Little, Brown and Company, London 1996)
241. E.g., FIRST data base operated by Fitch Risk under the label of OpVantage, http://www.fitchrisk.com
242. S. A. Klugman, H. H. Panjer, and G. E. Willmot: *Loss Models - From Data to Decisions,* (Wiley, New York 1998)
243. P. Neu and R. Kühn: cond-mat/0204368
244. C. Cornalba and P. Giudici: Physica A **338**, 166 (2004)
245. D. Duffie and J. Pan: J. Derivatives, Spring 1997, p. 7

246. P. Jorion: *Value at Risk: the New Benchmark for Measuring Financial Risk* (Mc Graw-Hill, New York 2001)

247. P. Artzner, F. Delbaen, J.-M. Eber, and D. Heath: Risk **10**, 68 (1997)

248. P. Artzner, F. Delbaen, J.-M. Eber, and D. Heath: Mathematical Finance **9**, 203 (1999)

249. U. Gaumert and G. Stahl: in *Handwörterbuch des Bank- und Finanzwesens,* ed. by W. Gerke and M. Steiner (Schäffer-Poeschel Verlag, Stuttgart 2001)

250. C. Acerbi, C. Nordio, and C. Sirtori: cond-mat/0102304

251. C. Acerbi and D. Tasche: J. Bank. Fin. **26**, 1487 (2002)

252. C. Acerbi and D. Tasche: cond-mat/0105191

253. H. Markowitz: *Portfolio Selection* (Basil Blackwell, Oxford 1991)

254. High-level information on risk management in two German blue chip companies, Lufthansa German Airlines and Bayer corporation, can be found in two articles in Deutsches Risk **4**, Winter 2004, p. 12 and 16

255. H. P. Deutsch: *Derivatives and Internal Models* (Palgrave MacMillan, 2002)

256. T. W. Koch and S. S. MacDonald: *Bank Management* (Thomson South-Western, Mason 2004)

257. P. Nakade and J. Kapitan: The RMA Journal, March 2004, p. 2

258. *International Convergence of Capital Measurement and Capital Standards,* The Basel Committee for Banking Supervision, Bank of International Settlements, Basel (1988), http://www.bis.org

259. *Amendment to the Capital Accord to Incorporate Market Risks,* The Basel Committee for Banking Supervision, Bank of International Settlements, Basel (1996), http://www.bis.org

260. http://www.standardandpoors.com

261. M. Böcker and H. Eckelmann: Betriebswirtschaftliche Blätter **51**, 168 (2002)

Index